D0712636

Probability and Its Applications

Published in association with the Applied Probability Trust

Editors: J. Gani, C.C. Heyde, P. Jagers, T.G. Kurtz

Probability and Its Applications

Richard Durrett

Probability Models for DNA Sequence Evolution

Second Edition

 Springer

Richard Durrett
Department of Mathematics
Cornell University
Ithaca, NY 14853-4201
USA

Series Editors

J. Gani
Stochastic Analysis Group, CMA
Australian National University
Canberra ACT 0200
Australia

C.C. Heyde
Stochastic Analysis Group, CMA
Australian National University
Canberra ACT 0200
Australia

P. Jagers
Mathematical Statistics
Chalmers University of Technology
SE-412 96 Göteberg
Sweden

T.G. Kurtz
Department of Mathematics
University of Wisconsin
480 Lincoln Drive
Madison, WI 53706
USA

ISBN: 978-0-387-78168-6   e-ISBN: 978-0-387-78169-3
DOI: 10.1007/978-0-387-78168-6

Library of Congress Control Number: 2008929756

Printed on acid-free paper

9 8 7 6 5 4 3 2 1

springer.com

Preface

"Mathematics seems to endow one with something like a new sense."
Charles Darwin

The goal of population genetics is to understand how genetic variability is shaped by natural selection, demographic factors, and random genetic drift. The stochastic evolution of a DNA segment that experiences recombination is a complex process, so many analyses are based on simulations or use heuristic methods. However, when formulas are available, they are preferable because, when simple, they show the dependence of observed quantities on the underlying parameters and, even when complicated, can be used to compute exact answers in a much shorter time than simulations can give good approximations.

The goal of this book is to give an account of useful analytical results in population genetics, together with their proofs. The latter are omitted in many treatments, but are included here because the derivation often gives insight into the underlying mechanisms and may help others to find new formulas. Throughout the book, the theoretical results are developed in close connection with examples from the biology literature that illustrate the use of these results. Along the way, there are many numerical examples and graphs to illustrate the main conclusions. To help the reader navigate the book, we have divided the sections into a large number of subsections listed in the index, and further subdivided the text with bold-faced headings (as in this Preface).

This book is written for mathematicians and for biologists alike. With mathematicians in mind, we assume no knowledge of concepts from biology. Section 1.1 gives a rapid introduction to the basic terminology. Other explanations are given as concepts arise. For biologists, we explain mathematical notation and terminology as it arises, so the only *formal* prerequisite for biologists reading this book is a one-semester undergraduate course in probability and some familiarity with Markov chains and Poisson processes will be very useful. We have emphasized the word *formal* here, because to read and understand all of the proofs will require more than these simple prerequisites. On the other hand, the book has been structured so that proofs can be omitted.

What is in this book?

Chapter 1 begins with the theory of neutral evolution in a homogeneously mixing population of constant size. We introduce and study the discrete-time Wright-Fisher model, the continuous-time Moran model, the coalescent, which describes the genealogy of a nonrecombining segment of DNA, and two simplified models of mutation: the infinite alleles and infinite sites models. Based on these results, Chapter 2 introduces the problem of testing to see if observed DNA sequences are consistent with the assumptions of the "null model" underlying the theory developed in Chapter 1.

Chapters 3 through 6 confront the complications that come from relaxing the assumptions of the models in Chapter 1. This material, which filled two chapters in the first edition, has doubled in size and contains many results from the last five years. Chapter 3 introduces the ancestral recombination graph and studies the effect of recombination on genetic variability and the problem of estimating the rate at which recombination occurs. Chapter 4 investigates the influence of large family sizes, population size changes, and population subdivision in the form of island models on the genealogy of a sample. Chapter 5 concerns the more subtle behavior of the stepping stone model, which depicts a population spread across a geographical range, not grouped into distinct subpopulations. Finally, Chapter 6 considers various forms of natural selection: directional selection and hitchhiking, background selection and Muller's ratchet, and balancing selection.

Chapters 7 and 8, which are new in this edition, treat the previous topics from the viewpoint of diffusion processes, continuous stochastic processes that arise from letting the population size $N \to \infty$ and at the same time running time at rate $O(N)$. A number of analytical complications are associated with this approach, but, at least in the case of the one-dimensional processes considered in Chapter 7, the theory provides powerful tools for computing fixation probabilities, expected fixation time, and the site frequency spectrum. In contrast, the theory of multidimensional diffusions described in Chapter 8 is more of an art than a science. However, it offers significant insights into recombination, Hill-Robertson interference, and gene duplication.

Chapter 9 tackles the relatively newer, and less well-developed, study of the evolution of whole genomes by chromosomal inversions, reciprocal translocations, and genome duplication. This chapter is the least changed from the previous edition but has new results about when the parsimony method is effective, Bayesian estimation of genetic distance, and the midpoint problem.

In addition to the three topics just mentioned, there are a number of results covered here that do not appear in most other treatments of the subject (given here with the sections in which they appear): Fu's covariance matrix for the site frequency spectrum (2.1), the sequentially Markovian coalescent (3.4), the beta coalescent for large family sizes (4.1), Malécot's recursion for

Contents

1

Basic Models

"All models are wrong, but some are useful." George Box

1.1 ATGCs of life

Before we can discuss modeling the evolution of DNA sequences, the reader needs a basic knowledge of the object being modeled. Biologists should skip this very rapid introduction, the purpose of which is to introduce some of the terminology used in later discussions. Mathematicians should concentrate here on the description of the genetic code and the notion of recombination. An important subliminal message is that DNA sequences are not long words randomly chosen from a four-letter alphabet; chemistry plays an important role as well.

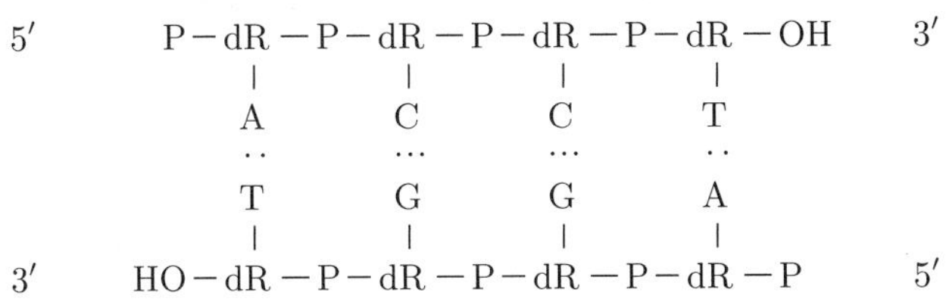

$$
\begin{array}{llll}
5' & \text{P}-\text{dR}-\text{P}-\text{dR}-\text{P}-\text{dR}-\text{P}-\text{dR}-\text{OH} & 3' \\
 & \quad | \qquad\quad | \qquad\quad | \qquad\quad | \\
 & \quad \text{A} \qquad\; \text{C} \qquad\;\; \text{C} \qquad\;\; \text{T} \\
 & \quad .. \qquad\;\; ... \qquad\; ... \qquad\;\; .. \\
 & \quad \text{T} \qquad\;\; \text{G} \qquad\; \text{G} \qquad\;\; \text{A} \\
 & \quad | \qquad\quad | \qquad\quad | \qquad\quad | \\
3' & \text{HO}-\text{dR}-\text{P}-\text{dR}-\text{P}-\text{dR}-\text{P}-\text{dR}-\text{P} & 5'
\end{array}
$$

Fig. 1.1. Structure of DNA.

The hereditary information of most living organisms is carried by deoxyribonucleic acid (DNA) molecules. DNA usually consists of two complementary

R. Durrett, *Probability Models for DNA Sequence Evolution*,
DOI: 10.1007/978-0-387-78168-6_1, © Springer Science+Business Media, LLC 2008

chains twisted around each other to form a double helix. As drawn in the figure, each chain is a linear sequence of four *nucleotides*: adenine (A), guanine (G), cytosine (C), and thymine (T). Adenine pairs with thymine by means of two hydrogen bonds, while cytosine pairs with guanine by means of three hydrogen bonds. The $A = T$ bond is weaker than the $C \equiv G$ one and separates more easily. The backbone of the DNA molecule consists of sugars (deoxyribose, dR) and phosphates (P) and is oriented. There is a phosphoryl radical (P) on one end (the $5'$ end) and a hydroxyl (OH) on the other ($3'$ end). By convention, DNA sequences are written in the order in which they are transcribed from the $5'$ to the $3'$ end.

The structure of DNA guarantees that the overall frequencies of A and T are equal and that the frequencies of C and G are equal. Indeed, this observation was one of the clues to the structure of DNA. If DNA sequences were constructed by rolling a four-sided die, then all four nucleotides (which are also called *base pairs*) would have a frequency near $1/4$, but they do not. If one examines the 12 million nucleotide sequence of the yeast genome, which consists of the sequence of one strand of each of its 16 chromosomes, then the frequencies of the four nucleotides are

$$A = 0.3090 \qquad T = 0.3078 \qquad C = 0.1917 \qquad G = 0.1913$$

Watson and Crick (1953a), in their first report on the structure of DNA, wrote: "It has not escaped our attention that the specific [nucleotide base] pairing we have postulated immediately suggests a possible copying mechanism of the genetic material." Later that year at a Cold Spring Harbor meeting, Watson and Crick (1953b) continued: "We should like to propose ... that the specificity of DNA replication is accomplished without recourse to specific protein synthesis and that each of our complimentary DNA chains serves as a template or mould for the formation onto itself of a new companion chain." This picture turned out to be correct. When DNA is ready to multiply, its two strands pull apart, along each one a new strand forms in the only possible way, and we wind up with two copies of the original. The precise details of the replication process are somewhat complicated, but are not important for our study.

Much of the sequence of the 3 billion nucleotides that make up the human genome apparently serves no function, but embedded in this long string are about 30,000 protein-coding genes. These genes are *transcribed* into ribonucleic acid (RNA), so-called messenger RNA (mRNA), which subsequently is *translated* into proteins. RNA is usually a single-stranded molecule and differs from DNA by having ribose as its backbone sugar and by using the nucleotide uracil (U) in place of thymine (T).

Amino acids are the basic structural units of proteins. All proteins in all organisms, from bacteria to humans, are constructed from 20 amino acids. The next table lists them along with their three-letter and one-letter abbreviations.

Ala	A	Alanine	Leu	L	Leucine
Arg	R	Arginine	Lys	K	Lysine
Asn	N	Asparagine	Met	M	Methionine
Asp	D	Aspartic acid	Phe	F	Phenylalanine
Cys	C	Cysteine	Pro	P	Proline
Gly	G	Glycine	Ser	S	Serine
Glu	E	Glutamic acid	Thr	T	Threonine
Gln	Q	Glutamine	Trp	W	Tryptophan
His	H	Histidine	Tyr	Y	Tyrosine
Ile	I	Isoleucine	Val	V	Valine

Amino acids are coded by triplets of adjacent nucleotides called *codons*. Of the 64 possible triplets, 61 code for amino acids, while 3 are stop codons, which terminate transcription. The correspondence between triplets of RNA nucleotides and amino acids is given by the following table. The first letter of the codon is given on the left edge, the second on the top, and the third on the right. For example, CAU codes for Histidine.

	U	C	A	G	
U	Phe	Ser	Tyr	Cys	U
	"	"	"	"	C
	Leu	"	Stop	Stop	A
	"	"	"	Trp	G
C	Leu	Pro	His	Arg	U
	"	"	"	"	C
	"	"	Gln	"	A
	"	"	"	"	G
A	Ile	Thr	Asn	Ser	U
	"	"	"	"	C
	"	"	Lys	Arg	A
	Met	"	"	"	G
G	Val	Ala	Asp	Gly	U
	"	"	"	"	C
	"	"	Glu	"	A
	"	"	"	"	G

Note that in 8 of 16 cases, the first two nucleotides determine the amino acid, so a mutation that changes the third base does not change the amino acid that is coded for. Mutations that do not change the amino acid are called *synonymous substitutions*; the others are *nonsynonymous*. For example, a change at the second position always changes the amino acid coded for, except for $UAA \rightarrow UGA$, which are both stop codons.

In DNA, adenine and guanine are *purines* while cytosine and thymine are *pyrimidines*. A substitution that preserves the type is called a *transition*; the others are called *transversions*. As we will see later in this chapter, transitions occur more frequently than transversions.

Most of the genes in our bodies reside on DNA in the nucleus of our cells and are organized into chromosomes. Lower organisms such as bacteria are *haploid*. They have one copy of their genetic material. Most higher organisms are *diploid* (i.e., have two copies). However, some plants are *tetraploid* (four copies), *hexaploid* (six copies, e.g., wheat), or *polyploid* (many copies, e.g., sorghum, which has more than 100 chromosomes of 8 basic types). Sex chromosomes in diploids are an exception to the two-copy rule. In humans, females have two X chromosomes, while males have one X and one Y. In birds, males have two Z chromosomes, while females have one Z and one W.

When haploid individuals reproduce, there is one parent that passes copies of its genetic material to its offspring. When diploid individuals reproduce, there are two parents, each of which contributes one of each of its pairs of chromosomes. Actually, one parent's contribution may be a combination of its two chromosomes, since *homologous* pairs (e.g., the two copies of human chromosome 14) undergo recombination, a reciprocal exchange of genetic material that may be diagrammed as follows:

Fig. 1.2. Recombination between homologous chromosomes.

As we will see in Chapter 3, recombination will greatly complicate our analysis. Two cases with no recombination are the Y chromosome, which except for a small region near the tip does not recombine, and the mitochondrial DNA (mtDNA), a circular double-stranded molecule about 16,500 base pairs in length that exist in multiple identical copies outside the nucleus and are inherited from the maternal parent. mtDNA, first sequenced by Anderson et al. (1981), contains genes that code for 13 proteins, 22 tRNA genes, and 2 rRNA genes. It is known that nucleotide substitutions in mtDNA occur at about 10 times the rate for nuclear genes. One important part of the molecule is the control region (sometimes referred to as the D loop), which is about 1,100 base pairs in length and contains promoters for transcription and the origin of replication for one of the DNA strands. It has received particular attention since it has an even higher mutation rate, perhaps an order of magnitude larger than the rest of the mtDNA.

These definitions should be enough to get the reader started. We will give more explanations as the need arises. Readers who find our explanations of the background insufficient should read the *Cartoon Guide to Genetics* by Gonick and Wheelis (1991) or the first chapter of Li's (1997) *Molecular Evolution*.

1.2 Wright-Fisher model

We will begin by considering a genetic locus with two alleles A and a that have the same fitness in a diploid population of constant size N with nonoverlapping generations that undergoes random mating. The first thing we have to do is to explain the terms in the previous sentence.

A *genetic locus* is just a location in the genome of an organism. A common example is the sequence of nucleotides that makes up a gene.

The two *alleles, A and a,* could be the "wrinkled" and "round" types of peas in Mendel's experiments. More abstractly, alleles are just different versions of the genetic information encoded at the locus.

The *fitness* of an individual is a measure of the individual's ability to survive and to produce offspring. Here we consider the case of *neutral evolution* in which the mutation changes the DNA sequence but this does not change the fitness.

Diploid individuals have two copies of their genetic material in each cell. In general, we will treat the N individuals as $2N$ copies of the locus and not bother to pair the copies to make individuals. Note: It may be tempting to set $M = 2N$ and reduce to the case of M haploid individuals, but that makes it harder to compare with formulas in the literature.

To explain the terms *nonoverlapping generations* and *random mating*, we use a picture.

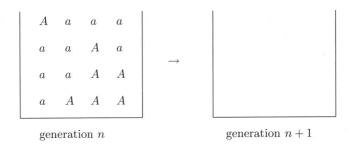

generation n generation $n + 1$

Fig. 1.3. Wright-Fisher model.

In words, we can represent the state of the population in generation n by an "urn" that contains $2N$ balls: i with A's on them and $2N - i$ with a's. Then, to build up the $(n + 1)$th generation, we choose at random from the urn $2N$ times with replacement.

Let X_n be the number of A's in generation n. It is easy to see that X_n is a Markov chain, i.e., given the present state, the rest of the past is irrelevant for predicting the future. Remembering the definition of the binomial distribution, it is easy to see that the probability there are j A's at time $n+1$ when there are i A's in the urn at time n is

$$p(i, j) = \binom{2N}{j} p_i^j (1 - p_i)^{2N-j} \tag{1.1}$$

Here $p_i = i/2N$ is the probability of drawing an A on one trial when there are i in the urn, and the *binomial coefficient*

$$\binom{2N}{j} = \frac{(2N)!}{j!(2N - j)!}$$

is the number of ways of choosing j things out of $2N$, where $j! = 1 \cdot 2 \cdots j$ is "j factorial."

Fixation probability

The long-time behavior of the Wright-Fisher model is not very exciting. Since we are, for the moment, ignoring mutation, eventually the number of A's in the population, X_n, will become 0, indicating the loss of the A allele, or $2N$, indicating the loss of a. Once one allele is lost from the population, it never returns, so the states 0 and $2N$ are *absorbing states* for X_n. That is, once the chain enters one of these states, it can never leave. Let

$$\tau = \min\{n : X_n = 0 \text{ or } X_n = 2N\}$$

be the *fixation time*; that is, the first time that the population consists of all a's or all A's.

We use P_i to denote the probability distribution of the process X_n starting from $X_0 = i$, and E_i to denote expected value with respect to P_i.

Theorem 1.1. *In the Wright-Fisher model, the probability of fixation in the all A's state,*

$$P_i(X_\tau = 2N) = \frac{i}{2N} \tag{1.2}$$

Proof. Since the number of individuals is finite, and it is always possible to draw either all A's or all a's, fixation will eventually occur. Let X_n be the number of A's at time n. Since the mean of the binomial in (1.1) is $2Np$, it follows that

$$E(X_{n+1}|X_n = i) = 2N \cdot \left(\frac{i}{2N}\right) = i = X_n \tag{1.3}$$

Taking expected value, we have $EX_{n+1} = EX_n$. In words, the average value of X_n stays constant in time.

Intuitively, the last property implies

$$i = E_i X_\tau = 2N P_i(X_\tau = 2N)$$

and this gives the desired formula. To prove this, we note that since $X_n = X_\tau$ when $n > \tau$,

$$i = E_i X_n = E_i(X_\tau; \tau \le n) + E_i(X_n; \tau > n)$$

where $E(X; A)$ is short for the expected value of X over the set A. Now let $n \to \infty$ and use the fact that $|X_n| \le 2N$ to conclude that the first term converges to $E_i X_\tau$ and the second to 0. $\qquad\square$

From (1.2) we get a famous result of Kimura:

Theorem 1.2. *Under the Wright-Fisher model, the rate of fixation of neutral mutations in a population of size N is the mutation rate μ.*

Proof. To see this note that mutations occur to some individual in the population at rate $2N\mu$ and go to fixation with probability $1/2N$.

Heterozygosity

To get an idea of how long fixation takes to occur, we will examine the *heterozygosity*, which we define here to be the probability that two copies of the locus chosen (without replacement) at time n are different:

$$H_n^o = \frac{2X_n(2N - X_n)}{2N(2N - 1)}$$

Theorem 1.3. *Let $h(n) = EH_n^o$ be the average value of the heterozygosity at time n. In the Wright-Fisher model*

$$h(n) = \left(1 - \frac{1}{2N}\right)^n \cdot h(0) \qquad (1.4)$$

Proof. It is convenient to number the $2N$ copies of the locus $1, 2, \ldots 2N$ and refer to them as individuals. Suppose we pick two individuals numbered $x_1(0)$ and $x_2(0)$ at time n. Each individual $x_i(0)$ is a descendant of some individual $x_i(1)$ at time $n - 1$, who is a descendant of $x_i(2)$ at time $n - 2$, etc. $x_i(m)$, $0 \le m \le n$ describes the lineage of $x_i(0)$, i.e., its ancestors working backwards in time.

If $x_1(m) = x_2(m)$, then we will have $x_1(\ell) = x_2(\ell)$ for $m < \ell \le n$. If $x_1(m) \ne x_2(m)$, then the two choices of parents are made independently, so $x_1(m+1) \ne x_2(m+1)$ with probability $1-(1/2N)$. In order for $x_1(n) \ne x_2(n)$, different parents must be chosen at all times $1 \le m \le n$, an event with probability $(1 - 1/2N)^n$. When the two lineages avoid each other, $x_1(n)$ and $x_2(n)$ are two individuals chosen at random from the population at time 0, so the probability that they are different is $H_0^o = h(0)$. $\qquad\square$

A minor detail. If we choose with replacement above, then the statistic is

$$H_n = \frac{2X_n(2N - X_n)}{(2N)^2} = \frac{2N - 1}{2N} H_n^o$$

and we again have $EH_n = (1 - 1/2N)^n \cdot H_0$. This version of the statistic is more commonly used, but is not very nice for the proof given above.

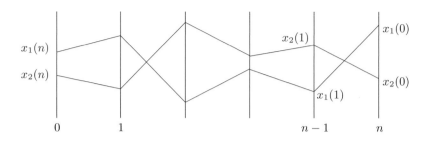

Fig. 1.4. A pair of genealogies.

1.2.1 The coalescent

When x is small, we have $(1 - x) \approx e^{-x}$. Thus, when N is large, (1.4) can be written as

$$h(n) \approx e^{-n/(2N)}h(0)$$

If we sample k individuals, then the probability that two will pick the same parent from the previous generation is

$$\approx \frac{k(k-1)}{2} \cdot \frac{1}{2N}$$

where the first factor gives the number of ways of picking two of the k individuals and the second the probability they will choose the same parent. Here we are ignoring the probability that two different pairs choose the same parents on one step or that three individuals will all choose the same parent, events of probability of order $1/N^2$.

Theorem 1.4. *When measured in units of $2N$ generation, the amount of time during which there are k lineages, t_k, has approximately an exponential distribution with mean $2/k(k-1)$.*

Proof. By the reasoning used above, the probability that the k lineages remain distinct for the first n generations is (when the population size N is large)

$$\approx \left(1 - \frac{k(k-1)}{2} \cdot \frac{1}{2N}\right)^n \approx \exp\left(-\frac{k(k-1)}{2} \cdot \frac{n}{2N}\right)$$

Recalling that the exponential distribution with rate λ is defined by

$$P(T > t) = e^{-\lambda t}$$

and has mean $1/\lambda$, we see that if we let the population size $N \to \infty$ and express time in terms of $2N$ generations, that is, we let $t = n/(2N)$, then the time to the first collision converges to an exponential distribution with mean $2/k(k-1)$. Using terminology from the theory of continuous-time Markov chains, k lineages coalesce to $k-1$ at rate $k(k-1)/2$. Since this reasoning applies at any time at which there are k lineages, the desired result follows. $\square$

The limit of the genealogies described in Theorem 1.4 is called the *coalescent*. Letting T_j be the first time that there are j lineages, we can draw a picture of what happens to the lineages as we work backwards in time:

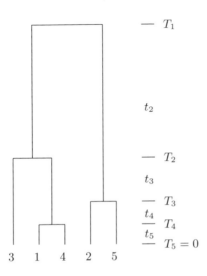

Fig. 1.5. A realization of the coalescent for a sample of size 5.

For simplicity, we do not depict how the lineages move around in the set before they collide, but only indicate when the coalescences occur. To give the reader some idea of the relative sizes of the coalescent times, we have made the t_k proportional to their expected values, which in this case are

$$Et_2 = 1, \quad Et_3 = 1/3, \quad Et_4 = 1/6, \quad Et_5 = 1/10$$

T_1 is the time of the *most recent common ancestor (MRCA)* of the sample. For a sample of size n, $T_1 = t_n + \cdots + t_2$, so the mean

$$ET_1 = \sum_{k=2}^{n} \frac{2}{k(k-1)} = 2 \sum_{k=2}^{n} \left(\frac{1}{k-1} - \frac{1}{k} \right) = 2 \cdot \left(1 - \frac{1}{n} \right)$$

This quantity converges to 2 as the sample size $n \to \infty$, but the time, t_2, at which there are only two lineages has $Et_2 = 1$, so the expected amount of time spent waiting for the last coalescence is always at least half of the expected total coalescence time.

Simulating the coalescent

It is fairly straightforward to translate the description above into a simulation algorithm, but for later purposes it is useful to label the internal nodes of the tree. The following picture should help explain the procedure

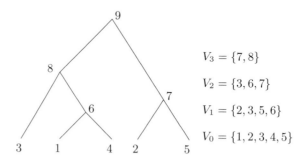

$$V_3 = \{7, 8\}$$

$$V_2 = \{3, 6, 7\}$$

$$V_1 = \{2, 3, 5, 6\}$$

$$V_0 = \{1, 2, 3, 4, 5\}$$

Fig. 1.6. Notation for the coalescent simulation algorithm.

For a sample of size n, we begin with $V_0 = \{1, 2, \dots n\}$ and $T_n = 0$.
For $k = 0, 1, \dots n - 2$ do

- Pick two numbers i_k and j_k from V_k.
- Let $V_{k+1} = V_k - \{i_k, j_k\} \cup \{n + k + 1\}$.
- In the tree connect $i_k \to n + k + 1$ and $j_k \to n + k + 1$.
- Let t_{n-k} be an independent exponential with mean $\binom{n-k}{2}^{-1}$.
- Let $T_{n-k-1} = T_{n-k} + t_{n-k}$.

To implement this in a computer, one can let $t_{n-k} = \binom{n-k}{2}^{-1} \log(1/U_k)$, where the U_k are independent uniform$(0, 1)$. From the construction it should be clear that the sequence of coalescence events is independent of the sequence of times of interevent times $t_n, \dots t_2$.

In the next two sections, we will introduce mutations. To do this in the computer, it is convenient to define the ancestor function in the third step of the algorithm above so that $\mathrm{anc}[i_k] = n + k + 1$ and $\mathrm{anc}[j_k] = n + k + 1$. For example, $\mathrm{anc}[2] = 7$ and $\mathrm{anc}[5] = 7$. One can then label the branches by the smaller number $1 \le i \le 2n - 2$ on the lower end and, if mutations occur at rate μ per generation and $\theta = 4N\mu$, introduce a Poisson number of mutations on branch i with mean

$$\frac{\theta}{2} \cdot (T_{\mathrm{anc}[i]} - T_i)$$

The reason for the definition of θ will become clear in Section 1.4. Before we move on, the reader should note that we first generate the genealogy and then introduce mutations.

1.2.2 Shape of the genealogical tree

The state of the coalescent at any time can then be represented as a *partition*, $A_1, \ldots A_m$, of $\{1, 2, \ldots n\}$. That is, $\cup_{i=1}^m A_i = \{1, 2, \ldots n\}$, and if $i \neq j$ the sets A_i and A_j are disjoint. In words, each A_i consists of one subset of lineages that have coalesced. To explain this notion, we will use the example that appears in the two previous figures. In this case, as we work backwards in time, the partitions are

$$
\begin{array}{cl}
T_1 & \{1, 2, 3, 4, 5\} \\
T_2 & \{1, 3, 4\} \quad \{2, 5\} \\
T_3 & \{1, 4\} \quad \{2, 5\} \quad \{3\} \\
T_4 & \{1, 4\} \quad \{2\} \quad \{3\} \quad \{5\} \\
\text{time } 0 & \{1\} \quad \{2\} \quad \{3\} \quad \{4\} \quad \{5\}
\end{array}
$$

Initially, the partition consists of five singletons since there has been no coalescence. After 1 and 4 coalesce at time T_4, they appear in the same set. Then 2 and 5 coalesce at time T_3, etc. Finally, at time T_1 we end up with all the labels in one set.

Let $\mathcal{E}_n$ be the collection of partitions of $\{1, 2, \ldots n\}$. If $\xi \in \mathcal{E}_n$, let $|\xi|$ be the number of sets that make up ξ, i.e., the number of lineages that remain in the coalescent. If, for example, $\xi = \{\{1, 4\}, \{2, 5\}, \{3\}\}$, then $|\xi| = 3$. Let ξ_i^n, $i = n, n-1, \ldots 1$ be the partition of $\{1, 2, \ldots n\}$ at time T_i, the first time there are i lineages. Kingman (1982a) has shown

Theorem 1.5. *If ξ is a partition of $\{1, 2, \ldots n\}$ with $|\xi| = i$, then*

$$
P(\xi_i^n = \xi) = c_{n,i} \, w(\xi)
$$

Here the weight $w(\xi) = \lambda_1! \cdots \lambda_i!$, where $\lambda_1, \ldots \lambda_i$ are the sizes of the i sets in the partition and the constant

$$
c_{n,i} = \frac{i!}{n!} \cdot \frac{(n-i)!(i-1)!}{(n-1)!}
$$

is chosen to make the sum of the probabilities equal to 1.

The proof of Theorem 1.6 will give some insight into the form of the constant. The weights $w(\xi)$ favor partitions that are uneven. For example, if $n = 9$ and $i = 3$, the weights based on the sizes of the sets in the partition are as follows:

3-3-3	4-3-2	5-2-2	4-4-1	5-3-1	6-2-1	7-1-1
216	288	480	576	720	1440	5040

Proof. We proceed by induction working backwards from $i = n$. When $i = n$, the partition is always $\{1\}, \ldots \{n\}$, all the $\lambda_i = 1$, and $c_{n,n} = 1$ (by definition, $0! = 1$). To begin the induction step now, write $\xi < \eta$ (and say ξ is finer than η) if $|\xi| = |\eta| + 1$ and η is obtained by combining two of the sets in ξ. For example, we might have

$$\xi = \{\{1,4\}, \{2,5\}, \{3\}\} \quad \text{and} \quad \eta = \{\{1,3,4\}, \{2,5\}\}$$

When $\xi < \eta$ and $|\xi| = i$, there is exactly one of the $\binom{i}{2}$ coalescence events that will turn ξ into η, so

$$P(\xi_{i-1}^n = \eta | \xi_i^n = \xi) = \begin{cases} \frac{2}{i(i-1)} & \text{if } \xi < \eta \\ 0 & \text{otherwise} \end{cases} \tag{1.5}$$

and we have

$$P(\xi_{i-1}^n = \eta) = \frac{2}{i(i-1)} \sum_{\xi < \eta} P(\xi_i^n = \xi) \tag{1.6}$$

If $\lambda_1, \ldots \lambda_{i-1}$ are the sizes of the sets in η, then for some ℓ with $1 \le \ell \le i-1$ and some ν with $1 \le \nu < \lambda_\ell$, the sets in ξ have sizes

$$\lambda_1, \ldots \lambda_{\ell-1}, \nu, \lambda_\ell - \nu, \lambda_{\ell+1}, \ldots \lambda_{i-1}$$

Using the induction hypothesis, the right-hand side of (1.6) is

$$= \frac{2}{i(i-1)} \sum_{\ell=1}^{i-1} \sum_{\nu=1}^{\lambda_\ell - 1} c_{n,i} \, w_{\ell,\nu} \binom{\lambda_\ell}{\nu} \cdot \frac{1}{2}$$

where the weight

$$w_{\ell,\nu} = \lambda_1! \cdots \lambda_{\ell-1}! \, \nu! \, (\lambda_\ell - \nu)! \, \lambda_{\ell+1}! \cdots \lambda_{i-1}!$$

and $\binom{\lambda_\ell}{\nu} \cdot \frac{1}{2}$ gives the number of ways of picking $\xi < \eta$ with the ℓth set in η subdivided into two pieces of size ν and $\lambda_\ell - \nu$. (We pick ν of the elements to form a new set but realize that we will generate the same choice again when we pick the $\lambda_\ell - \nu$ members of the complement.)

It is easy to see that $w_{\ell,\nu}\binom{\lambda_\ell}{\nu} = w(\eta)$ so the sum above is

$$= w(\eta) \frac{c_{n,i}}{i(i-1)} \sum_{\ell=1}^{i-1} \sum_{\nu=1}^{\lambda_\ell - 1} 1$$

The double sum $= \sum_{\ell=1}^{i-1} (\lambda_\ell - 1) = n - (i-1)$. The last detail to check is that

$$\frac{c_{n,i}}{i(i-1)} \cdot (n-i+1) = c_{n,i-1} \quad \text{or} \quad \frac{c_{n,i}}{c_{n,i-1}} = \frac{i(i-1)}{n-i+1} \tag{1.7}$$

but this is clear from the definition. $\qquad \square$

To write the partition ξ_i^n, it is natural, as we have done in the example above, to order the sets so that $\xi_{i,1}^n$ is the set containing 1, $\xi_{i,2}^n$ contains the smallest number not in $\xi_{i,1}^n$, etc. However, to compute the distribution of the sizes of sets in the coalescent, it is useful to put the sets in the partition into a randomly chosen order.

Theorem 1.6. *Let π be a randomly chosen permutation of $\{1, 2, \ldots i\}$ and let $\lambda_j = |\xi_{i,\pi(j)}^n|$ be the size of the jth set in ξ_i^n when they are rearranged according to π. $(\lambda_1, \lambda_2, \ldots \lambda_i)$ is uniformly distributed over the vectors of positive integers that add to n.*

Tajima (1983) proved this in the case $i = 2$. In words, if we pick one of the two sets in ξ_2^n at random, its size is uniformly distributed on $1, 2, \ldots n - 1$.

Proof. If we randomly order the sets in ξ, then each ordered arrangement has probability $c_{n,i}w(\xi)/i!$. If we only retain information about the sizes, then by considering the number of collections of sets that can give rise to the vector $(\lambda_1, \ldots \lambda_i)$, we see that it has probability:

$$\frac{c_{n,i}w(\xi)}{i!} \cdot \frac{n!}{\lambda_1!\lambda_2!\cdots\lambda_i!} = \frac{(n-i)!(i-1)!}{(n-1)!} = 1 \bigg/ \binom{n-1}{i-1}$$

Since the final quantity depends only on n and i and not on the actual vector, we have shown that the distribution is uniform. To see that the denominator of the last fraction gives the number of vectors of positive integers of length i that add up to n, imagine n balls separated into i groups by $i - 1$ pieces of cardboard. For example, if $n = 10$ and $i = 4$, we might have

$$O\,O\,O|O|O\,O\,O\,O|O\,O$$

Our $i - 1$ pieces of cardboard can go in any of the $n - 1$ spaces between balls, so there are $\binom{n-1}{i-1}$ possible vectors $(j_1, \ldots, j_i)$ of positive integers that add up to n. $\square$

As a consequence of Theorem 1.6, we get the following amusing fact.

Theorem 1.7. *The probability that the most recent common ancestor of a sample of size n is the same as that of the population converges to $(n-1)/(n+1)$ as the population size tends to ∞.*

When $n = 10$, this is $9/11$.

Proof. Consider the first split in the coalescent tree of the population. Let X be the limiting proportion of lineages in the left half of the tree, and recall that X is uniformly distributed on $(0, 1)$. In order for the MRCA of the sample to come before that of the population either all of the n lineages must be in the left half or all n in the right half. Thus, the probability the MRCAs coincide is

$$1 - \int_0^1 x^n + (1-x)^n \, dx = 1 - \frac{2}{n+1}$$

and this gives the desired result. $\square$

Our final result in this section gives a dynamic look at the coalescent process running backwards.

Theorem 1.8. *To construct the partition ξ_i^n from $\xi_{i-1}^n = \{A_1, \ldots A_{i-1}\}$, we pick a set at random, picking A_j with probability $(\lambda_j - 1)/(n - i - 1)$, where $\lambda_j = |A_j|$, and then split A_j into two sets with sizes k and $\lambda_j - k$, where k is uniform on $1, 2, \ldots \lambda_j - 1$.*

Proof. By elementary properties of conditional probability,

$$P(\xi_i^n = \xi | \xi_{i-1}^n = \eta) = \frac{P(\xi_{i-1}^n = \eta | \xi_i^n = \xi) P(\xi_i^n = \xi)}{P(\xi_{i-1}^n = \eta)}$$

Suppose $\xi < \eta$ are partitions of the correct sizes, and ξ is obtained from η by splitting A_j into two sets with sizes k and $\lambda_j - k$. It follows from Theorem 1.5 and (1.5) that

$$\frac{P(\xi_i^n = \xi)}{P(\xi_{i-1}^n = \eta)} = \frac{c_{n,i}}{c_{n,i-1}} \cdot \frac{k!(\lambda_j - k)!}{\lambda_j!} = \frac{i(i-1)}{n-i+1} \cdot \frac{k!(\lambda_j - k)!}{\lambda_j!}$$

Using (1.7), now we have

$$P(\xi_i^n = \xi | \xi_{i-1}^n = \eta) = \frac{1}{n-i+1} \cdot \frac{k!(\lambda_j - k)!}{\lambda_j!} \cdot 2$$

The first factor corresponds to picking A_j with probability $(\lambda_j - 1)/(n-i-1)$ and then picking k with probability $1/(\lambda_j - 1)$. Getting the correct division of A_j to produce ξ has probability $1/\binom{\lambda_j}{k}$. The final 2 takes into account the fact that we can also generate ξ by choosing $\lambda_j - k$ instead of k. □

1.3 Infinite alleles model

In this section, we will consider the *infinite alleles model*. As the name should suggest, we assume that there are so many alleles that each mutation is always to a new type never seen before. To explain the reason for this assumption, Kimura (1971) argued that if a gene consists of 500 nucleotides, the number of possible DNA sequences is

$$4^{500} = 10^{500 \log 4 / \log 10} = 10^{301}$$

For any of these, there are $3 \cdot 500 = 1500$ sequences that can be reached by single base changes, so the chance of returning where one began in two mutations is $1/1500$ (assuming an equal probability for all replacements). Thus, the total number of possible alleles is essentially infinite.

The infinite alleles model arose at a time when one had to use indirect methods to infer diferences between individuals. For example, Coyne (1976) and Singh, Lewontin, and Felton (1976) studied *Drosophila* by performing electrophoresis under various conditions. Coyne (1976) found 23 alleles in 60 family lines at the xanthine dehydrogenase locus of *Drosophila persimilis* that displayed the following pattern, which we call the *allelic partition*:

$$a_1 = 18,\, a_2 = 3,\, a_4 = 1,\, a_{32} = 1$$

That is, there were 18 unique alleles, 3 alleles had 2 representatives, 1 had 4, and 1 had 32. Singh, Lewontin, and Felton (1976) found 27 alleles in 146 genes from the xanthine dehydrogenase locus of *D. pseudoobscura* with the following pattern:

$$a_1 = 20,\, a_2 = 3,\, a_3 = 7,\, a_5 = 2,\, a_6 = 2,\, a_8 = 1,\, a_{11} = 1,\, a_{68} = 1$$

The infinite alleles model is also relevant to DNA sequence data when there is no recombination. Underhill et al. (1997) studied 718 Y chromosomes. They found 22 nucleotides that were *polymorphic* (i.e., not the same in all of the individuals). The sequence of nucleotides at these variable positions gives the *haplotype* of the individual. In the sample, there were 20 distinct haplotypes. The sequences can be arranged in a tree in which no mutation occurs more than once, so it is reasonable to assume that the haplotypes follow the infinite alleles model. The allelic partition has

$$a_1 = 7,\, a_2 = a_3 = a_5 = a_6 = a_8 = a_9 = a_{26} = a_{36} = a_{37} = 1,$$
$$a_{82} = 2,\, a_{149} = 1,\, a_{266} = 1.$$

After looking at the data, the first obvious question is: What do we expect to see? The answer to this question is given by *Ewens' sampling formula*. This section is devoted to the derivation of the formula and the description of several perspectives from which one can approach it. At the end of this section, we will lapse into a mathematical daydream about the structure of a randomly chosen permutation.

1.3.1 Hoppe's urn, Ewens' sampling formula

The genealogical process associated with the infinite alleles version of the Wright-Fisher model is a coalescent with killing. When there are k lineages, coalescence and mutation occur on each step with probability

$$\frac{k(k-1)}{2} \cdot \frac{1}{2N}$$

as before, but now killing of one of the lineages occurs with probability $k\mu$ because if a mutation is encountered, we know the genetic state of that individual and all of its descendants in the sample. Speeding up the system by running time at rate $2N$, the rates become $k(k-1)/2$ and $k\theta/2$, where $\theta = 4N\mu$.

Turning the coalescent with killing around backwards leads to *Hoppe's (1984) urn model.* This urn contains a black ball with mass θ and various colored balls with mass 1. At each time, a ball is selected at random with a probability proportional to its mass. If a colored ball is drawn, that ball and another of its color are returned to the urn. If the black ball is chosen, it is returned to the urn with a ball of a new color that has mass 1. The choice of the black ball corresponds to a new mutation and the choice of a colored ball corresponds to a coalescence event. A simulation should help explain the definition. Here a black dot indicates that a new color was added at that time step.

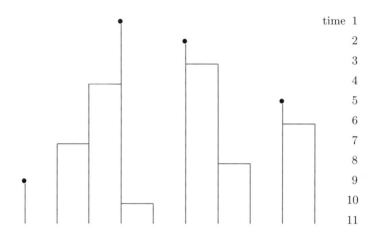

Fig. 1.7. A realization of Hoppe's urn.

As we go backwards from time $k+1$ to time k in Hoppe's urn, we encounter a mutation with probability $\theta/(\theta+k)$ and have a coalescence with probability $k/(\theta+k)$. Since in the coalescent there are $k+1$ lineages that are each exposed to mutations at total rate $k\theta/2$ and collisions occur at rate $(k+1)k/2$, this is the correct ratio. Since by symmetry all of the coalescence events have equal probability, it follows that

Theorem 1.9. *The genealogical relationship between k lineages in the coalescent with killing can be simulated by running Hoppe's urn for k time steps.*

This observation is useful in computing properties of population samples under the infinite alleles model. To illustrate this, let K_n be the random variable that counts the number of different alleles found in a sample of size n. Here and throughout the book, log is the "natural logarithm" with base e.

Theorem 1.10 (Watterson (1975)). *For fixed θ, as the sample size $n \to \infty$,*

$$EK_n \sim \theta \log n \quad and \quad var\,(K_n) \sim \theta \log n$$

where $a_n \sim b_n$ means that $a_n/b_n \to 1$ as $n \to \infty$. In addition, the central limit theorem holds. That is, if χ has the standard normal distribution, then

$$P\left(\frac{K_n - EK_n}{\sqrt{var\,(K_n)}} \leq x\right) \to P(\chi \leq x)$$

Proof. Let $\eta_i = 1$ if the ith ball added to Hoppe's urn is a new type and 0 otherwise. It is clear from the definition of the urn scheme that $K_n = \eta_1 + \cdots + \eta_n$ and $\eta_1, \ldots \eta_n$ are independent with

$$P(\eta_i = 1) = \theta/(\theta + i - 1) \tag{1.8}$$

To compute the asymptotic behavior of EK_n, we note that (1.8) implies

$$EK_n = \sum_{i=1}^{n} \frac{\theta}{\theta + i - 1} \tag{1.9}$$

Viewing the right-hand side as a Riemann sum approximating an integral,

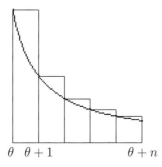

it follows that

$$\sum_{i=1}^{n} \frac{1}{\theta + i - 1} \sim \int_{\theta}^{n+\theta} \frac{1}{x}\,dx = \log(n + \theta) - \log(\theta) \sim \log n \tag{1.10}$$

From this, the first result follows. To prove the second, we note that

$$var\,(K_n) = \sum_{i=1}^{n} var\,(\eta_i) = \sum_{i=2}^{n} \frac{\theta(i - 1)}{(\theta + i - 1)^2} \tag{1.11}$$

As $i \to \infty$, $(i - 1)/(\theta + i - 1) \to 1$, so

$$\text{var}\,(K_n) \sim \sum_{i=2}^{n} \frac{\theta}{\theta + i - 1} \sim \theta \log n$$

by the reasoning for EK_n. Since the η_i are independent, the final claim follows from the triangular array form of the central limit theorem. See, for example, (4.5) in Chapter 2 of Durrett (2005). □

An immediate consequence of Theorem 1.10 is that $K_n / \log n$ is an asymptotically normal estimator of the scaled mutation rate θ. However, the asymptotic standard deviation of $K_n / \log n$ is quite large, namely of order $1/\sqrt{\log n}$. Thus, if the true $\theta = 1$ and we want to estimate θ with a standard error of 0.1, a sample of size e^{100} is required. Given this depressingly slow rate of convergence, it is natural to ask if there is another way to estimate θ from the data. The answer is NO, however. As we will see in Theorem 1.13 below, K_n is a sufficient statistic. That is, it contains all the information in the sample that is useful for estimating θ.

The last result describes the asymptotic behavior of the number of alleles. The next one, due to Ewens (1972), deals with the entire distribution of the sample under the infinite alleles model.

Theorem 1.11 (Ewens' sampling formula). *Let a_i be the number of alleles present i times in a sample of size n. When the scaled mutation rate is $\theta = 4N\mu$,*

$$P_{\theta,n}(a_1, \ldots a_n) = \frac{n!}{\theta_{(n)}} \prod_{j=1}^{n} \frac{(\theta/j)^{a_j}}{a_j!}$$

where $\theta_{(n)} = \theta(\theta + 1) \cdots (\theta + n - 1)$.

The formula may look strange at first, but it becomes more familiar if we rewrite it as

$$c_{\theta,n} \prod_{j=1}^{n} e^{-\theta/j} \frac{(\theta/j)^{a_j}}{a_j!}$$

where $c_{\theta,n}$ is a constant that depends on θ and n and guarantees the sum of the probabilities is 1. In words, if we let $Y_1, \ldots Y_n$ be independent Poisson random variables with means $EY_j = \theta/j$, then the allelic partition $(a_1, a_2, \ldots a_n)$ has the same distribution as

$$\left(Y_1, Y_2, \ldots, Y_n \,\middle|\, \sum_m m Y_m = n \right)$$

One explanation of this can be found in Theorem 1.19.

Proof. In view of Theorem 1.9, it suffices to show that the distribution of the colors in Hoppe's urn at time n is given by Ewens' sampling formula. We proceed by induction. When $n = 1$, the partition $a_1 = 1$ has probability 1 so the result is true. Suppose that the state at time n is $a = (a_1, \ldots, a_n)$ and let $\bar{a} = (\bar{a}_1, \ldots, \bar{a}_{n-1})$ be the state at the previous time. There are two cases to consider.

(i) $\bar{a}_1 = a_1 - 1$, i.e., a new color was just added. In this case, the transition probability for Hoppe's urn is

$$p(\bar{a}, a) = \frac{\theta}{\theta + n - 1}$$

and the ratio of the probabilities in Ewens' sampling formula is

$$\frac{P_{\theta,n}(a)}{P_{\theta,n-1}(\bar{a})} = \frac{n}{\theta + n - 1} \cdot \frac{\theta}{a_1}$$

(ii) For some $1 \le j < n$, we have $\bar{a}_j = a_j + 1$, $\bar{a}_{j+1} = a_{j+1} - 1$, i.e., an existing color with j representatives was chosen and increased to size $j + 1$. In this case, the transition probability is

$$p(\bar{a}, a) = \frac{j \bar{a}_j}{\theta + n - 1}$$

and the ratio of the probabilities in Ewens' sampling formula is

$$\frac{P_{\theta,n}(a)}{P_{\theta,n-1}(\bar{a})} = \frac{n}{\theta + n - 1} \cdot \frac{j \bar{a}_j}{(j + 1) a_{j+1}}$$

To complete the proof now, we observe

$$\sum_{\bar{a}} \frac{P_{\theta,n-1}(\bar{a})}{P_{\theta,n}(a)} p(\bar{a}, a) = \frac{\theta}{\theta + n - 1} \cdot \frac{\theta + n - 1}{n} \cdot \frac{a_1}{\theta}$$

$$+ \sum_{j=1}^{n-1} \frac{j \bar{a}_j}{\theta + n - 1} \cdot \frac{\theta + n - 1}{n} \cdot \frac{(j + 1) a_{j+1}}{j \bar{a}_j}$$

Cancelling on the right-hand side, we have

$$= \frac{a_1}{n} + \sum_{j=1}^{n-1} \frac{(j + 1) a_{j+1}}{n} = 1$$

since $\sum_k k a_k = n$. Rearranging, we have

$$\sum_{\bar{a}} P_{\theta,n-1}(\bar{a}) p(\bar{a}, a) = P_{\theta,n}(a)$$

Since the distribution of Hoppe's urn also satisfies this recursion with the same initial condition, the two must be equal.

Small sample sizes

To get a feeling for what Theorem 1.11 says, we will consider some small values of n.

$n = 2$. The factor in front is $2/\theta(\theta + 1)$. There are two possible partitions: $(a_1, a_2) = (0, 1)$ or $(2, 0)$. The products in these cases are $(\theta/2)^1/1!$ and $(\theta/1)^2/2!$, so the probabilities of the two partitions are $1/(\theta+1)$ and $\theta/(\theta+1)$. From this it follows that the probability that two randomly chosen individual are identical (also known as the *homozygosity*) is $1/(\theta + 1)$. This is easy to see directly. Two lineages coalesce with probability $1/2N$ per generation and experience mutation with probability $2u$, so the probability of coalescence before mutation is

$$\frac{1/2N}{2u + 1/2N} = \frac{1}{1 + \theta} \tag{1.12}$$

$n = 3$. The factor in front is $6/D$ where $D = \theta(\theta + 1)(\theta + 2)$. There are three possible partitions. The next table gives the value of the product in this case and the probability.

(a_1, a_2, a_3)	product	probability
$(0, 0, 1)$	$\theta/3$	$2\theta/D$
$(1, 1, 0)$	$(\theta/1)(\theta/2)$	$3\theta^2/D$
$(3, 0, 0)$	$\theta^3/3!$	θ^3/D

The probabilities add to 1, since $D = \theta^3 + 3\theta^2 + 2\theta$.

$n = 4$. The factor in front is $24/D$, where $D = \theta(\theta + 1)(\theta + 2)(\theta + 3)$. There are five possible partitions, listed below so that the largest set is decreasing.

(a_1, a_2, a_3, a_4)	product	probability
$(0, 0, 0, 1)$	$\theta/4$	$6\theta/D$
$(1, 0, 1, 0)$	$(\theta/1)(\theta/3)$	$8\theta^2/D$
$(0, 2, 0, 0)$	$(\theta/2)^2/2$	$3\theta^2/D$
$(2, 1, 0, 0)$	$(\theta/1)^2(\theta/2)/2$	$6\theta^3/D$
$(4, 0, 0, 0)$	$\theta^4/4!$	θ^4/D

The probabilities add to 1, since

$$D = (\theta^3 + 3\theta^2 + 2\theta)(\theta + 3) = \theta^4 + 6\theta^3 + 11\theta^2 + 6\theta$$

1.3.2 Chinese restaurants and sufficient statistics

Joyce and Tavaré (1987) enriched Hoppe's urn by adding bookkeeping to keep track of the history of choices in the process as a permutation written in its cycle decomposition. To explain, consider the following permutation of eight objects, which mathematically is a mapping π from $\{1, 2, \ldots 8\}$ onto $\{1, 2, \ldots 8\}$.

$$i \quad 1\ 2\ 3\ 4\ 5\ 6\ 7\ 8$$
$$\pi(i)\ 7\ 8\ 2\ 1\ 6\ 5\ 4\ 3$$

To construct the cycle decomposition of the permutation, we note that

$$1 \to 7 \to 4 \to 1 \qquad 2 \to 8 \to 3 \to 2 \qquad 5 \to 6 \to 5$$

and write $(174)(283)(56)$. To reconstruct the permutation from the short form, each integer is mapped to the one to its right, except for the one adjacent to the right parenthesis, which is mapped to the one at the beginning of the cycle. Because of this the same permutation can be written as $(741)(832)(65)$.

Numbering the balls according to the order that they enter the urn, the rules are simple. A new color starts a new cycle. If ball k's color was determined by choosing ball j, it is inserted to the left of j. For an example, consider the first eight events in the realization from the beginning of this section.

(1)	1 is always a new color
(1)(2)	2 is a new color
(1)(32)	3 is a child of 2
(41)(32)	4 is a child of 1
(41)(32)(5)	5 is a new color
(41)(32)(65)	6 is a child of 5
(741)(32)(65)	7 is a child of 4
(741)(832)(65)	8 is a child of 3

This scheme is also known as the "Chinese restaurant process"; see Aldous (1985). In that formulation, one thinks of the numbers as successive arrivals to a restaurant and the groups as circular tables. The nth person sits at a new table with probability $\theta/(\theta + n - 1)$ and sits to the left of $1 \le j \le n - 1$ with probability $1/(\theta + n - 1)$. The cycles indicate who is sitting at each table and in what order.

Let Π_n be the permutation when there are n individuals. An important property of this representation is that given a permutation, the path to it is unique. Indeed, one can compute the path by successively deleting the largest number that remains in the permutation. This property is the key to the following result.

Theorem 1.12. *If π is a permutation with k cycles, then in the Joyce-Tavaré process,*

$$P_\theta(\Pi_n = \pi) = \frac{\theta^k}{\theta_{(n)}}$$

When $\theta = 1$, the right-hand side is $1/n!$, i.e., all permutations are equally likely.

Proof. In our example, $P_\theta\{\Pi_8 = (741)(832)(65)\}$ is

$$\frac{\theta}{\theta} \cdot \frac{\theta}{\theta+1} \cdot \frac{1}{\theta+2} \cdot \frac{1}{\theta+3} \cdot \frac{\theta}{\theta+4} \cdot \frac{1}{\theta+5} \cdot \frac{1}{\theta+6} \cdot \frac{1}{\theta+7}$$

where the θ's in the numerators of the first, second, and fifth fractions are due to individuals sitting at a new table at those steps and the 1's in the other numerators come from the fact that the other individuals have exactly one place they can sit. Generalizing from the example, we see that if a permutation of $\{1, 2, \ldots n\}$ has k cycles, then the numerator is always θ^k and the denominator is $\theta_{(n)}$.   □

Comparing with Ewens' sampling formula in Theorem 1.11 and noting that $k = \sum a_j$ is the number of cycles, we see that the number of permutations of $\{1, 2, \ldots n\}$ with a_1 cycles of size 1, a_2 cycles of size 2, etc., is

$$\frac{n!}{\prod_{j=1}^{n}(j^{a_j}(a_j)!)}$$

Let S_n^k be the set of permutations of $\{1, 2, \ldots n\}$ with exactly k cycles, and let $|S_n^k|$ be the number of elements of S_n^k. Letting K_n be the number of alleles in a sample of size n, it follows from Theorem 1.12 that

$$P_\theta(K_n = k) = \frac{\theta^k}{\theta_{(n)}} \cdot |S_n^k| \tag{1.13}$$

The $|S_n^k|$ are called the *Stirling numbers of the first kind*. In order to compute them, it is enough to know that they satisfy the relationship

$$|S_n^k| = (n-1)|S_{n-1}^k| + |S_{n-1}^{k-1}|$$

In words, we can construct a $\pi \in S_n^k$ from a member of S_{n-1}^{k-1} by adding (n) as a new cycle, or from a $\sigma \in S_{n-1}^k$ by picking an integer $1 \le j \le n-1$ and setting $\pi(j) = n$, $\pi(n) = \sigma(j)$.

Combining (1.13) with Ewens' sampling formula in Theorem 1.11, we see that

$$P_\theta(a_1, \ldots, a_n | K_n = k) = \frac{n!}{|S_n^k|} \prod_{j=1}^{n} \left(\frac{1}{j}\right)^{a_j} \frac{1}{a_j!} \tag{1.14}$$

Since the conditional distribution of the allelic partition $(a_1, a_2, \ldots a_n)$ does not depend on θ,

Theorem 1.13. K_n *is a sufficient statistic for estimating* θ.

To develop an estimate of θ based on K_n, we will use maximum-likelihood estimation. Let

$$L_n(\theta, k) = \frac{\theta^k}{\theta_{(n)}} \cdot |S_n^k|$$

be the likelihood of observing $K_n = k$ when the true parameter is θ. The maximum-likelihood estimator finds the value of θ that maximizes the probability of the observed value of k. To find the maximum value, we compute

$$\frac{\partial}{\partial\theta} L_n(\theta, k) = |S_n^k| \frac{k\theta^{k-1}\theta_{(n)} - \theta^k \theta'_{(n)}}{(\theta_{(n)})^2} = \frac{\theta^k |S_n^k|}{\theta_{(n)}} \cdot \left[\frac{k}{\theta} - \frac{\theta'_{(n)}}{\theta_{(n)}}\right]$$

Setting the right-hand side equal to 0 and solving for k yields

$$k = \theta \frac{\theta'_{(n)}}{\theta_{(n)}} = \theta \frac{d}{d\theta} \log \theta_{(n)}$$

Recalling $\theta_{(n)} = \theta(\theta + 1) \cdots (\theta + n - 1)$, we need to compute

$$\frac{d}{d\theta} \sum_{i=1}^{n} \log(\theta + i - 1) = \sum_{i=1}^{n} \frac{1}{\theta + i - 1}$$

and thus

$$k = \frac{\theta}{\theta} + \frac{\theta}{\theta + 1} + \cdots + \frac{\theta}{\theta + n - 1} = EK_n$$

In words, the maximum-likelihood estimator $\hat{\theta}_{MLE}$ is the θ that makes the mean number of alleles equal to the observed number.

The theory of maximum-likelihood estimation tells us that asymptotically $E\hat{\theta}_{MLE} = \theta$ and $\text{var}(\hat{\theta}_{MLE}) = 1/I(\hat{\theta}_{MLE})$, where $I(\theta)$ is the Fisher information

$$I(\theta) = E\left(\frac{\partial}{\partial\theta} \log L_n(\theta, k)\right)^2$$

This is easy to compute.

$$\frac{\partial}{\partial\theta} \log L_n(\theta, k) = \frac{\partial}{\partial\theta} \log \frac{\theta^k |S_n^k|}{\theta_{(n)}} = \frac{k}{\theta} - \frac{\partial}{\partial\theta} \log \theta_{(n)}$$

$$= \frac{k}{\theta} - \sum_{i=1}^{n} \frac{1}{\theta + i - 1} = \frac{1}{\theta}\left(k - \sum_{i=1}^{n} \frac{\theta}{\theta + i - 1}\right)$$

Since $EK_n = \sum_{i=1}^{n} \frac{\theta}{\theta+i-1}$, it follows that

$$I(\theta) = \frac{1}{\theta^2} \text{var}(K_n)$$

Theorem 1.10 implies that $\text{var}(K_n) \sim \theta \log n$, so $\text{var}(\hat{\theta}_{MLE}) \to 0$ as the sample size $n \to \infty$ but rather slowly.

Sample homozygosity

Letting $\delta_{ij} = 1$ if individuals i and j have the same allele and 0 otherwise,

$$\hat{F}_n = \binom{n}{2}^{-1} \sum_{i<j} \delta_{ij}$$

is the *sample homozygosity*. It is a random variable that gives the probability that in the given sample two randomly chosen members are identical. As (1.12) shows,

$$E\hat{F}_n = \frac{1}{1+\theta}$$

We will now show

$$\begin{aligned}
\text{var}\left(\hat{F}_n\right) = \binom{n}{2}^{-1} \Big(&\frac{\theta}{(1+\theta)^2} + 2(n-2) \cdot \frac{\theta}{(1+\theta)^2(2+\theta)} \\
&+ \binom{n-2}{2} \cdot \frac{2\theta}{(1+\theta)^2(2+\theta)(3+\theta)} \Big)
\end{aligned} \qquad (1.15)$$

Since $\binom{n-2}{2} \big/ \binom{n}{2} \to 1$ as $n \to \infty$, it follows that

$$\text{var}\left(\hat{F}_n\right) \to \frac{2\theta}{(1+\theta)^2(2+\theta)(3+\theta)}$$

as found by Stewart (1976). In principle, one could estimate θ by

$$\bar{\theta} = \frac{1}{\hat{F}_n} - 1$$

but (i) the last calculation shows that $\bar{\theta} \not\to \theta$ as $n \to \infty$, and (ii) $1/x$ is convex so Jensen's inequality implies that

$$E\bar{\theta} > \frac{1}{E\hat{F}_n} - 1 = \theta$$

i.e., the estimator is biased.

Proof. To compute the variance of $\hat{F}_n$, we begin by observing that

$$\hat{F}_n^2 = \binom{n}{2}^{-2} \sum_{\{i_1,j_1\}} \sum_{\{i_2,j_2\}} \delta_{i_1,j_1} \delta_{i_2,j_2}$$

where the sum is over all subsets of size 2. There are three types of terms:

(i) $|\{i_1, j_1, i_2, j_2\}| = 2$. Since $i_1 = i_2$, $j_1 = j_2$, there are $\binom{n}{2}$ of these.

(ii) $|\{i_1, j_1, i_2, j_2\}| = 3$. There are $\binom{n}{2} 2(n-2)$ of these.

(iii) $\{i_1, j_1\} \cap \{i_2, j_2\} = \emptyset$. There are $\binom{n}{2}\binom{n-2}{2}$ of these.

To check the combinatorics, note that

$$1 + 2(n-2) + \frac{(n-2)(n-3)}{2} = \frac{2 + 4n - 8 + n^2 - 5n + 6}{2} = \frac{n^2 - n}{2} = \binom{n}{2}$$

Sorting out the three types of terms

$$E(\hat{F}_n^2) = \binom{n}{2}^{-1}\left(E(\delta_{ij}^2) + 2(n-2)E(\delta_{ij}\delta_{ik}) + \binom{n-2}{2}E(\delta_{ij}\delta_{k\ell})\right) \quad (1.16)$$

where the indices i, j, k, ℓ are different. Since δ_{ij} only takes the values 1 and 0, $E(\delta_{ij}^2) = E(\delta_{ij}) = 1/(1+\theta)$. $\delta_{ij}\delta_{ik} = 1$ only if i, j, k are identical, i.e., the allelic partition $a_3 = 1$, which by our discussion of small sample sizes after Theorem 1.11 is $2/(\theta+1)(\theta+2)$. $\delta_{ij}\delta_{k\ell} = 1$ can only happen if the partition is $a_4 = 1$ or $a_2 = 2$. It always occurs in the first case and with probability $1/3$ in the second, so by the small sample size discussion,

$$E(\delta_{ij}\delta_{k\ell}) = \frac{6\theta + \theta^2}{\theta(\theta+1)(\theta+2)(\theta+3)}$$

Combining the results in the previous paragraph

$$E(\hat{F}_n^2) = \binom{n}{2}^{-1}\left(\frac{1}{1+\theta} + \frac{2(n-2)\cdot 2}{(1+\theta)(2+\theta)}\right. \quad (1.17)$$

$$\left. + \binom{n-2}{2}\cdot\frac{(6+\theta)}{(1+\theta)(2+\theta)(3+\theta)}\right) \quad (1.18)$$

Subtracting

$$\frac{1}{(1+\theta)^2} = \binom{n}{2}^{-1}\left(1 + 2(n-2) + \binom{n-2}{2}\right)\frac{1}{(1+\theta)^2}$$

gives the desired result.   □

Testing the infinite alleles model

Returning to the motivating question from the beginning of the section, we would like to know if the observations are consistent with Ewens' sampling formula. Since the conditional distribution of the allelic partition given the number of alleles given in (1.14) does not depend on θ, one can use any function of the allele frequencies to test for departures from the infinite alleles model. Watterson (1977) suggested the use of the following slightly different version of the sample homozygosity:

$$F_n = \sum_{i=1}^{k} x_i^2 = \sum_{j=1}^{n} a_j\left(\frac{j}{n}\right)^2$$

where x_i is the frequency of the ith allele, and a_j is the number of alleles with j representatives. In words, F_n is the probability that two individuals chosen with replacement are the same, while the version $\hat{F}_n$ discussed above is the probability that two individuals chosen without replacement are the same.

For the Coyne data, $F_n = 0.2972$. For that of Singh, Lewontin, and Felton, $F_n = 0.2353$. To assess the significance of the value for the Coyne data, Watterson (1977) generated 1000 samples of size 60 with 23 alleles. No sample had a value of $\hat{F}$ larger than 0.2972. For the Singh, Lewontin, and Felton data, 2000 samples of size 146 with 27 alleles produced only 8 values larger than 0.2353. We may thus conclude that both data sets depart significantly from the neutral alleles distribution in the direction of excess homozygosity.

1.3.3 Branching process viewpoint

In addition to introducing the bookkeeping described above, Joyce and Tavaré (1987) related Hoppe's urn to the *Yule process with immigration*. In this process, immigrants enter the population at times of a Poisson process with rate θ, and each individual in the population follows the rules of the *Yule process*. That is, they never die and they give birth to new individuals at rate 1. If we only look at the process when the number of individuals increases, then we get a discrete-time process in which when there are k individuals, a new type is added with probability $\theta/(k + \theta)$ and a new individual with a type randomly chosen from the urn is added with probability $k/(k + \theta)$. From this description, it should be clear that

Theorem 1.14. *If each immigrant is a new type and offspring are the same type as their parents, then the sequence of states the branching process with immigration moves through has the same distribution as those generated by Hoppe's urn.*

This result becomes very useful when combined with the fact, see page 109 of Athreya and Ney (1972), that

Theorem 1.15. *Starting from a single individual, the number of individuals in the Yule process at time t has a geometric distribution with success probability $p = e^{-t}$.*

In this way, we obtain a simple proof of Theorem 1.6.

Theorem 1.16. *Consider the coalescent starting with ℓ lineages and stop when there are k. Let $J_1, \ldots J_k$ be the number of lineages in the k elements of the partition when they are labeled at random. $(J_1, \ldots J_k)$ is uniformly distributed over the vectors of positive integers that add up to ℓ, and hence*

$$P(J_i = m) = \binom{\ell - m - 1}{k - 2} \Big/ \binom{\ell - 1}{k - 1} \tag{1.19}$$

for $1 \le i \le k$ and $1 \le m \le \ell - k + 1$.

Proof. Let Z_t^i, $1 \leq i \leq k$, be independent copies of the Yule process. If $j_1, \ldots j_k$ are positive integers that add up to ℓ then by Lemma 1.15

$$P(Z_t^1 = j_1, \ldots Z_t^k = j_k) = (1 - p)^{\ell-k} p^k \quad \text{where} \quad p = e^{-t}$$

Since the right-hand side only depends on the sum ℓ and the number of terms k, all of the possible vectors have the same probability. As shown in the proof of Theorem 1.6, there are $\binom{\ell-1}{k-1}$ possible vectors $(j_1, \ldots, j_k)$ of positive integers that add up to ℓ. From this it follows that

$$P\left(Z_t^1 = j_1, \ldots Z_t^k = j_k \middle| \sum_{j=1}^{k} Z_t^j = \ell \right) = 1 \middle/ \binom{\ell-1}{k-1}$$

that is, the conditional distribution is uniform over the set of possible vectors. Since the number of vectors $(j_2, \ldots, j_k)$ of positive integers that add up to $\ell - m$ is $\binom{\ell-m-1}{k-2}$,

$$P\left(Z_t^1 = m \middle| \sum_{j=1}^{k} Z_t^j = \ell \right) = \binom{\ell-m-1}{k-2} \middle/ \binom{\ell-1}{k-1}$$

which completes the proof. □

The passage from the discrete-time urn model to the continuous-time branching process with immigration is useful because it makes the growth of the various families independent. This leads to some nice results about the asymptotic behavior of Ewens' sampling distribution when n is large. The proofs are somewhat more sophisticated than the others in this section and we will not use the results for applications, so if the discussion becomes confusing, the reader should feel free to move on to the next section. Let $s_j(n)$ be the size of the jth family when there are n individuals. Donnelly and Tavaré (1986) have shown in their Theorem 6.1 that

Theorem 1.17. *For $j = 1, 2, \ldots$,* $\lim_{n \to \infty} s_j(n)/n = P_j$ *with*

$$P_j = Z_j \prod_{i=1}^{j-1} (1 - Z_i)$$

where the Z_i are independent and have a beta$(1, \theta)$ density: $\theta(1 - z)^{\theta-1}$.

An immediate consequence of this result is

Theorem 1.18. *The sample homozygosity $\hat{F}_n$ converges in distribution to $\sum_{j=1}^{\infty} P_j^2$ as $n \to \infty$.*

Note that when $\theta = 1$, the Z_i are uniform. By our discussion of the Chinese restaurant process, Theorem 1.17 describes the "cycle structure" of a random permutation. To be precise, given a permutation, we start with 1 and follow the successive iterates $\pi(1)$, $\pi(\pi(1))$ until we return to 1. This is the first cycle. Now pick the smallest number not in the first cycle and repeat the procedure to construct the second cycle, etc. This gives us a size-biased look at the cycle structure, i.e., cycles are chosen with probability proportional to their length.

Sketch of proof of Theorem 1.17. We content ourselves to explain the main ideas, referring the reader to Tavaré (1987) for details. Let $X(t)$ be the number of individuals at time t in the Yule process. It follows from Theorem 1.15 that as $t \to \infty$, $X(t)/e^t$ converges to a limit we will denote by $\mathcal{E}$. From Lemma 1.15 it follows that $\mathcal{E}$ has an exponential distribution. Let $X_i(t)$ be the number of individuals in the ith family at time t. From the previous result, it follows that

$$e^{-t}(X_1(t), X_2(t), \ldots) \to (e^{-T_1}\mathcal{E}_1, e^{-T_2}\mathcal{E}_2, \ldots)$$

where $T_1, T_2, \ldots$ are the arrival times of the rate θ Poisson process and $\mathcal{E}_1, \mathcal{E}_2, \ldots$ are independent mean 1 exponentials. Let $I(t) = X_1(t) + X_2(t) + \cdots$ be the total number of individuals at time t. Leaving the boring details of justifying the interchange of sum and limit to the reader, we have

$$e^{-t}I(t) \to \sum_{i=1}^{\infty} e^{-T_i}\mathcal{E}_i$$

A little calculation shows that the sum has a gamma(θ,1) distribution. From the last two results, it follows that

$$\frac{X_1(t)}{I(t)} \to \frac{e^{-T_1}\mathcal{E}_1}{\sum_{i=1}^{\infty} e^{-T_i}\mathcal{E}_i} = \frac{\mathcal{E}_1}{\mathcal{E}_1 + \sum_{i=2}^{\infty} e^{-(T_i - T_1)}\mathcal{E}_i} \equiv Z_1$$

where the last three-lined equality ($\equiv$) indicates we are making a definition.

Writing $\overset{d}{=}$ to indicate that two things have the same distribution,

$$\sum_{i=2}^{\infty} e^{-(T_i - T_1)}\mathcal{E}_i \overset{d}{=} \sum_{i=1}^{\infty} e^{-T_i}\mathcal{E}_i$$

has a gamma(θ,1) distribution and is independent of $\mathcal{E}_1$, which has an exponential distribution, so the ratio has a beta($1, \theta$) distribution. To get the result for $i = 2$, we note that

$$\frac{X_2(t)}{I(t)} \to \frac{e^{-T_2}\mathcal{E}_2}{\sum_{i=1}^{\infty} e^{-T_i}\mathcal{E}_i}$$

$$= (1 - Z_1) \cdot \frac{e^{T_2}\mathcal{E}_2}{\sum_{i=2}^{\infty} e^{-T_i}\mathcal{E}_i} \equiv (1 - Z_1)Z_2$$

Similar algebra leads to the result for $i > 2$. $\qquad\qquad\qquad\square$

Theorem 1.17 gives the limiting distribution of the sizes of the partition when they are ordered by their ages. In words, the oldest part of the partition contains a fraction Z_1, the next a fraction Z_2 of what is left, etc. The amount that is not in the oldest j part of the partition is $\prod_{i=1}^{j}(1 - Z_i) \to 0$ exponentially fast, so for large n most of the sample can be found in the oldest parts of the partition. The ages of the alleles are, of course, not visible in the data, but this result translates immediately into a limit theorem for the sizes in decreasing order: $\hat{P}_1 \geq \hat{P}_2 \geq \hat{P}_3 \ldots$. The limit is called the Poisson Dirichlet distribution.

Somewhat surprisingly, it has a simple description. Let $y_1 > y_2 > y_3 \ldots$ be the points of a Poisson process with intensity $\theta e^{-x}/x$, i.e., the number of points in the interval (a, b) is Poisson with mean $\int_a^b \theta e^{-x}/x\, dx$ and the number of points in disjoint intervals are independent. $\hat{P}_1 \geq \hat{P}_2 \geq \hat{P}_3 \ldots$ has the same distribution as $y_1/y, y_2/y, y_3, \ldots$, where $y = \sum_{i=1}^{\infty} y_i$.

Theorem 1.17 concerns the large parts of the partition. At the other end of the size spectrum, Arratia, Barbour, and Tavaré (1992) have studied $A_j(n)$, the number of parts of size j, and showed

Theorem 1.19. *As $n \to \infty$, $(A_1(n), A_2(n), \ldots)$ converges to $(Y_1, Y_2, \ldots)$, where the Y_i are independent Poissons with mean θ/i.*

Proof (Sketch). Let $T_{\ell,m} = (\ell+1)Y_\ell + \cdots + mY_m$. By the remark after (3.5), if $z = y_1 + 2y_2 + \cdots + \ell y_\ell$, then

$$P(A_1(n) = y_1, \ldots A_\ell(n) = y_\ell) = \frac{P(Y_1 = y_1, \ldots Y_\ell = y_\ell)P(T_{\ell,n} = n - z)}{P(T_{0,n} = n)}$$

so it suffices to show that for each fixed z and ℓ,

$$P(T_{\ell,n} = n - a)/P(T_{0,n} = n) \to 1$$

When $\theta = 1$, this can be done using the local central limit theorem but for $\theta \neq 1$ this requires a simple large deviations estimate, see Arratia, Barbour, and Tavaré (1992) for details. $\square$

The book by Arratia, Barbour, and Tavaré (2003) contains a thorough account of these results.

1.4 Infinite sites model

The infinite alleles model arose in an era where the information about the genetic state of an individual was obtained through indirect means. With the availability of DNA sequence data, it became more natural to investigate the *infinite sites model* of Kimura (1969) in which mutations always occur at distinct sites. In this section, we will study several aspects of this model. To motivate the developments and help illustrate the concepts we will consider data from Ward et al. (1991), who sequenced 360 nucleotides in the D loop of mitochondria in 63 humans.

```
       1111111222222222333333
       6890246699013455677900 0134
       9816492604093715715612 4994

       TCCGCTCTGTCCCCGCCCTGTTCTTA
    1  .......CA.T....T.......... 3
    2  .......A.T....T.......... 2
    3  ..........T....T.......... 1
    4  ..........T....T.......C. 1
    5  .T.A..T...T.....T..A....C. 2
    6  .T.A......T........A....C. 2
    7  CT.A......T..T.....A....C. 1
    8  .T.A......T.....T..A....C. 2
    9  CT........T.....T..A....C. 2
   10  .T........T.....T..A....CG 1
   11  .T........T.....T..A....C. 5
   12  .T............T..A....C. 9
   13  .T......A.......T..A....C. 1
   14  .T........T....TT..A....C. 1
   15  .T........T....TT..AC...C. 2
   16  ..........TT..........T.C. 1
   17  ....T.....T.........C...C. 1
   18  ....T.....T.........C..C. 2
   19  ..T.......T......T...C..C. 1
   20  .....C....T...A......C..C. 3
   21  ..........T.........C..C. 3
   22  C.........T.........C.... 3
   23  ..........TT......C..C.... 1
   24  ..........T.......C..CT... 7
   25  ..........TT......C..CTC.. 3
   26  ...........T......C..CTC.. 1
   27  .......C.C................ 1
   28  .......C.C..T............. 1
```

We ignore the positions at which all of the sequences are the same. The sequence at the top is the published human reference sequence from Anderson et al. (1981). The numbers at the top indicate the positions where the mutations occurred in the fragment sequenced. To make it easier to spot mutations, the others sequences have dots where they agree with the reference sequence. The numbers at the right indicate how many times each pattern, or *haplotype*, was observed. If we were to examine this data from the view point of the infinite alleles model, it would reduce to the allelic partition:

$$a_1 = 13, \ a_2 = 7, \ a_3 = 5, \ a_5 = 1, \ a_7 = 1, \ a_9 = 1$$

1.4.1 Segregating sites

We begin by considering the number of *segregating sites*, S_n, in the sample of size n, that is, the number of positions where some pair of sequences differs. In the example above, $n = 63$ and $S_{63} = 26$.

Theorem 1.20. *Let u be the mutation rate for the locus being considered and $\theta = 4Nu$. Under the infinite sites model, the expected number of segregating sites is*

$$ES_n = \theta h_n \quad where \quad \sum_{i=1}^{n-1} \frac{1}{i} \tag{1.20}$$

Here h stands for *harmonic series*. The reader should note that although the subscript is n to indicate the sample size, the last term has $i = n - 1$.

Proof. Let t_j be the amount of time in the coalescent during which there are j lineages. In Theorem 1.4, we showed that if N is large and time is measured in units of $2N$ generations, then t_j has approximately an exponential distribution with mean $2/j(j-1)$. The total amount of time in the tree for a sample of size n is

$$T_{tot} = \sum_{j=2}^{n} j t_j$$

Taking expected values, we have

$$ET_{tot} = \sum_{j=2}^{n} j \cdot \frac{2}{j(j-1)} = 2 \sum_{j=2}^{n} \frac{1}{j-1} \tag{1.21}$$

Since mutations occur at rate $2Nu$, $ES_n = 2Nu \cdot ET_{tot}$, and the desired result follows. $\square$

To understand the variability in S_n, our next step is to show that

Theorem 1.21. *Under the infinite sites model, the number of segregating sites S_n has*

$$var(S_n) = \theta h_n + \theta^2 g_n \quad where \quad g_n = \sum_{i=1}^{n-1} \frac{1}{i^2} \tag{1.22}$$

Proof. Let s_j be the number of segregating sites created when there were j lineages. While there are j lineages, we have a race between mutations happening at rate $2Nuj$ and coalescence at rate $j(j-1)/2$. Mutations occur before coalescence with probability

$$\frac{2Nuj}{2Nuj + j(j-1)/2} = \frac{4Nu}{4Nu + j - 1}$$

so we have

$$P(s_j = k) = \left(\frac{\theta}{\theta + j - 1}\right)^k \frac{j - 1}{\theta + j - 1} \quad \text{for } k = 0, 1, 2, \ldots$$

This is a shifted geometric distribution (i.e., the smallest value is 0 rather than 1) with success probability $p = (j - 1)/(\theta + j - 1)$, so

$$\begin{aligned}
\text{var}(s_j) &= \frac{1 - p}{p^2} = \frac{\theta}{\theta + j - 1} \cdot \frac{(\theta + j - 1)^2}{(j - 1)^2} \\
&= \frac{\theta^2 + (j - 1)\theta}{(j - 1)^2} = \frac{\theta}{j - 1} + \frac{\theta^2}{(j - 1)^2}
\end{aligned}$$

Summing from $j = 2$ to n and letting $i = j - 1$ gives the result. □

The last proof is simple, but one can obtain additional insight and generality by recalling the simulation algorithm discussed earlier. First, we generate the tree, and then we put mutations on each branch according to a Poisson process with rate $\theta/2$. This implies that the distribution of $(S_n | T_{tot} = t)$ is Poisson with mean $t\theta/2$. From this it follows that

$$\begin{aligned}
E(S_n) &= \frac{\theta}{2} E T_{tot} \\
\text{var}(S_n) &= E\{\text{var}(S_n | T_{tot})\} + \text{var}\{E(S_n | T_{tot})\} \quad (1.23) \\
&= \frac{\theta}{2} E T_{tot} + \left(\frac{\theta}{2}\right)^2 \text{var}(T_{tot})
\end{aligned}$$

The last result is valid for any genealogy. If we now use the fact that in the Wright-Fisher model $T_{tot} = \sum_{j=2}^{n} j t_j$, where the t_j are exponential with mean $2/j(j - 1)$ and variance $4/j^2(j - 1)^2$, then (1.22) follows and we see the source of the two terms.

- The first, $E\{\text{var}(S_n | T_{tot})\}$ is called the *mutational variance* since it is determined by how many mutations occur on the tree.
- The second, $\text{var}\{E(S_n | T_{tot})\}$, is called the *evolutionary variance* since it is due to the fluctuations in the genealogy.

(1.20) and (1.22) show that

Theorem 1.22. *Watterson's estimator* $\theta_W = S_n/h_n$ *has* $E(\theta_W) = \theta$ *and*

$$\text{var}(\theta_W) = \theta \frac{1}{h_n} + \theta^2 \frac{g_n}{h_n^2} \quad (1.24)$$

Example 1.1. In the Ward et al. (1991) data set, there are $n = 63$ sequences and 26 segregating sites. Summing the series gives $h_{63} = 4.7124$, so our estimate based on the number of segregating sites S_{63} is

$$\theta_W = 26/4.7124 = 5.5173$$

Dividing by 360 gives an estimate of $\theta_b = 0.015326$ per nucleotide.

Remark. In the derivation of (1.10) we observed that $h_n = \sum_{i=1}^{n-1} 1/i >$ $\log n$. The difference increases with n and converges to Euler's constant $\gamma = 0.577215$, so if you don't want to sum the series then you can use $h_{63} \approx \gamma + \log 63 = 4.7203$ (instead of the exact answer of 4.7124).

Figure 1.8 shows the distribution of S_n when $\theta = 3$ and $n = 10, 30, 100$. It follows from (1.20) and (1.22) that the mean ES_n and standard deviation $\sigma(S_n)$ are given by:

n	10	30	100
ES_n	8.49	11.88	15.53
$\sigma(S_n)$	4.73	5.14	5.49

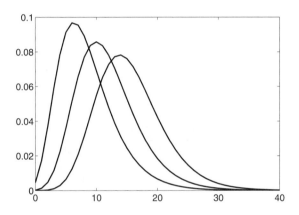

Fig. 1.8. The distribution of S_n is approximately normal.

Theorem 1.23. *Under the infinite sites model, Watterson's estimate θ_W has*

$$P\left(\frac{\theta_W - \theta}{\sqrt{var(\theta_W)}} \le x\right) \to P(\chi \le x)$$

where χ has the standard normal distribution.

Proof. The proof of (1.22) shows that the number of segregating sites in the infinite sites model $S_n = \sum_{j=2}^{n} s_j$, where the s_j are independent shifted geometrics with success probabilities $p = (j-1)/(\theta+j-1)$. Using the triangular array form of the central limit theorem, see (4.5) on page 116 of Durrett (2005), we have that as $n \to \infty$,

$$P\left(\frac{S_n - ES_n}{\sqrt{var(S_n)}} \le x\right) \to P(\chi \le x)$$

Dividing by h_n in the numerator and denominator gives the desired result. □

(1.24) implies

$$\text{var}\,(\theta_W) = \theta\frac{1}{h_n} + \theta^2\frac{g_n}{h_n^2}$$

Since $h_n = \sum_{i=1}^{n-1} 1/i \sim \log n$ and $g_n = \sum_{i=1}^{n-1} 1/i^2 \to g_\infty < \infty$ as $n \to \infty$, $\text{var}\,(\theta_W) \sim \theta/\log n$, so as in Theorem 1.10, the rate of convergence to the limiting expected value $\theta = E\hat{\theta}$ is painfully slow.

Can we do better?

Fu and Li (1993) had a clever idea that allows us to use statistical theory to get lower bounds on the variance of unbiased estimates of θ.

Theorem 1.24. *Any unbiased estimator of θ has variance at least*

$$J(\theta) = \theta \left/ \sum_{k=1}^{n-1} \frac{1}{k + \theta} \right.$$

Proof. The proof of (1.22) shows that the number of segregating sites in the infinite sites model $S_n = \sum_{j=2}^{n} s_j$, where the s_j are independent shifted geometrics with success probabilities $p = (j - 1)/(\theta + j - 1)$. So the likelihood

$$L_n(\theta) = \prod_{j=2}^{n} \left(\frac{\theta}{\theta + j - 1}\right)^{s_j} \frac{j - 1}{\theta + j - 1}$$

$$= (n - 1)!\theta^{S_n} \prod_{j=2}^{n} (\theta + j - 1)^{-(s_j+1)}$$

$\log L_n = \log((n - 1)!) + S_n \log \theta - \sum_{j=2}^{n}(s_j + 1)\log(\theta + j - 1)$, so

$$\frac{d}{d\theta}\log L_n = \frac{S_n}{\theta} - \sum_{j=2}^{n} \frac{s_j + 1}{\theta + j - 1}$$

and the maximum-likelihood estimate, θ_B, where B is for bound, is obtained by solving

$$\theta = S_n \left/ \sum_{j=2}^{n} \frac{s_j + 1}{\theta + j - 1} \right.$$

The estimator θ_B is not relevant to DNA sequence data. The next picture explains why. At best we can only reconstruct the branches on which mutations occur and even if we do this, we would not have enough information needed to compute θ_B, since moving the mutations on the branches will change the values of the s_j.

The reason for pursuing the likelihood calculation is that the Cramér-Rao lower bound implies that all unbiased estimators have variance larger than $J(\theta) = 1/I(\theta)$, where $I(\theta)$ is the Fisher information:

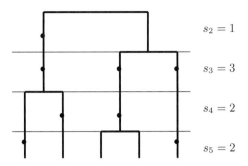

Fig. 1.9. The variables s_j cannot be determined from the data.

$$E\left(\frac{d^2}{d\theta^2}\log L_n\right)^2 = -E\frac{d^2}{d\theta^2}\log L_n.$$

The equality is a well-known identity, see e.g., Casella and Berger (1990) p. 312, which can be established by integrating by parts. To compute the lower bound, we differentiate again:

$$\frac{d^2}{d\theta^2}\log L_n = -\frac{S_n}{\theta^2} + \sum_{j=2}^{n}\frac{s_j+1}{(\theta+j-1)^2}$$

Taking expected value and recalling $E(s_j+1) = (\theta+j-1)/(j-1)$, we get

$$-E\frac{d^2}{d\theta^2}\log L_n = \frac{h_n}{\theta} + \sum_{j=2}^{n}\frac{1}{(j-1)(\theta+j-1)}$$

$$= \frac{1}{\theta}\sum_{k=1}^{n-1}\left(\frac{1}{k} - \frac{\theta}{k(\theta+k)}\right) = \frac{1}{\theta}\sum_{k=1}^{n-1}\frac{1}{k+\theta}$$

which completes the proof. □

To compare with Watterson's estimator, we recall

$$\operatorname{var}(\theta_W) = \theta\frac{1}{h_n} + \theta^2\frac{g_n}{h_n^2}$$

where $h_n = \sum_{i=1}^{n-1}1/i$ and $g_n = \sum_{i=1}^{n-1}1/i^2$. If we fix θ and let $n \to \infty$, then since the series in the denominator of the lower bound $J(\theta)$ is the harmonic series with a few terms left out at the beginning and a few added at the end, $J(\theta) \sim \theta/h_n$ and hence

$$\frac{\text{var}(\theta_W)}{J(\theta)} \to 1$$

On the other hand, if we fix n and let $\theta \to \infty$ then $J(\theta) \sim \theta^2/(n-1)$ and

$$\frac{\text{var}(\theta_W)}{J(\theta)} \to \frac{h_n^2}{g_n} \geq \frac{(\log n)^2}{\pi^2/6}$$

As Tavaré (2004) says in his St. Flour notes, this suggests that we explore estimation of θ using the likelihoods formed from the full data rather than their summary statistics, a computationally intensive approach that we will not consider in this book.

Exact distribution of S_n

Breaking things down according to whether the first event is a mutation or coalescence, we conclude

$$P(S_n = k) = \frac{\theta}{n-1+\theta} P(S_n = k-1) + \frac{n-1}{n-1+\theta} P(S_{n-1} = k)$$

Tavaré (1984) used this approach to obtain explicit expressions for the distribution of S_n:

$$P(S_n = k) = \frac{n-1}{\theta} \sum_{i=1}^{n-1} (-1)^{i-1} \binom{n-2}{i-1} \left(\frac{\theta}{\theta+i}\right)^{k+1} \tag{1.25}$$

The result is very simple for $k = 0$. In that case,

$$P(S_n = 0) = P(S_{n-1} = 0) \cdot \frac{n-1}{\theta+n-1}$$

Since $P(S_1 = 0) = 1$, iterating gives

$$P(S_n = 0) = \frac{(n-1)!}{(\theta+1)\cdots(\theta+n-1)} \tag{1.26}$$

Example 1.2. Dorit, Akashi, and Gilbert (1995) examined a 729 base pair intron between the third and fourth exons of the ZFY gene in 38 individuals and found no variation. Whether or not this is unusual depends on the value of θ. A simple bisection search shows that when $\theta = 0.805$, $P(S_{38} = 0) = 0.049994$, so if $\theta > 0.805$, no variation has a probability smaller than 0.05. Dividing by 729 translates this into a per base pair value of $\theta_b = 0.0011$. As we will see in a moment, this number is exactly Li and Sadler's value for the nucleotide diversity. However, females have no Y chromosome and males one, so if N_m is the number of males, then for the Y chromosome $\theta_b = 2N_m u$, and if we assume males are half the population, this is $1/4$ of the value of $\theta = 4Nu$ for autosomes. If we use $\theta = 729(0.0011)/4 = 0.2$, we find $P(S_{38} = 0) = 0.443$, and the data does not look unusual at all.

Segregating sites vs. haplotypes

To compare the number of segregating sites S_n with the number of alleles (haplotypes) K_n, we note that $S_n \geq K_n - 1$ since a new mutation is needed to increase the number of haplotypes. We have computed in (1.20) and (1.9) that

$$ES_n = \sum_{j=1}^{n-1} \frac{\theta}{j} \quad \text{and} \quad E(K_n - 1) = \sum_{j=1}^{n-1} \frac{\theta}{\theta + j}$$

Taking the difference, we see that

$$0 \leq E\{S_n - (K_n - 1)\} = \sum_{j=1}^{n-1} \frac{\theta^2}{j(\theta + j)}$$

If $\theta = 1$, this is

$$\sum_{j=1}^{n-1} \frac{1}{j(j+1)} = \sum_{j=1}^{n-1} \frac{1}{j} - \frac{1}{(j+1)} = 1 - \frac{1}{n}$$

so on the average there is one more segregating site than haplotype. The last computation generalizes to show that if $\theta = k$ is an integer, then

$$E\{S_n - (K_n - 1)\} = k \sum_{j=1}^{n-1} \left(\frac{1}{j} - \frac{1}{(j+k)} \right) \approx k \sum_{j=1}^{k} \frac{1}{j} = k h_{k+1}$$

so if $\theta = k$ is large, there are many more segregating sites than haplotypes.

Example 1.3. At this point, some readers may recall that in the Ward et al. (1991) data set, there are $S_n = 26$ segregating sites but $K_n = 28$ haplotypes. The reason for this is that the data does not satisfy the assumptions of the infinite sites model. To explain this, consider the following example genealogy.

The sets of individuals affected by mutations A, B, and C are $S_A = \{1, 2\}$, $S_B = \{5, 6\}$, and $S_C = \{5, 6, 7\}$. Note that $S_B \subset S_C$ while S_C and S_A are disjoint. A little thought reveals that any two mutations on this tree must obey this pattern. That is, their sets of affected individuals must be nested (i.e., one a subset of the other) or disjoint. If we introduce lowercase letters a, b, c to indicate the absence of the corresponding capital-letter mutations, then this can be reformulated as saying that if all loci share the same genealogy, it is impossible to observe all four combinations AB, Ab, aB, and aa in one data set. Returning to the original data set of Ward et al. (1991), we see that all four combinations occur in columns 69 and 88, 190 and 200, 255 and 267, 296 and 301, and 302 and 304, so at least five sites have been hit twice by mutation.

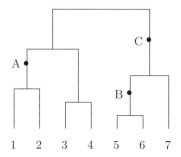

Fig. 1.10. It is is impossible to observe all four combinations of alleles at two sites without repeat mutations (or recombination).

Birthday problem

At first, it may seem surprising that if there are 26 mutations in a 360 nucleotide region, then there will be more than one at some site. However, this is just the classical *birthday problem* of probability theory. The probability that no site is hit more than once is

$$\left(1 - \frac{1}{360}\right)\left(1 - \frac{2}{360}\right)\cdots\left(1 - \frac{25}{360}\right) = 0.396578$$

Using $\prod_{i=1}^{25}(1 - x_i) \approx \exp(-\sum_{i=1}^{25} x_i)$ and $\sum_{i=1}^{25} i = (26 \cdot 25)/2$, the above is

$$\approx \exp\left(-\frac{26 \cdot 25}{2} \cdot \frac{1}{360}\right) = 0.40544$$

Thus, we should not be surprised to find one site hit more than once. Generalizing, we see that if there are m mutations in a region of L nucleotides, the probability is

$$\approx \exp\left(-\frac{m(m-1)}{2L}\right)$$

which becomes < 1 when $m \approx \sqrt{L}$.

On the other hand, it is very unusual in this example to have five sites hit twice. If we suppose that there are 31 mutations in a 360 nucleotide region, then the expected number of sites that are hit more than once is

$$\frac{31 \cdot 30}{2} \cdot 1360 = 1.2916$$

Since we have a lot of events with small probability of success and these events are almost independent, the number of double hits will have roughly a Poisson distribution with mean 1.2916. (This intuition can be made precise using the Chen-Stein method in the form of Arratia, Gordon, and Goldstein 1989.) For the Poisson, the probability of exactly five double hits is

$$e^{-1.2916} \frac{(1.2916)^5}{5!} = 0.00823$$

Using a computer, we find

$$\sum_{k=5}^{\infty} e^{-1.2916} \frac{(1.2916)^k}{k!} = 0.010393$$

or only 1% of the time are there 5 a more double hits.

1.4.2 Nucleotide diversity

The probability that two nucleotides differ in two randomly chosen individuals is called the *nucleotide diversity* and is denoted by π.

Theorem 1.25. *Let μ be the mutation rate per nucleotide per generation and $\theta = 4N\mu$. Under the infinite sites model,*

$$E\pi = \frac{\theta}{1+\theta} \approx \theta \qquad (1.27)$$

since in most cases $4N\mu$ is small.

Proof. In each generation two lineages coalesce with probability $1/2N$ and mutate with probability 2μ, so the probability of mutation before coalescence is

$$\frac{2\mu}{2\mu + 1/2N} = \frac{4N\mu}{1 + 4N\mu} \qquad \square$$

Li and Sadler (1991) estimated π for humans by examining 49 genes. At *four-fold degenerate sites* (i.e., where no substitution changes the amino acid) they found $\pi = 0.11\%$ (i.e., $\pi = 0.0011$). At *two-fold degenerate sites* (i.e., where only one of the three possible changes is synonymous) and nondegenerate sites, the values were 0.06% and 0.03%, respectively. More recent studies have confirmed this. Harding et al. (1997) sequenced a 3 kb stretch including the β-globin gene in 349 chromosomes from nine populations in Africa, Asia, and Europe, revealing an overall nucleotide diversity of $\pi = 0.18\%$. Clark et al. (1998) and Nickerson et al. (1998) sequenced a 9.7 kb region near the lipoprotein lipase gene in 142 chromosomes, finding an average nucleotide diversity of 0.2%.

In contrast, data compiled by Aquadro (1991) for various species of *Drosophila* give the following estimates for π:

	D. pseudoobscura	D. simulans	D. melanogaster
Adh	0.026	0.015	0.006
Amy	0.019		0.008
rosy	0.013	0.018	0.005

Since $\pi = 4N\mu$, the differences in the value of π can be due to differences in N or in μ. Here N is not the physical population size, i.e., 6 billion for humans or an astronomical number for *Drosophila*, but instead is the "effective population size." To explain the need for this concept, we note that in the recent past the human population has grown exponentially and *Drosophila* populations undergo large seasonal fluctuations, so neither fits our assumption of a population of constant size. The effective population size will be defined precisely in Section 3.1. To illustrate its use in the current example, we note that current estimates (see Drake et al. 1998) of the mutation rate in humans are $\mu = 10^{-8}$ per nucleotide per generation. Setting $1.1 \times 10^{-3} = 4N \cdot 10^{-8}$ and solving gives $N_e = 27,500$.

1.4.3 Pairwise differences

Given two DNA sequences of length L, let Δ_2 be the number of *pairwise differences*. For example, the two sequences

$$AATCGCTTGATACC$$
$$A\underline{C}TCGC\underline{C}TGATA\underline{A}C$$

have three pairwise differences at positions 2, 7, and 13. Given n DNA sequences, let Δ_{ij} be the number of pairwise differences between the ith and jth sequences, and define the average number of pairwise differences by

$$\Delta_n = \binom{n}{2}^{-1} \sum_{\{i,j\}} \Delta_{ij}$$

where the sum is over all pairs $\{i,j\} \subset \{1,2,\ldots n\}$ with $i \neq j$.

Theorem 1.26. *Let u be the mutation rate for the locus and let $\theta = 4Nu$. Under the infinite sites model,*

$$E\Delta_n = \theta \tag{1.28}$$

In words, Δ_n is an unbiased estimator of θ. In situations where we want to emphasize that Δ_n is an estimator of θ, we will call it θ_π, i.e., the estimate of θ based on the nucleotide diversity π.

Proof. To compute the distribution of Δ_2, note that (1.12) implies that the probability of coalescence before mutation is $1/(1 + 4Nu)$. If, however, mutation comes before coalescence, we have an equal chance of having another mutation before coalescence, so letting $\theta = 4Nu$, we have

$$P(\Delta_2 = k) = \left(\frac{\theta}{\theta+1}\right)^k \frac{1}{\theta+1} \quad \text{for } k = 0, 1, 2, \ldots \qquad (1.29)$$

To compute the mean of Δ_2 now, we recall that the shifted geometric distribution $P(X = k) = (1-p)^k p$ for $k = 0, 1, 2, \ldots$ has mean $(1/p) - 1$, so $E\Delta_2 = \theta$. Since there are $\binom{n}{2}$ pairs, and $E\Delta_{ij} = E\Delta_2$, the desired result follows. □

Example 1.4. Ward et al. (1991). To compute Δ_n, it is useful to note that if k_m is the number of columns in which the less frequent allele has m copies, then

$$\Delta_n = \binom{n}{2}^{-1} \sum_{ij} \Delta_{ij} = \sum_{m=1}^{n/2} k_m \cdot m(n-m)$$

In the example under consideration, $k_1 = 6$, $k_2 = 2$, $k_3 = 3$, $k_4 = 1$, $k_6 = 4$, $k_7 = 1$, $k_{10} = 1$, $k_{12} = 2$, $k_{13} = 1$, $k_{23} = 1$, $k_{24} = 1$, $k_{25} = 1$, and $k_{28} = 2$, so our formula gives $\Delta_{63} = 5.2852$ compared to Watterson's estimate $\theta_W = 5.5173$ based on segregating sites.

Tajima (1983) has shown that

$$\text{var}\,(\Delta_n) = \frac{n+1}{3(n-1)}\theta + \frac{2(n^2+n+3)}{9n(n-1)}\theta^2 \qquad (1.30)$$

Note that when $n = 2$, this says $\text{var}\,(\Delta_2) = \theta + \theta^2$, which follows from (1.29). We will postpone the proof to Chapter 3. See (3.8).

As in the case of Watterson's estimator,

- the term with θ is the *mutational variance* due to the placement of mutations on the tree;
- the term with θ^2 is the *evolutionary variance* due to fluctuations in the shape of the tree.

To prepare for future estimates of θ, we will prove a general result. Let η_k be the number of sites in the sample where k individuals have the mutant allele, and consider an estimate of the form

$$\hat{\theta} = \sum_{k=1}^{n-1} c_{n,k}\eta_k$$

where $c_{n,k}$ are constants. In Section 2.2, we will see that many estimates of θ have this form. So far we have seen two: Δ_n corresponds to $c_{n,k} = \frac{2}{n(n-1)}k(n-k)$ and Watterson's estimate θ_W to $c_{n,k} = 1/h_n$. Adapting the argument in (1.23) we will now show that

Theorem 1.27. *There are constants a_n and b_n, which depend on the $c_{n,k}$ so that*

$$\text{var}\,(\hat{\theta}) = a_n\theta + b_n\theta^2$$

Proof. Let L_k be the total length of branches in the tree with k descendants and let $\mathcal{L} = (L_1, \ldots, L_{n-1})$. Since mutations occur on the tree according to a Poisson process with rate $\theta/2$,

$$\text{dist}(\hat{\theta}|\mathcal{L}) = \sum_{k=1}^{n-1} c_{n,k} \, \text{Poisson}(\theta L_k/2)$$

where $\text{Poisson}(\lambda)$ is shorthand for Poisson with mean λ and the different Poisson variables are independent. Since the mean and variance of the $\text{Poisson}(\lambda)$ distribution are both equal to λ, it follows that

$$E(\hat{\theta}|\mathcal{L}) = \sum_{k=1}^{n-1} c_{n,k} \cdot \theta L_k/2 \qquad \text{var}(\hat{\theta}|\mathcal{L}) = \sum_{k=1}^{n-1} c_{n,k}^2 \cdot \theta L_k/2$$

From this we can conclude that

$$E\{\text{var}(\Delta_n|\mathcal{L})\} = \theta a_n \quad \text{where} \quad a_n = \sum_{k=1}^{n-1} c_{n,k}^2 \cdot EL_k/2$$

$$\text{var}\{E(\Delta_n|\mathcal{L})\} = \theta^2 b_n \quad \text{where}$$
$$b_n = (1/4) \sum_{1 \le j,k \le n-1} c_{n,j} c_{n,k} \, \text{cov}(L_j, L_k)$$

which proves the desired result. □

Letting $n \to \infty$ in (1.30) we see that

$$\text{var}(\Delta_n) \to \frac{\theta}{3} + \frac{2}{9}\theta^2$$

so the variance of Δ_n does not go to 0. This is due to the fact that the value of Δ_n is determined mostly by mutations on a few braches near the top of the tree. The next result will make this precise and determine the limiting distribution of Δ_n.

Theorem 1.28. *Let $U_1, U_2, \ldots$ be independent uniform on $(0,1)$. For fixed m let $U_{m,1} < U_{m,2} < \ldots < U_{m,m}$ be the first m of the U_i arranged in increasing order, and let $U_{m,0} = 0$, $U_{m,m+1} = 1$. Let $V_{m,j} = U_{m,j} - U_{m,j-1}$ be the spacing in between the order statistics $U_{m,i}$ for $1 \le j \le m+1$. Let t_m be independent exponentials with mean $2/m(m+1)$ and, conditional on t_m, let $Z_{m,0}, \ldots Z_{m,m}$ be independent Poisson with mean $\theta t_m/2$. As $n \to \infty$, Δ_n converges in distribution to*

$$\sum_{m=1}^{\infty} \sum_{j=0}^{m} 2V_{m,j}(1 - V_{m,j})Z_{n,j}$$

Proof. We look at the coalescent tree starting from the most common ancestor and working down. Theorem 1.8 shows that if λ_j^k, $1 \leq j \leq k$ are the number of members of the sample that are descendants of the jth lineage when there are k lineages left, then λ_j^k is constructed from λ_j^{k-1} by picking λ_j^{k-1} with probability $(\lambda_j^{k-1} - 1)/(n - k - 1)$ and then splitting λ_j^{k-1} into i and $\lambda_j^{k-1} - i$ where i is uniform on $1, 2, \ldots \lambda_j^{k-1} - 1$.

When $k = 1$, $\lambda_1^1 = n$, i.e., all individuals are descendants of the most recent common ancestor. When $k = 2$, λ_1^2 is uniform on $1, 2, \ldots n - 1$, so as $n \to \infty$, $(\lambda_1^2/n, \lambda_2^2/n)$ converges in distribution to $(V_{1,0}, V_{1,1}) = (U_1, 1 - U_1)$. Continuing we see that the fractions of descendants $(\lambda_1^k/n, \ldots \lambda_k^k/n)$ converge in distribution to $(V_{k-1,1}, \ldots V_{k-1,k})$. The first index of V is $k - 1$ because $k - 1$ uniforms will break the interval into k pieces.

The amount of time in the coalescent when there are $m + 1$ lineages is exponential with mean $2/m(m + 1)$. Conditional on this time, the number of mutations on the $m + 1$ lineages are independent Poisson with mean $\theta t_m/2$. If we number the lineages by $0 \leq j \leq m$ then a mutation on the jth produces approximately $V_{m,j}(1 - V_{m,j})n^2$ pariwise differences. Dividing by $\binom{n}{2}$, we end up with $2V_{m,j}(1 - V_{m,j})$. □

As a check on our derivation, we will show that the limit has mean θ. It is known and not hard to check that $V_{m,j}$ has a beta(1,m) distribution, i.e., the density function is $m(1 - x)^{m-1}$. Integrating by parts, we find

$$E(V_{m,j}) = \int_0^1 x \cdot m(1 - x)^{m-1} \, dx = \int_0^1 (1 - x)^m \, dx = \frac{1}{m + 1}$$

which is intuitive since we are breaking the interval into $m + 1$ pieces.

$$E(V_{m,j}^2) = \int_0^1 x^2 \cdot m(1 - x)^{m-1} \, dx = \int_0^1 2x(1 - x)^m \, dx$$

$$= \frac{2}{m + 1} \int_0^1 (1 - x)^{m+1} \, dx = \frac{2}{(m + 1)(m + 2)}$$

Using this we have

$$E\Delta_n \to \sum_{m=1}^\infty (m + 1)2E(V_{m,j} - V_{m,j}^2) \cdot \frac{\theta}{2} \cdot \frac{2}{m(m + 1)}$$

$$= 2\theta \sum_{m=1}^\infty \left(1 - \frac{2}{m + 2}\right) \cdot \frac{1}{m(m + 1)}$$

$$= 2\theta \sum_{m=1}^\infty \frac{1}{(m + 1)(m + 2)} = 2\theta \sum_{m=1}^\infty \frac{1}{m + 1} - \frac{1}{m + 2} = \theta$$

1.4.4 Folded site frequency spectrum

For motivation, consider the following data set.

Example 1.5. Aquadro and Greenberg (1983) studied data on 900 nucleotides in 7 mtDNA sequences. We ignore the positions at which all of the sequences are the same, and three positions where there has been an insertion or deletion. As before, to make it easier to spot mutations, the last six sequences have dots where they agree with the first one.

```
1 GTCTCATATGCGAGGATCTAAAGAGAAGTTGAGTAGAGGAGTGGC
2 AC.C.....................G.............G.....
3 ...CTGC.C.T......TC..G..AG.......C.AGCAGA.A.T
4 ..TCT...GA.AGA..C..GG.A..G.ACCA....AG.A..C...
5 ...C........G..GC..G.....G........G..........
6 ...C...G....G.A.C..G...G.......GA..........A.
7 ...C...........C..G..........A..............
```

In this data set, we have no way of telling what is the ancestral nucleotide, so all we can say is that there are m of one nucleotide and $n - m$ of the other. This motivates the definition of the *folded site frequency spectrum*. Let $\tilde{\eta}_n$ be the number of sites where the less frequent nucleotide appears m times.

The next result tells us what to expect.

Theorem 1.29. *Consider a segregating site where two different nucleotides appear in the sample. The probability the less frequent nucleotide has m copies is*

$$\frac{1}{h_n}\left(\frac{1}{m} + \frac{1}{n-m}\right) \quad \text{if } m < n/2$$

$$\frac{1}{h_n}\left(\frac{1}{m}\right) \quad \text{if } m = n/2$$

Proof. To derive this result, we use Ewens' sampling formula to compute the conditional distribution of the number of individuals with the two alleles given that there was one mutation. Consider first the situation in which one allele has m and the other $n - m > m$ representatives. Let $a^{m,n-m}$ be the allelic partition with $a_m^{m,n-m} = 1$ and $a_{n-m}^{m,n-m} = 1$. Writing $q(m, n-m)$ as shorthand for $P_{\theta,n}(a^{m,n-m})$, Theorem 1.11 gives

$$q(m, n-m) = \frac{n!}{\theta_{(n)}} \cdot \frac{\theta}{m^1 \cdot 1!} \cdot \frac{\theta}{(n-m)^1 \cdot 1!} = \frac{n!\theta^2}{\theta_{(n)}} \cdot \frac{1}{m(n-m)}$$

In the exceptional case that $n - m = m$, the allelic partition has $a_m = 2$, so

$$q(m, m) = \frac{n!\theta^2}{\theta_{(n)}} \cdot \frac{1}{m^2 \cdot 2!}$$

We are interested in the conditional probability:

$$p_m = q(m, n-m) \bigg/ \sum_{1 \leq m \leq n/2} q(m, n-m)$$

To compute this, we note that since $n!\theta^2/\theta_{(n)}$ is common to both formulas, there is a constant c so that

$$
p_m = \begin{cases} \frac{c}{n}\left(\frac{1}{m} + \frac{1}{n-m}\right) & \text{if } m < n/2 \\ \frac{c}{n}\frac{1}{m} & \text{if } m = n/2 \end{cases}
$$

where in the second formula we have substituted $m = n/2$. Since the probabilities have to add to 1, $c = 1/h_n$ and the desired result follows. □

Example 1.6. In the case of $n = 7$ individuals, the probabilities are

$$
1/6 : \frac{c_7}{1 \cdot 6} \qquad 2/5 : \frac{c_7}{2 \cdot 5} \qquad 3/4 : \frac{c_7}{3 \cdot 4}
$$

where c_7 is the constant needed to make the sum of the probabilities equal to 1. A little arithmetic shows that $c_7 = 60/21$, so the relative frequencies are

$$
1/6 : \frac{10}{21} = 0.476 \qquad 2/5 : \frac{6}{21} = 0.286 \qquad 3/4 : \frac{5}{21} = 0.238
$$

Ignoring the ninth position in the Aquadro and Greenberg data set, which has had two mutations, we find an excess of $1/6$ splits over what is expected:

partition	observed	expected
3/4	4	10.48
2/5	6	12.58
1/6	34	20.94

While we have the data in front of us, we would like to note that there are 44 segregating sites and

$$
h_7 = 1 + \frac{1}{2} + \frac{1}{3} + \frac{1}{4} + \frac{1}{5} + \frac{1}{6}
$$

so $\theta_W = 44/2.45 = 17.96$. Using the table,

$$
\Delta_7 = 34 \cdot \frac{(1 \cdot 6)}{21} + 6 \cdot \frac{(2 \cdot 5)}{21} + 4 \cdot \frac{(3 \cdot 4)}{21} = \frac{312}{21} = 14.86
$$

which is a considerably smaller estimate of θ. Is the difference of the two estimates large enough to make us doubt the assumptions underlying the Wright-Fisher model (a homogeneously mixing population of constant size with neutral mutations)? Chapter 2 will be devoted to this question.

Singletons

Taking the viewpoint of Hoppe's urn, the expected number of colors that have exactly one ball in the urn is

$$Ea_1 = \sum_{i=1}^{n} \frac{\theta}{\theta + i - 1} \prod_{k=i+1}^{n} \left(1 - \frac{1}{\theta + k - 1}\right)$$

since the color must be introduced at some time i and then not chosen after that. The right-hand side simplifies nicely:

$$= \sum_{i=1}^{n} \frac{\theta}{\theta + i - 1} \prod_{k=i+1}^{n} \frac{\theta + k - 2}{\theta + k - 1} = \theta \frac{n}{\theta + n - 1}$$

In contrast, it follows from the folded site frequency spectrum described in Theorem 1.29 and the formula for the expected number of segregating sites (1.20) that the number of sites $\tilde{\eta}_m$, where there are $m < n/2$ of the less frequent allele, has (for $n > 2$)

$$E\tilde{\eta}_m = ES_n \cdot \frac{1}{2h_n} \cdot 2 \left(\frac{1}{m} + \frac{1}{n - m}\right) = \theta \frac{n}{m(n - m)} \qquad (1.31)$$

When $m = 1$ and $n > 2$, we have

$$E\tilde{\eta}_1 = \frac{n\theta}{n - 1}$$

A haplotype that occurs in only one individual must have at least one nucleotide that appears in only one individual so $\tilde{\eta}_1 \geq a_1$. The difference

$$E\tilde{\eta}_1 - Ea_1 = \frac{n\theta^2}{(n - 1)(\theta + n - 1)}$$

is the number of extra singletons in unique haplotypes. This is small if n is larger than θ^2.

1.5 Moran model

The Wright-Fisher model considers nonoverlapping generations, as one has in annual plants and the deer herds of upstate New York. However, in many species, e.g., humans, *Drosophila*, and yeast, the generations are not synchronized and it is convenient to use a model due to Moran (1958) in which only one individual changes at a time. In the Wright-Fisher model, the $2N$ copies of our locus could either come from N diploid individuals or $2N$ haploid individuals (who have one copy of their genetic material in each cell). In our formulation of the *Moran model* we think in terms of $2N$ haploid individuals.

- Each individual is replaced at rate 1. That is, individual x lives for an exponentially distributed amount with mean 1 and then is "replaced."
- To replace individual x, we choose an individual at random from the population (including x itself) to be the parent of the new individual.

One could make a version of the Moran model for N diploid individuals by replacing two copies of the locus at once. This should not change the dynamics very much, but it would eliminate one property that is very important for the analysis. The number of copies of an allele never changes by more than one.

Even though the Wright-Fisher model and the Moran model look much different going forward, they are almost identical going backwards.

Theorem 1.30. *When time is run at rate N, the genealogy of a sample of size n from the Moran model converges to Kingman's coalescent.*

Thus, when we reverse time, the only difference is that the Moran model coalesces twice as fast, so if we want to get the same answers we have to double the mutation rate in the Moran model.

Proof. If we look backwards in time, then when there are k lineages, each replacement leads to a coalescence with probability $(k-1)/2N$. If we run time at rate N, then jumps occur at rate $N \cdot k/2N = k/2$, so the total rate of coalescence is $k(k-1)/2$, the right rate for Kingman's coalescent.   □

Returning to the forward perspective, suppose that we have two alleles A and a, and let X_t be the number of copies of A. The transition rates for X_t are

$$i \to i+1 \quad \text{at rate} \quad b_i = (2N - i) \cdot \frac{i}{2N}$$

$$i \to i-1 \quad \text{at rate} \quad d_i = i \cdot \frac{2N - i}{2N}$$

where b is for birth and d is for death. In words, a's are selected for possible replacement at total rate $2N - i$. The number of A's will increase if an A is chosen to be the parent of the new individual, an event of probability $i/2N$. The reasoning is similar for the second rate. Note that the two rates are equal.

1.5.1 Fixation probability and time

The next result is the same as the one for the Wright-Fisher model given in (1.2). Let $\tau = \min\{t : X_t = 0 \text{ or } X_t = 2N\}$ be the fixation time.

Theorem 1.31. *In the Moran model, the probability that A becomes fixed when there are initially i copies is $i/2N$.*

Proof. The rates for up and down jumps are the same, so $(d/dt)E_i X_t = 0$, and hence $E_i X_t$ is constant in time. As in the Wright-Fisher model, this implies

$$i = E_i X_\tau = 2N P_i(X_\tau = 2N)$$

and the desired result follows.   □.

Writing $\bar{E}_i \tau = E_i(\tau | T_{2N} < T_0)$, we can state

Theorem 1.32. *Let $p = i/2N$. In the Moran model*

$$\bar{E}_i \tau \approx -\frac{2N(1-p)}{p} \log(1-p) \tag{1.32}$$

Recalling that the Moran model coalesces twice as fast as the Wright-Fisher model, this agrees with the classic result due to Kimura and Ohta (1969a). As $p \to 0$, $-\log(1-p)/p \to 1$, so

$$\bar{E}_i \tau \to 2N \tag{1.33}$$

In particular, this gives the expected fixation time of a new mutation, conditional on its fixation.

Proof. Let S_j be the amount of time spent at j before time τ and note that

$$E_i \tau = \sum_{j=1}^{2N-1} E_i S_j \tag{1.34}$$

Let N_j be the number of visits to j. Let $q(j) = j(2N-j)/2N$ be the rate at which the chain leaves j. Since each visit to j lasts for an exponential amount of time with mean $1/q(j)$, we have

$$E_i S_j = \frac{1}{q(j)} E_i N_j \tag{1.35}$$

If we let $T_j = \min\{t : X_t = j\}$ be the first hitting time of j, then

$$P_i(N_j \geq 1) = P_i(T_j < \infty)$$

Letting $T_j^+ = \min\{t : X_t = j \text{ and } X_s \neq j \text{ for some } s < t\}$ be the time of the first return to j, we have for $n \geq 1$

$$P_i(N_j \geq n+1 | N_j \geq n) = P_j(T_j^+ < \infty)$$

The last formula shows that, conditional on $N_j \geq 1$, N_j has a geometric distribution with success probability $P_j(T_j^+ = \infty)$. Combining this with our formula for $P_i(N_j \geq 1)$, we have

$$E_i N_j = \frac{P_i(T_j < \infty)}{P_j(T_j^+ = \infty)} \tag{1.36}$$

Since the average value of X_t is constant in time, the martingale argument shows that for $0 \leq i \leq j$

$$i = j P_i(T_j < T_0) + 0 \cdot [1 - P_i(T_j < T_0)]$$

and solving gives

$$P_i(T_j < T_0) = \frac{i}{j} \qquad P_i(T_j > T_0) = \frac{j - i}{j} \qquad (1.37)$$

Similar reasoning shows that for $j \leq i \leq 2N$,

$$i = jP_i(T_j < T_{2N}) + 2N[1 - P_i(T_j < T_{2N})]$$

and solving gives

$$P_i(T_j < T_{2N}) = \frac{2N - i}{2N - j} \qquad P_i(T_j > T_{2N}) = \frac{i - j}{2N - j} \qquad (1.38)$$

When the process leaves j, it goes to $j - 1$ or $j + 1$ with equal probability, so

$$P_j(T_j^+ = \infty) = \frac{1}{2} \cdot P_{j+1}(T_j > T_{2N}) + \frac{1}{2} \cdot P_{j-1}(T_j > T_0)$$

$$= \frac{1}{2} \cdot \frac{1}{2N - j} + \frac{1}{2} \cdot \frac{1}{j} = \frac{2N}{2j(2N - j)}$$

Putting our results into (1.36) gives

$$E_i N_j = \begin{cases} \frac{i}{j} \cdot \frac{2j(2N-j)}{2N} & i \leq j \\ \frac{2N-i}{2N-j} \cdot \frac{2j(2N-j)}{2N} & j \leq i \end{cases}$$

Since $q(j) = 2j(2N - j)/2N$, (1.35) gives

$$E_i S_j = \begin{cases} \frac{i}{j} & i \leq j \\ \frac{2N-i}{2N-j} & j \leq i \end{cases} \qquad (1.39)$$

If we let $h(i) = P_i(T_{2N} < T_0)$ and let $p_t(i, j)$ be the transition probability for the Moran model, then it follows from the definition of conditional probability and the Markov property that

$$\bar{p}_t(i, j) = \frac{P_i(X_t = j, T_{2N} < T_0)}{P_i(T_{2N} < T_0)} = p_t(i, j) \cdot \frac{h(j)}{h(i)}$$

Integrating from $t = 0$ to ∞, we see that the conditioned chain has

$$\bar{E}_i S_j = \int_0^\infty \bar{p}_t(i, j)\, dt = \frac{h(j)}{h(i)} E_i S_j \qquad (1.40)$$

$h(i) = i/2N$, so using the formula for $E_i S_j$ given in (1.39), we have

$$\bar{E}_i S_j = \begin{cases} 1 & i \leq j \\ \frac{2N-i}{i} \cdot \frac{j}{2N-j} & j \leq i \end{cases} \qquad (1.41)$$

By the reasoning that led to (1.34),

$$\bar{E}_i\tau = \sum_{j=1}^{2N-1} \bar{E}_i S_j = \sum_{j=i}^{2N-1} 1 + \frac{2N-i}{i} \cdot \sum_{j=1}^{i-1} \frac{j}{2N-j}$$

The first sum is $2N - i$. For the second we note that

$$\sum_{j=1}^{i-1} \frac{j}{2N-j} = 2N \sum_{j=1}^{i-1} \frac{j/2N}{1-j/2N} \cdot \frac{1}{2N} \approx 2N \int_0^p \frac{u}{1-u}\, du$$

where $p = i/2N$. To evaluate the integral, we note that it is

$$= \int_0^p -1 + \frac{1}{1-u}\, du = -p - \log(1-p)$$

Combining the last three formulas gives

$$\bar{E}_i\tau \approx 2N(1-p) + \frac{2N(1-p)}{p}(-p - \log(1-p))$$
$$= -\frac{2N(1-p)}{p} \log(1-p)$$

which gives (1.32).   □

1.5.2 Site frequency spectrum mean

Consider now the Moran model with infinite sites mutations.

Theorem 1.33. *Suppose the per locus mutation rate is u, and we have a sample of size n. Let η_m be the sites where m individuals in the sample have the mutant nucleotide. For the Moran model, $E\eta_m = 2Nu/m$. For the Wright-Fisher model, $E\eta_m = 4Nu/m$.*

Proof. It suffices to prove the result for the Moran model, because to pass from the Moran model to the Wright-Fisher model, we only have to replace u by $2u$. If we are to find a mutation in k individuals at time 0, then it must have been introduced into one individual at some time t units in the past and risen to frequency k, an event of probability $p_t(1,k)$, where p_t is the transition probability for the Moran model. Integrating over time gives that the expected number of mutations with frequency k in the population is

$$2Nu \int_0^\infty p_t(1,k)\, dt$$

The quantity $G(i,j) = \int_0^\infty p_t(i,j)\, dt$ is called the *Green's function* for the Markov chain and represents the expected value of S_j, the time spent at j for a process starting at i. By (1.39),

$$G(i, j) = E_i S_j = \begin{cases} \frac{i}{j} & i \le j \\ \frac{2N-i}{2N-j} & j \le i \end{cases}$$

When $i = 1$ we only have the first case, so $G(1, k) = 1/k$, and the expected number of mutations with k copies in the population is $2Nu/k$. If there are k copies of a mutant in the population and we take a sample of size n from the population (with replacement), then the number of mutants in the sample is binomial$(n, k/2N)$ and hence the expected number of mutations with m copies in the sample is

$$\sum_{k=1}^{2N-1} \frac{2Nu}{k} \binom{n}{m} (k/2N)^m (1 - k/2N)^{n-m}$$

If the sample size n is much smaller than the population size N, this will be a good approximation for sampling without replacement. To evaluate this sum, we write it as

$$2Nu \cdot \frac{1}{2N} \sum_{k=1}^{2N-1} \frac{2N}{k} \binom{n}{m} (k/2N)^m (1 - k/2N)^{n-m}$$

and change variables $y = k/2N$ to see that this is approximately

$$\frac{2N\mu}{m} \int_0^1 dy \binom{n}{m} m y^{m-1} (1 - y)^{n-m}$$

The term in front is the answer, so we want to show that the integral is 1. To do this, we integrate by parts (differentiating y^{m-1} and integrating $(1 - y)^{n-m}$) to conclude that if $2 \le m < n$,

$$\frac{n(n - 1) \cdots (n - m + 1)}{(m - 1)!} \int_0^1 y^{m-1}(1 - y)^{n-m} \, dy$$
$$= \frac{n(n - 1) \cdots (n - m + 2)}{(m - 2)!} \int_0^1 y^{m-2}(1 - y)^{n-m+1} \, dy$$

Iterating, we finally reach $\int_0^1 n y^{n-1} \, dy = 1$ when $m = 2$. $\qquad\square$

Example 1.7. Single nucleotide polymorphisms (abbreviated SNPs and pronounced "snips") are nucleotides that are variable in a population but in which no allele has a frequency of more than 99%. If we restrict our attention to nucleotides for which there has been only one mutation, then our previous result can be used to estimate the number of SNPs in the human genome. The probability that the less frequent allele is present in at least a fraction x of the population is then

$$\theta \sum_{m=xn}^{(1-x)n} \frac{1}{m} = \theta \int_x^{1-x} \frac{1}{y} \, dy = \theta \log \left(\frac{1 - x}{x} \right)$$

in agreement with a computation of Kruglyak and Nickerson (2001). If we use their figure for the nucleotide diversity of $\theta = 1/1331$ and take $x = 0.01$ then we see the density of SNPs is $\log(99) = 4.59$ times θ or 1 every 289 bp, or more than 10 million in the human genome.

2

Estimation and Hypothesis Testing

"It is easy to lie with statistics. It is hard to tell the truth without it." Andrejs Dunkels

2.1 Site frequency spectrum covariance

Assume the Wright-Fisher model with infinitely many sites, and let η_i be the number of sites where the mutant (or derived) allele has frequency i in a sample of size n. In this section, we will describe work of Fu (1995), which allows us to compute $\text{var}(\eta_i)$ and $\text{cov}(\eta_i, \eta_j)$. We begin with the new proof of the result for the mean given in Theorem 1.33.

Theorem 2.1. $E\eta_i = \theta/i$.

Proof. We say that a time t is at level k if there are k sample lineages in the coalescent at that time. The key to the proof is to break things down according to the level at which mutations occur.

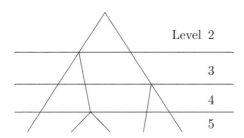

Let L_m be the total length of branches with m descendants. Let J_ℓ^k be the number of sampled individuals that are descendants of edge ℓ at level k.

R. Durrett, *Probability Models for DNA Sequence Evolution*,
DOI: 10.1007/978-0-387-78168-6_2, © Springer Science+Business Media, LLC 2008

Theorem 1.6 implies that $(J_1^k, \ldots, J_k^k)$ is uniformly distributed over the vectors of k positive integers that add up to n. Recalling that the number of such vectors is $\binom{n-1}{k-1}$, it follows that for $1 \le i \le n-k+1$

$$P(J_\ell^k = i) = \binom{n-i-1}{k-2} / \binom{n-1}{k-1} \tag{2.1}$$

since the numerator gives the number of ways of breaking the remaining $n-i$ individuals into $k-1$ nonempty groups.

Since level k lasts for an amount of time with mean $2/k(k-1)$ and there are k edges on level k,

$$EL_i = \sum_{k=2}^{n} \frac{2}{k(k-1)} \cdot k \cdot \frac{\binom{n-i-1}{k-2}}{\binom{n-1}{k-1}} \tag{2.2}$$

Since mutations occur on the tree at rate $\theta/2$, it suffices to show $EL_i = 2/i$. To evaluate the sum, we need the following identity:

$$\frac{\binom{n-i-1}{k-2}}{\binom{n-1}{k-1}} \frac{1}{k-1} = \frac{\binom{n-k}{i-1}}{\binom{n-1}{i}} \frac{1}{i} \tag{2.3}$$

Skipping the proof for the moment and using this in (2.2) gives

$$EL_i = \frac{2}{i} \sum_{k=2}^{n} \frac{\binom{n-k}{i-1}}{\binom{n-1}{i}} = \frac{2}{i}$$

To see the last equality, note that if we are going to pick i things out of $n-1$ and the index of the first one chosen is $k-1$, with $2 \le k \le n$, then we must choose $i-1$ from the last $(n-1) - (k-1)$ items.

The last detail is to prove (2.3). Recalling the definition of the binomial coefficients,

$$\frac{\binom{n-i-1}{k-2}}{\binom{n-1}{k-1}} \frac{1}{k-1} = \frac{1}{k-1} \frac{(n-i-1)!}{(k-2)!(n-i-k+1)!} \cdot \frac{(k-1)!(n-k)!}{(n-1)!}$$

Cancelling the $(k-1)!$, swapping the positions of $(n-i-1)!$ and $(n-k)!$, and then multiplying by $i!/(i-1)!i$, the above becomes

$$= \frac{(n-k)!}{i(i-1)!(n-k-(i-1))!} \cdot \frac{i!(n-i-1)!}{(n-1)!} = \frac{\binom{n-k}{i-1}}{\binom{n-1}{i}} \frac{1}{i}$$

and the proof is complete. □

To state Fu's (1995) result for the covariances, we recall $h_n = \sum_{i=1}^{n-1} 1/i$ and let

$$\beta_n(i) = \frac{2n}{(n-i+1)(n-i)}(h_{n+1} - h_i) - \frac{2}{n-i}$$

Theorem 2.2. $var(\eta_i) = \theta/i + \theta^2\sigma_{ii}$ and for $i \neq j$, $cov(\eta_i, \eta_j) = \sigma_{ij}\theta^2$. The diagonal entries σ_{ii} are given by

$$
\begin{cases}
\beta_n(i+1) & i < n/2 \\[2mm]
2\dfrac{h_n - h_i}{n - i} - \dfrac{1}{i^2} & i = n/2 \\[3mm]
\beta_n(i) - \dfrac{1}{i^2} & i > n/2
\end{cases}
\tag{2.4}
$$

while σ_{ij} with $i > j$ are given by

$$
\begin{cases}
\dfrac{\beta_n(i+1) - \beta_n(i)}{2} & i + j < n \\[3mm]
\dfrac{h_n - h_i}{n - i} + \dfrac{h_n - h_j}{n - j} - \dfrac{\beta_n(i) + \beta_n(j+1)}{2} - \dfrac{1}{ij} & i + j = n \\[3mm]
\dfrac{\beta_n(j) - \beta_n(j+1)}{2} - \dfrac{1}{ij} & i + j > n
\end{cases}
\tag{2.5}
$$

To help understand the formulas, we begin by computing σ_{ij} in the special case $n = 8$.

	$j = 1$	2	3	4	5	6	7
$i = 1$	**0.3211**	-0.0358	-0.0210	-0.0141	-0.0103	-0.0079	**0.1384**
2	-0.0358	**0.2495**	-0.0210	-0.0141	-0.0103	**0.1328**	-0.0356
3	-0.0210	-0.0210	**0.2076**	-0.0141	**0.1283**	-0.0346	-0.0356
4	-0.0141	-0.0141	-0.0141	**0.3173**	-0.0359	-0.0275	-0.0267
5	-0.0103	-0.0103	**0.1283**	-0.0359	**0.1394**	-0.0230	-0.0216
6	-0.0079	**0.1328**	-0.0346	-0.0275	-0.0230	**0.1310**	-0.0183
7	**0.1384**	-0.0356	-0.0267	-0.0216	-0.0183	-0.0159	**0.1224**

The numbers on the diagonal must be positive since $var(\eta_i) > 0$. All of the off-diagonal elements are negative numbers, except for the σ_{ij} with $i + j = n$. Intuitively, these are positive due the fact that the first split in the tree may have i lineages on the left and $n - i$ on the right, and this event increases both η_i and η_{n-i}. The negative off-diagonal elements are small, but since there are $O(n^2)$ of them, their sum is significant.

The next two figures give the values of the covariance matrix when $N = 25$. The first gives the values on the diagonal $i = j$ and the anti-diagonal $i + j = 25$. Since 25 is odd, these do not intersect. The bump in the middle of the graph of the diagonal covariances may look odd, but when $i \geq 13$, there cannot be two edges in the tree that produce these mutations. To better see the off-diagonal entries in the covariance matrix in the second figure, we have plotted $-\sigma_{ij}$ and inserted 0's on the two diagonals. The largest value is at σ_{12}. Note the jump in the size of the correlations when we enter the region $i + j > 25$.

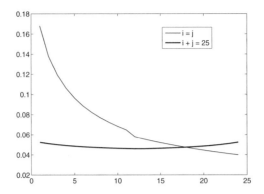

Fig. 2.1. Diagonals of the covariance matrix when $N = 25$.

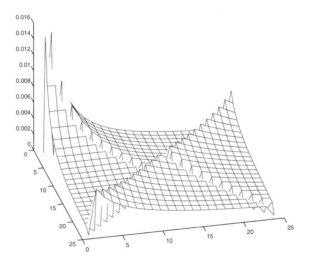

Fig. 2.2. Off-diagonal entries (times -1) of the covariance matrix when $N = 25$.

Proof. Let $\nu_{\ell,k}$ = the number of mutations that occur on edge ℓ at level k. To compute variances and covariances, we use the identity

$$\eta_i = \sum_{k=2}^{n} \sum_{\ell=1}^{k} 1_{(J_\ell^k = i)} \nu_{\ell,k}$$

where the indicator function $1_{(J_\ell^k = i)}$ is equal to 1 if $J_\ell^k = i$ and 0 otherwise. Multiplying two copies of the sum and sorting the terms, we conclude that

$$E(\eta_i \eta_j) = 1_{(i=j)} \sum_{k=2}^{n} kP(J_1^k = i)E\nu_{1,k}^2$$

$$+ \sum_{k=2}^{n} k(k-1)P(J_1^k = i, J_2^k = j)E(\nu_{1,k}\nu_{2,k}) \qquad (2.6)$$

$$+ \sum_{k \neq h} khP(J_1^k = i, J_1^h = j)E(\nu_{1,k}\nu_{1,h})$$

To evaluate the last expression, we have to compute (i) the expected values of the $\nu_{\ell,k}$, and (ii) the joint distribution of the J_ℓ^k. Readers who lose interest in the details can skip to the result given in Theorem 2.2.

The expected values of the $\nu_{\ell,k}$ are easy to compute. Let t_k be the amount of time at level k. Conditioning on the value of t_k,

$$(\nu_{\ell,k}|t_k = t) = \text{Poisson}(\theta t/2)$$

Since t_k is exponential with mean $2/k(k-1)$,

$$E\nu_{\ell,k} = \frac{\theta}{k(k-1)} \qquad (2.7)$$

Reasoning as in the derivation of (1.23), recalling Poisson($\theta t/2$) has variance $\theta t/2$, and using var$(t_k) = 4/(k^2(k-1)^2)$, we have

$$\text{var}(\nu_{\ell,k}) = E(\text{var}(\nu_{\ell,k}|t_k)) + \text{var}(E(\nu_{\ell,k}|t_k))$$
$$= \frac{\theta}{2}E(t_k) + \frac{\theta^2}{4}\text{var}(t_k))$$
$$= \frac{\theta}{k(k-1)} + \frac{\theta^2}{k^2(k-1)^2}$$

Adding the square of the mean we have

$$E\nu_{\ell,k}^2 = \frac{\theta}{k(k-1)} + \frac{2\theta^2}{k^2(k-1)^2} \qquad (2.8)$$

Conditioning on the values of t_k and t_h, and using $E(t_k^2) = 8/(k^2(k-1)^2)$,

$$\text{for } \ell \neq m \quad E(\nu_{\ell,k}\nu_{m,k}) = \frac{\theta^2}{4}Et_k^2 = \frac{2\theta^2}{k^2(k-1)^2} \qquad (2.9)$$

$$\text{for } k \neq h \quad E(\nu_{\ell,k}\nu_{m,h}) = \frac{\theta^2}{4}Et_k Et_h = \frac{\theta^2}{k(k-1)h(h-1)} \qquad (2.10)$$

The joint probabilities for pairs of edges are more difficult to compute. The easiest situation occurs when both edges are on the same level. When $k = 2$, Tajima's result, Theorem 1.6, tells us that

$$P(J_1^2 = i, J_2^2 = n - i) = \frac{1}{n - 1} \qquad (2.11)$$

Extending the reasoning for (2.1) shows that for $k \geq 3$ and $i + j < n$

$$P(J_\ell^k = i, J_m^k = j) = \frac{\binom{n-i-j-1}{k-3}}{\binom{n-1}{k-1}} \qquad (2.12)$$

since the remaining $n - (i + j)$ sampled individuals are divided between the other $k - 2$ lineages. Similar reasoning shows that if $2 \leq t \leq k - 1$ and $j < i < n$,

$$P\left(J_1^k = j, \sum_{\ell=2}^{t} J_\ell^k = i - j \right) = \frac{\binom{i-j-1}{t-2}\binom{n-i-1}{k-t-1}}{\binom{n-1}{k-1}} \qquad (2.13)$$

Suppose now that $k < h$ and let $D_{\ell,k}^h$ be the number of descendants on level h of edge (ℓ, k), i.e., edge ℓ on level k. Writing $(m, h) < (\ell, k)$ for (m, h) is a descendant of (ℓ, k), there are two cases to consider: $(m, h) < (\ell, k)$ and $(m, h) \not< (\ell, k)$. In each case there is the general situation and a degenerate subcase in which one group of vertices is empty.

Case 1.

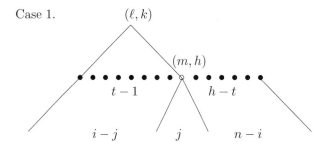

Here we have not drawn the tree, but have indicated the sizes of the various sets of descendants. Recalling that edges on each level are randomly labeled, we see that if $j < i < n$ and $t \geq 2$,

$$P(D_{\ell,k}^h = t, (m, h) < (\ell, k), J_\ell^k = i, J_m^h = j)$$
$$= \frac{\binom{h-t-1}{k-2}}{\binom{h-1}{k-1}} \cdot \frac{t}{h} \cdot \frac{\binom{i-j-1}{t-2}\binom{n-i-1}{h-t-1}}{\binom{n-1}{h-1}} \qquad (2.14)$$

The first factor is $P(D_{\ell,k}^h = t)$, computed from (2.1). The second is $P((m, h) < (\ell, k)|D_{\ell,k}^h = t)$. When the first two events occur, we need $J_m^h = j$ and the other $t - 1$ descendants of (ℓ, k) on level h to have a total of $i - j$ descendants, so the desired result follows from (2.13).

If $i = j$, then we must have $t = 1$ and

$$P(D^h_{\ell,k} = 1, (m,h) < (\ell,k), J^k_\ell = i, J^h_m = i)$$
$$= \frac{\binom{h-2}{k-2}}{\binom{h-1}{k-1}} \cdot \frac{1}{h} \cdot \frac{\binom{n-i-1}{h-2}}{\binom{n-1}{h-1}} \quad (2.15)$$

If one is willing to adopt the somewhat strange convention that $\binom{-1}{-1} = 1$, this can be obtained from the previous formula by setting $t = 1$ and $i - j = 0$.

Case 2.

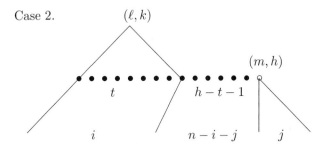

Reasoning similar to Case 1 shows that if $t \leq h - 2$ and $i + j < n$,

$$P(D^h_{\ell,k} = t, (m,h) \not< (\ell,k), J^k_\ell = i, J^h_m = j)$$
$$= \frac{\binom{h-t-1}{k-2}}{\binom{h-1}{k-1}} \cdot \frac{h-t}{h} \cdot \frac{\binom{i-1}{t-1}\binom{n-(i+j)-1}{h-t-2}}{\binom{n-1}{h-1}} \quad (2.16)$$

If $i + j = n$, then we must have $t = h - 1$ and $k = 2$. (We may have $h > 2$.)

$$P(D^h_{\ell,k} = h-1, (m,h) \not< (\ell,k), J^k_\ell = i, J^h_m = n - i)$$
$$= \frac{1}{h-1} \cdot \frac{1}{h} \cdot \frac{\binom{i-1}{h-2}}{\binom{n-1}{h-1}} \quad (2.17)$$

Again, if $\binom{-1}{-1} = 1$, this can be obtained from the previous formula by setting $t = h - 1$ and $n - (i+j) = 0$.

At this point we have everything we need to compute the variances and covariances. The expressions in (2.14)–(2.17) are complicated and two of them must be summed over t. Remarkably, Fu (1995) was able to obtain the given analytical formulas for the quantities of interest. Many details need to end up with these results, so we refer the reader to the original article for details.

2.2 Estimates of θ

Let η_i be the number of sites where the mutant allele is present i times in a sample of size n. There are a number of ways of estimating $\theta = 4Nu$, where u

is the mutation rate for the locus, using linear functions of the site frequency spectrum, that is, estimators of the form

$$\hat{\theta} = \sum_{i=1}^{n-1} c_{n,i}\eta_i.$$

Fu and Li (1993). $\theta_{FL} = \eta_1$.

Watterson (1975). $\theta_W = h_n^{-1}\sum_{i=1}^{n-1}\eta_i$, where $h_n = \sum_{i=1}^{n-1} 1/i$.

Zeng, Fu, Shi, and Wu (2006). $\theta_L = \frac{1}{n-1}\sum_{i=1}^{n-1} i\eta_i$

Tajima (1983). $\theta_\pi = \frac{2}{n(n-1)}\sum_{i=1}^{n-1} i(n-i)\eta_i$.

Fay and Wu (2000). $\theta_H = \frac{2}{n(n-1)}\sum_{i=1}^{n-1} i^2\eta_i = 2\theta_L - \theta_\pi$.

To check that these are unbiased estimators, we note that $E\hat{\theta} = \sum_{i=1}^{n-1} c_{n,i}\theta/i$. The fact that $E\hat{\theta} = \theta$ in the first three cases is easy to see. For the fourth and fifth, we note that

$$\sum_{i=1}^{n-1} i = \frac{n(n-1)}{2} = \sum_{i=1}^{n-1}(n-i)$$

It is much more difficult to compute the variances. Using $g_n = \sum_{i=1}^{n-1} 1/i^2$, we can write the answers as

$$\operatorname{var}\left(\theta_W\right) = \frac{\theta}{h_n} + \frac{g_n}{h_n^2}\theta^2 \tag{2.18}$$

$$\operatorname{var}\left(\eta_1\right) = \theta + 2\frac{nh_n - 2(n-1)}{(n-1)(n-2)}\theta^2 \tag{2.19}$$

$$\operatorname{var}\left(\theta_\pi\right) = \frac{n+1}{3(n-1)}\theta + \frac{2(n^2+n+3)}{9n(n-1)}\theta^2 \tag{2.20}$$

$$\operatorname{var}\left(\theta_L\right) = \frac{n}{2(n-1)}\theta + \left[2\left(\frac{n}{n-1}\right)^2 (g_{n+1}-1) - 1\right]\theta^2 \tag{2.21}$$

$$\operatorname{var}\left(\theta_H\right) = \theta + \frac{2[36n^2(2n+1)g_{n+1} - 116n^3 + 9n^2 + 2n - 3]}{9n(n-1)^2}\theta^2 \tag{2.22}$$

We have seen the first and third results in (1.22) and (1.30). The second is due to Fu and Li (1993). The fourth and fifth are from Zeng, Fu, Shi, and Wu (2006). Note that, as we proved in Theorem 1.27, each variance has the form $a_n\theta + b_n\theta^2$. The term with θ is the mutational variance due to the placement of mutations on the tree, while the term with θ^2 is the evolutionary variance due to fluctuations in the shape of the tree.

It is easy to visually compare the terms with θ. In the case of θ_W, $\theta/h_n \to 0$. η_1 and θ_H have θ, while for θ_π and θ_L the terms are $\approx \theta/3$ and $\approx \theta/2$

respectively. It is hard to understand the relationship between the variances by looking at the formulas. In the next table, we have evaluated the coefficient of θ^2 for the indicated values of n. The limits as $n \to \infty$ are 0, 0, 2/9, 0.289863, and 0.541123.

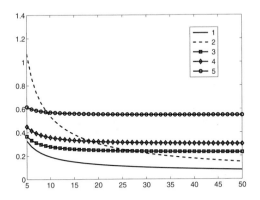

Fig. 2.3. Coefficient of θ^2 in the variance of 1. θ_W, 2. η_1, 3. θ_π, 4. θ_L, and 5. θ_H.

In addition to formulas for the variances, we will need the following co-variances:

$$\operatorname{cov}(\eta_k, S_n) = \frac{\theta}{k} + \frac{h_n - h_k}{n-k}\theta^2 \tag{2.23}$$

$$\operatorname{cov}(\theta_L, S_n) = \theta + \frac{ng_n - (n-1)}{(n-1)}\theta^2 \tag{2.24}$$

$$\operatorname{cov}(\theta_\pi, S_n) = \theta + \left(\frac{n+2}{2n}\right)\theta^2 \tag{2.25}$$

$$\operatorname{cov}(\theta_L, \theta_\pi) = \frac{n+1}{3(n-1)}\theta + \frac{7n^2 + 3n - 2 - 4n(n+1)g_{n+1}}{2(n-1)^2} \tag{2.26}$$

The first for $k = 1$ is formula (25) from Fu and Li (1993). We prove the more general result here. The third is (25) from Tajima (1989). The other two are in (A1) of Zeng, Fu, Shi, and Wu (2006).

Where do these formulas come from?

In principle, the problem is solved by Fu's result for the site freqeuncy spectrum covariance. Given $X_j = \sum_{i=1}^{n-1} c_{n,i}^j \eta_i$, $\operatorname{cov}(X_1, X_2) = a_n\theta + b_n\theta^2$, where

$$a_n = \sum_{i=1}^{n-1} c_{n,i}^1 c_{n,i}^2 / i \tag{2.27}$$

$$b_n = \sum_{i,j} c_{n,i}^1 \sigma_{ij} c_{n,j}^2 \tag{2.28}$$

Of course, $\operatorname{var}(X_1) = \operatorname{cov}(X_1, X_1)$.

If one wants numerical values for a fixed n, then (2.27), (2.28), and Theorem 2.2 lead easily to the desired answer. It is fairly straightforward to use (2.27) to algebraically compute the coefficients a_n, but the b_n are considerably more difficult. To illustrate this point, we will compute $\operatorname{cov}(\eta_k, S_n)$, which is the simplest of these tasks.

Proof of 2.23. In this case, $c_{n,i}^1 = 1$ if $i = k$ and $c_{n,i}^2 = 1$ for all i, so $a_n = 1/k$. Using (2.28) and Theorem 2.2,

$$b_n = \sum_j \sigma_{kj} = \beta_n(k) + \frac{\beta_n(1) - \beta_n(k)}{2}$$

$$+ (n-k-1)\frac{\beta_n(k) - \beta_n(k+1)}{2} - (n-k-1)\frac{\beta_n(k) - \beta_n(k+1)}{2}$$

$$- \frac{\beta_n(1) - \beta_n(n-k)}{2} - \frac{\beta_n(i) + \beta_n(n-k)}{2}$$

$$+ \frac{h_n - h_k}{n-k} + \frac{h_n - h_{n-k}}{k} - \frac{1}{k}\sum_{j=n-k}^{n-1} \frac{1}{j}$$

The terms involving β_n add up to 0, and the last two terms cancel, leaving us with

$$b_n = \sum_{j=1}^{n-1} \sigma_{kj} = \frac{h_n - h_k}{n-k} \tag{2.29}$$

which proves (2.23). $\qquad\square$

Using (2.23), one can, with some effort, compute $\operatorname{var}(S_n)$, $\operatorname{cov}(\theta_L, S_n)$, and $\operatorname{cov}(\theta_\pi, S_n)$, since they are all equal to

$$\sum_{k=1}^{n-1} c_{n,k}^1 \operatorname{cov}(\eta_k, S_n)$$

However, in the other cases, the terms involving β_n do not cancel, and we have not been able to derive the other formulas given above from Theorem 2.2. On the other hand, the results in this section allow one to easily compute the variances and covariances numerically, so perhaps such tedious algebraic manipulations are obsolete.

2.3 Hypothesis testing overview

One of the most important questions we face is:

Is the observed DNA sequence data consistent with neutral evolution in a homogeneously mixing population of constant size?

As the last sentence indicates, there are many assumptions that can be violated. Much of the rest of the book is devoted to investigating consequences of alternatives to this null hypothesis. In Chapters 4 and 5, we study population size changes and population structure. In Chapters 6 and 7, we study various types of fitness differences between alleles. Most of the alternative hypotheses can be grouped into two categories.

A. Those that tend to make a *star-shaped* genealogy:

Examples of this are:

- *Population bottleneck.* If, as we work backwards in time, there is a sudden decrease in the population size, then the coalescence rate will become large. Situations that can cause this are the founding of a new population by a small number of migrants or, what is essentially the same, improving a crop species by choosing a few individuals with desirable properties.
- *Selective sweep.* This term refers to the appearance of a favorable mutation that rises in frequency until it takes over the population. In the absence of recombination, this is a severe population bottleneck because the entire population will trace its ancestry to the individual with the favorable mutation.
- *Population growth.* In the human population, which has experienced a period of exponential growth, then the coalescence rate will be initially small, and the genealogical tree will have tips that are longer than usual.

B. At the other extreme from a star-shaped genealogy is a *chicken legs genealogy*. There are two long, skinny legs with feet on the ends.

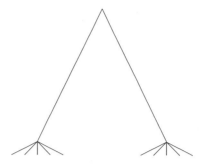

Examples of this are:

- *Population subdivision.* Imagine two isolated populations that exchange migrants infrequently. If we sample 10 individuals from each population, then the two subsamples will coalesce as usual and then eventually the their two ancestral lineages will coalesce.

- *Balancing selection.* This term refers to a situation in which the fitness of heterozygotes Aa is larger than that of homozygotes AA and aa. In this case, the population will settle into a equilibrium in which the two alleles A and a are each present with fixed frequencies. As we will see in Section 6.2, this is essentially a two-population model, with migration between chromosomes with A and chromosomes with a being caused by mutation or recombination.

Notice that in each case there are different explanations that produce the same effect. Thus, one of the problems we will face in hypothesis testing is that if we reject neutral evolution in a homogeneously mixing population of constant size, it will be difficult to say if this is due to natural selection or to demographic factors such as population structure or population size changes.

This is a serious problem for the difference statistics considered in the next section because the tests are performed by comparing the observation to the distribution of the statistic under the neutral model. The HKA test discussed in Section 2.5 avoids this problem by comparing patterns of variability at two regions in the same individuals. The McDonald-Kreitman test discussed in Section 2.6 compares the ratio of nonsynonymous to synonymous polymorphisms within species to the ratio of nonsynonymous to synonymous fixed differences between species, which should not be affected by the shape of the tree.

Here, we will content ourselves to describe the mechanics of the tests and give a few examples to illustrate their use. For more on the issues involved in the use of these and other tests, there are a number of excellent survey articles: Kreitman (2000), Nielsen (2001), Fay and Wu (2001), Bamshad and Wooding (2003), Nielsen (2005), and Sabeti et al. (2006).

As the reader will see from these articles, there are many tests that we have not discussed. An important omission is the method of demonstrating the presence of positive selection by comparing the number of nonsynonymous mutations per nonsynonymous site (d_N) to the number of nonsynonymous mutations per nonsynonymous site (d_S). Hughes and Nei (1988) showed that $\omega = d_N/d_S > 1$ for the antigen binding cleft of the Major Histocompatibility Complex. A statistical framework for making inferences regarding d_N and d_S was developed by Goldman and Yang (1994) and Muse and Gaut (1994). In this framework, the evolution of a gene is modeled as a continuous-time Markov chain with state space the 61 possible non-stop codons.

In general, testing $\omega < 1$ for an entire gene is a very conservative test of neutrality. Purifying selection often acts on large parts of genes to preserve their function. To address this, Nielsen and Yang (1998) developed a model in which there are three categories of sites: invariable sites ($\omega = 0$), neutral sites ($\omega = 1$), and positively selected sites ($\omega > 1$). Later, Yang et al. (2000) replaced the neutral class by constrained sites that have a distribution of ω values in $(0, 1)$. This test, which is implemented in the computer program PAML, has been used to provide evidence of positive selection in a number of cases; see Nielsen (2001) for some examples.

2.4 Difference statistics

Given two unbiased estimators of θ, we can subtract them to get a random variable with mean 0 that can be used for testing whether the data is consistent with our model of neutral mutations in a homogeneously mixing population of constant size.

2.4.1 Tajima's D

Tajima (1989) was the first to do this, taking the difference $d = \theta_W - \theta_\pi$. We have computed that

$$\text{var}(S_n) = a_1\theta + a_2\theta^2 \quad \text{where} \quad a_1 = \sum_{i=1}^{n-1} 1/i \qquad a_2 = \sum_{i=1}^{n-1} 1/i^2$$

$$\text{var}(\theta_\pi) = b_1\theta + b_2\theta^2 \quad \text{where} \quad b_1 = \frac{n+1}{3(n-1)} \qquad b_2 = \frac{2(n^2+n+3)}{9n(n-1)}$$

To compute $\text{var}(d)$, we need (2.25):

$$\text{cov}(S_n, \theta_\pi) = \theta + \left(\frac{1}{2} + \frac{1}{n}\right)\theta^2$$

Recalling $\theta_W = S_n/a_1$, we have

$$\text{var}\,(d) = c_1\theta + c_2\theta^2 \quad \text{where} \quad c_1 = b_1 - \frac{1}{a_1}, \quad c_2 = b_2 - \frac{n+2}{a_1 n} + \frac{a_2}{a_1^2}$$

To finish the definition of Tajima's statistic, we need estimators of θ and θ^2, so we can construct an estimator of $\text{var}\,(d)$. For the first we use $\hat{\theta} = S_n/a_1$. For the second we note that

$$E(S_n^2) - ES_n = \text{var}\,(S_n) + E(S_n)^2 - ES_n = (a_1^2 + a_2)\theta^2$$

so our estimate of $\text{var}\,(d)$ is $\hat{v}(d) = e_1 S_n + e_2 S_n(S_n - 1)$, where $e_1 = c_1/a_1$ and $e_2 = c_2/(a_1^2 + a_2)$ and Tajima's test statistic is

$$D_T = \frac{\theta_\pi - \theta_W}{\sqrt{e_1 S_n + e_2 S_n(S_n - 1)}} \tag{2.30}$$

The smallest value of D_T, call it u, occurs when the minor allele has frequency 1 at each segregating site. This happens when there is a *star-shaped* genealogy. In this case,

$$\theta_\pi - \theta_W = \left[\binom{n}{2}^{-1}(n-1) - \frac{1}{h_n}\right] S_n$$

If S_n is large, we have

$$u \approx \left(\frac{2}{n} - \frac{1}{h_n}\right) \Big/ \sqrt{e_2}$$

The largest value of D_T, call it v, occurs where the split between the two nucleotides at each site is as even as it can be. If n is even,

$$\theta_\pi - \theta_W = \left[\binom{n}{2}^{-1}(n/2)^2 - \frac{1}{h_n}\right] S_n$$

If S_n is large, we have

$$v \approx \left(\frac{1}{2} - \frac{1}{h_n}\right) \Big/ \sqrt{e_2}$$

Tajima argued on the basis of simulations that the density function of D_T is approximately a generalized beta distribution with range $[u, v]$:

$$f(D) = \frac{\Gamma(\alpha + \beta)(v - D)^{\alpha-1}(D - u)^{\beta-1}}{\Gamma(\alpha)\Gamma(\beta)(v - u)^{\alpha+\beta-1}}$$

where α and β are chosen to make the mean 0 and the variance 1:

$$\alpha = -\frac{(1 + uv)v}{v - u} \qquad \beta = \frac{(1 + uv)u}{v - u}$$

Table 2 in Tajima's paper gives 90%, 95%, 99%, and 99.9% confidence intervals for a variety of sample sizes.

Example 2.1. Aquadro and Greenberg (1983). In Section 1.4, we computed $\theta_W = 17.959184$ and $\theta_\pi = 14.857143$. Using a small computer program, one can compute

$$
\begin{aligned}
a_1 &= 2.450000 & a_2 &= 1.491389 \\
b_1 &= 0.444444 & b_2 &= 0.312169 \\
c_1 &= 0.036281 & c_2 &= 0.035849 \\
e_1 &= 0.014809 & e_2 &= 0.004784
\end{aligned}
$$

and $\sqrt{\widehat{\mathrm{var}}\,(d)} = 3.114888$, so

$$
D_T = \frac{14.857143 - 17.959184}{3.114888} = -0.995875
$$

The negative value of D_T is caused by an excess of rare alleles. However, from Table 2 on page 592 of Tajima (1989), we see that the 90% confidence interval for D_T in the case $n = 7$ is $(-1.498, 1.728)$, so this value of D_T is not very unusual.

2.4.2 Fu and Li's D

Fu and Li (1993) used the difference $\theta_W - \theta_{FL}$ as a test statistic, or what is essentially the same, $d = S_n - h_n \eta_1$, where $h_n = \sum_{i=1}^{n-1} 1/i$. Again, this statistic is normalized by dividing by the square root of an estimate of $\mathrm{var}\,(d)$. The ingredients necessary for the computation of $\mathrm{var}\,(d)$ are given in the previous section, but we skip the somewhat messy details of the derivation, which is similar to the computation of Tajima's denominator.

$$
D_{FL} = \frac{S_n - h_n \eta_1}{\sqrt{u_D S_n + v_D S_n^2}} \tag{2.31}
$$

where $u_D = h_n - 1 - v_D$, $g_n = \sum_{i=1}^{n-1} 1/i^2$,

$$
v_D = 1 + \frac{h_n^2}{g_n + h_n^2}\left(c_n - \frac{n+1}{n-1}\right) \quad \text{and} \quad c_n = \frac{2nh_n - 4(n-1)}{(n-1)(n-2)}
$$

To save you some arithmetic, Table 1 of Fu and Li (1993) gives the values of h_n (which they call a_n) and v_D.

Table 2 of Fu and Li (1993) gives cutoff values for their test, which are based on simulation. Since the cutoff values depend on θ, they chose to present conservative percentage points that are valid for $\theta \in [2, 20]$. To illustrate their method, we will consider a data set.

Example 2.2. Hamblin and Aquadro (1996) studied DNA sequence variation at the *glucose dehydrogenase (Gld)* locus in *Drosophila simulans*. The *Gld* locus is near the centromere of chromosome 3 in a region of low recombination.

Hamblin and Aquadro sequenced 970 nucleotides from exon 4 from 11 *D. simulans* chromosomes sampled from a population in Raleigh, N.C. These 11 sequences and that of one *D. melanogaster* individual are given in the table below. As usual, dots in rows 1–11 indicate that the sequence agrees with the *D. melanogaster* sequence.

These two *Drosophila* species diverged about 2.5 million years ago, which is about 25 million generations. Since a typical estimate of the *Drosophila* effective population size is one million, it seems likely that the most recent common ancestor of the 11 *D. simulans* individuals, which has mean roughly 4 million generations, will occur before coalescence with the *D. melanogaster* lineage. Thus, the *D. melanogaster* sequence gives us information about the state of the most recent common ancestor of the *D. simulans* individuals and allows us to conclude which nucleotides represent mutations. Note that in all cases but position 5413, the nucleotide in *D. melanogaster* agrees with one of the *D. simulans* individuals. In this case, the state of the most recent common ancestor is ambiguous, but it is clear from the data that the mutation is

```
     44444444455555555555555555
     66677789901111233344444555
     01925869913579056912568115
     92039316947867002235516024
mel  CCTTACCCGTGAAGTCCCCTGACCGG
  1  T.GT.G....AG.A....G.......
  2  T.G..G....AG.A....G.......
  3  T....G.....G.A....G.A...G.
  4  T....G....AG.A....G.A...G.
  5  T..CT.....AGGA...TA.....G.
  6  T..CT......G.A...TA.....G.
  7  T..CT.....AGGA.TTTA.....G.
  8  ...CTG.AAC.G.C.TTTGC...A..
  9  ...CTG.AACAGGC...TGC...A..
 10  .T...GA..CA...ATTTG.AT..GA
 11  .T...GA..CA...ATTTG.ATT.GA
```

There are 26 segregating sites, but since there are two mutations at 5197, the total number of mutations $\eta = 27$. This represents a deviation from the infinite sites model, but in light of the birthday problem calculations in Section 1.4 is not an unexpected one, since the expected number of nucleotides hit twice is $(27 \cdot 26)/(2 \cdot 970) = 0.35926$ and the probability of no double hit is $\approx \exp(-0.35962) = 0.6982$. However, as we will now see, it is very unusual for there to be only one external mutation at 5486. Table 2 of Fu and Li (1993) gives $h_{11} = 2.929$ and $v_D = 0.214$, so $u_D = h_{11} - 1 - v_D = 1.929 - 0.214 = 1.715$. The value of Fu and Li's statistic is thus

$$D_{FL} = \frac{27 - 1 \cdot 2.929}{\sqrt{(1.715)(27) + (0.214)(27)^2}} = 1.68$$

When $n = 11$, a 95% confidence interval for D_{FL} is $(-2.18, 1.57)$. Thus, there is a significant deficiency of external mutations.

To perform Tajima's test on these data, we ignore the column with two mutations so that there are 25 segregating sites. The folded site frequency spectrum is

$$
\begin{array}{cccccc}
m & 1 & 2 & 3 & 4 & 5 \\
\tilde{\eta}_m & 1 & 11 & 4 & 7 & 2
\end{array}
$$

Using the fact that $\binom{11}{2} = 55$, we can compute that

$$
\theta_\pi = \frac{1 \cdot 10 + 11 \cdot 18 + 4 \cdot 24 + 7 \cdot 28 + 2 \cdot 30}{55} = \frac{560}{55} = 10.181818
$$

Dividing θ_π by the 970 bp gives an estimate of π of 0.0105, which is consistent with the estimates for *Adh*, *Amy*, and *rosy* discussed in Section 1.4. In contrast, the estimate of θ based on the 25 segregating sites is $25/2.928968 = 8.535429$. Computing as in the previous example, we find

$$
\begin{array}{ll}
a_1 = 2.928968 & a_2 = 1.549768 \\
b_1 = 0.400000 & b_2 = 0.272727 \\
c_1 = 0.058583 & c_2 = 0.049884 \\
e_1 = 0.020001 & e_2 = 0.004925
\end{array}
$$

and $\sqrt{\widehat{\mathrm{var}}\,(d)} = 1.858779$, so Tajima's D is

$$
D_T = \frac{10.1818181 - 8.535429}{1.858779} = 0.885737
$$

Consulting Tajima's table, we see that a 90% confidence interval for D_T is $(-1.572, 1.710)$, so this value of D_T is far from significant.

p values. Tajima's use of the beta distribution and Fu and Li's choice of $\theta \in [2, 20]$ are somewhat arbitrary. To improve the computation of p values for these tests, Simonsen, Churchill, and Aquadro (1995) used (1.25) to construct a $1 - \beta$ confidence interval (θ_L, θ_U) for θ and then defined an interval for the test statistic which had probability $\geq 1-(\alpha-\beta)$ for a grid of values in $[\theta_L, \theta_U]$.

2.4.3 Fay and Wu's H

Fay and Wu (2000) considered the difference

$$
H = \theta_H - \theta_\pi
$$

In their original paper, they did not normalize the statistic to have variance approximately 1, and they determined the distribution of H using simulations of a neutral coalescence algorithm (without recombination) conditioning on the observed number of segregating sites. Their motivation for defining H, which we will not be able to explain until we consider hitchhiking in Section 6.5, and the site freqeuncy spectrum in Section 7.11, is that the fixation of advantageous alleles should result in an excess of high frequency derived alleles. Since

$$\theta_H = \frac{2}{n(n-1)} \sum_{i=1}^{n-1} \eta_i i^2 \quad \text{versus} \quad \theta_\pi = \frac{2}{n(n-1)} \sum_{i=1}^{n-1} \eta_i i(n-i)$$

and $i^2 > i(n-i)$ when $i > n/2$, this will produce positive values of H. For this reason they consider a one-sided test that rejects the null hypothesis if the observed value is too large.

In computing θ_H an outgroup is needed to infer the ancestral and derived states, but a mutation before the most recent common ancestor or in the outgroup lineage would lead to misinference of the ancestral state. To compensate for this, the probability of misinference was incorporated in the null distribution of the H statistic by exchanging the frequencies of the derived and the ancestral states with probability pd, where d is the net divergence or the average number of fixed differences per site between the two species. If all mutation rates are equal $p = 1/3$. For *Drosophila* data they use $p = 3/8$. This number is chosen based on the idea that in *Drosophila* transitions $A \leftrightarrow G$ and $C \leftrightarrow T$ occur at twice the rate of transversions (the other eight possible mutations). Since for any nucleotide there are two transversions and one transition, $1/2$ of the mutations are transitions and $1/2$ are transversions. Taking into account the rate of back mutations in the two cases, we get

$$\frac{1}{2} \cdot \frac{1}{2} + \frac{1}{2} \cdot \frac{1}{4} = \frac{3}{8}$$

Example 2.3. Accessory gland proteins. This is a group of specialized proteins in the seminal fluid of *Drosophila* that have been suggested to be involved in egg-laying stimulation, remating inhibition, and sperm competition. Among the *Acp* genes, *Acp26Aa* and nearby *Acp26Ab* have been extensively studied. Here we use data of Tsaur, Ting, and Wu (1998) who analyzed 49 sequences from five populations in four continents.

The site frequency spectrum for *Acp26Aa* given in their Table 2 for the 31 sites (out of 38) where the derived nucleotide could be unambiguously inferred with reference to the three outgroup species had $\eta_1 = 9$, $\eta_2 = 5$, $\eta_3 = 2$, $\eta_{46} = 3$, and $\eta_m = 1$ for $m = 6, 7, 11, 16, 21, 29, 31, 38, 39, 42, 45,$ and 47. There are 31 segregating sites and $h_{49} = 4.458797$, so

$$\theta_W = 31/4.458797 = 6.95248$$

Using the site frequency spectrum, one can compute

$$\theta_\pi = 5.265306 \qquad \theta_H = 15.359694$$

Fay and Wu's test can be run from

http://www.genetics.wustl.edu/jflab/htest.html

Because of results of Zeng, Shi, Fu, and Wu (2006) they now scale the statistic by the variance of the difference, so their $H = -4.369243$. Using 10,000 simulations and a back mutation probability of 0.03 produces a p value of 0.0254. If we grind through the details of computing Tajima's D, we find

$$a_1 = 4.458797 \quad a_2 = 1.624316$$
$$b_1 = 0.347222 \quad b_2 = 0.231765$$
$$c_1 = 0.122946 \quad c_2 = 0.080038$$
$$e_1 = 0.027574 \quad e_2 = 0.003722$$

and $\sqrt{\widehat{\mathrm{var}}\,(d)} = 2.077509$, so Tajima's D is

$$D_T = \frac{5.265306 - 6.95248}{2.077509} = -0.812146$$

Consulting Tajima's table, we see that a 90% confidence interval for D_T is $(-1.571, 1.722)$, so this value of D_T is far from significant. Indeed, the output from Fay's program says that the one-tailed p value is 0.229700.

One can, of course, also calculate Fu and Li's D_{FL}. To do this, we begin by computing $v_D = 0.137393$ and $u_D = 3.321404$. Since $\eta_1 = 9$,

$$D_{FL} = \frac{31 - 9 \cdot 4.458797}{\sqrt{(3.321404)(31) + (0.137393)(31)^2}} = -0.595524$$

When $n = 50$, a 95% confidence interval for D_{FL} is $(-1.96, 1.37)$, so again this value is far from significant.

Example 2.4. In the Hamblin and Aquadro (1996) data, throwing out the column in which the outgroup sequence differs from both nucleotides in *D. simulans*, we have 24 segregating sites and a site frequency spectrum of

m	1	2	3	4	5	6	7	8	9	10
η_m	1	10	1	4	2	0	3	2	1	0

Running Fay and Wu's test from the web page with 10,000 simulations and a back mutation probability of 0.03, gives a value for H of 0.117960 (scaled by the variance) and a one-tailed p value of 0.3343.

2.4.4 Conditioning on S_n

Since the number of segregating sites is observable and θ is not, it is natural, as Fay and Wu (2002) have done, to perform simulations of the coalescent conditional on the number of segregating sites $S_n = k$. An easy, but incorrect, way to do this is to generate a genealogy and then place k mutations on it at random. This method was suggested by Hudson (1993, page 27). However, discussion in his paper indicates he knew it was not correct. The reason is that for fixed values of θ and k not all genealogies have an equal probability of producing k segregating sites.

To see this, let τ_j be the amount of time there are j lineages:

$$P(\tau_2 = t_2, \ldots \tau_n = t_n) = \prod_{j=2}^{n} \binom{j}{2} \exp\left(-\binom{j}{2}t\right)$$

and note that if $\tau = 2\tau_2 + \cdots + n\tau_n$ is the total size of the tree,

$$P(S_n = k | \tau_2 = t_2, \ldots \tau_n = t_n) = e^{-\theta\tau}\frac{(\theta\tau)^k}{k!}$$

Combining the last two formulas and dividing by $P(S_n = k)$, we have

$$P(\tau_2 = t_2, \ldots \tau_n = t_n | S_n = k)$$
$$= c_{\theta,n,k}(2t_2 + \cdots + nt_n)^k \prod_{j=2}^{n} \binom{j}{2} \exp\left(-\left(\binom{j}{2} + j\theta\right)t_j\right)$$

where $c_{\theta,n,k}$ is a normalizing constant that depends on the indicated parameters. It is clear from the last formula that, for fixed k, as θ changes not only the total size of the tree τ changes but:

- Due to the $j\theta$ in the exponential, the relative sizes of $\tau_2, \ldots \tau_n$ change, so the distribution of test statistics conditional on $S_n = k$ will depend on θ.
- Due to the $(2t_2 + \cdots + nt_n)^k$, the τ_j are no longer independent, since the joint density function is not a product $f_2(t_2) \cdots f_n(t_n)$.

Markovstova, Marjoram, and Tavaré (2001) have shown for two tests of neutrality of Depaulis and Veuille (1998) that under the simple but incorrect algorithm of generating a tree and then putting a fixed number of mutations on it, the fraction of observations that fall in an interval claimed to have probability 0.95 can be very small for extreme values of θ. In the other direction, Wall and Hudson (2001) have shown that these problems do not occur if the true value of θ is near Watterson's estimator $\theta_W = S_n/h_n$.

2.5 The HKA test

Suppose now that we have a sample from one species and one sequence from another closely related species. The ratio of the number of segregating sites in one species to the amount of divergence between the two species is determined by the time since divergence of the two species, the effective population size, and the size of the sample, but does not depend on the mutation rate at the locus. Hence, these ratios should be similar for different loci, and sufficiently large differences provide evidence for nonneutral evolution.

Having explained the motivation behind the HKA test, we turn now to the mechanics. Consider data collected from two species, referred to as species A and species B, and from $L \geq 2$ regions of the genome referred to as locus 1 through locus L. Assume that a random sample of n_A gametes from species A have been sequenced at all L loci and n_B gametes from species B have been sequenced at the same loci. Let S_i^A denote the number of sites that are polymorphic at locus i in the sample from species A. Similarly, let S_i^B denote the number of sites that are polymorphic at locus i in the sample from species B. Let D_i denote the number of differences between a random gamete from species A and a random gamete from species B. The $3L$ observations S_i^A, S_i^B, and D_i constitute the data with which the test devised by Hudson, Kreitman, and Aguadé (1987) is carried out.

It is assumed that each locus evolves according to the standard Wright-Fisher infinite sites model: (1) generations are discrete, (2) all mutations are selectively neutral, (3) the number of sites at each locus is very large, so that each mutation occurs at a previously unmutated site, (4) in each generation, mutations occur independently in each gamete and at each locus, (5) at locus i, the number of mutations per gamete in each generation is Poisson distributed with mean u_i, and (6) no recombination occurs within the loci. In addition, we assume that (7) all loci are unlinked, (8) species A and B are at stationarity at the time of sampling with population sizes $2N$ and $2Nf$, respectively, and (9) the two species were derived T' generations ago from a single ancestral population with size $2N(1 + f)/2$ gametes.

Letting $\theta_i = 4Nu_i$ and $C(n) = \sum_{j=1}^{n-1} 1/j$, it follows from (1.20) that

$$E(S_i^A) = \theta_i C(n_A) \quad E(S_i^B) = f\theta_i C(n_B)$$

Using (1.22) in Chapter 1 and letting $C_2(n) = \sum_{j=1}^{n-1} 1/j^2$, we have

$$\mathrm{var}\,(S_i^A) = E(S_i^A) + \theta_i^2 C_2(n_A)$$
$$\mathrm{var}\,(S_i^B) = E(S_i^B) + (f\theta_i)^2 C_2(n_B)$$

To compute the expected value of D_i, we note that it is $2u_i$ times the expected coalescence time of two individuals: one chosen at random from A and one from B. Those two lineages must stay apart for T' units of time and then coalescence occurs as in a single population of size $2N(1 + f)/2$. Measured in units of $2N$ generations, the second phase takes an exponentially distributed amount of time with mean $(1 + f)/2$, so letting $T = T'/2N$,

$$ED_i = \theta_i(T + (1 + f)/2)$$

To compute the variance, we note that in the first phase, the number of mutations is Poisson with mean $2u_iT' = \theta_iT$. By (1.22) with $n = 2$, the number in the second phase has variance $\theta_i(1 + f)/2 + (\theta_i(1 + f)/2)^2$ and is independent of the number in the first phase, so

$$\operatorname{var}(D_i) = ED_i + (\theta_i(1+f)/2)^2$$

There are $L+2$ parameters. These can be estimated by solving the following $L+2$ equations:

$$\sum_{i=1}^{L} S_i^A = C(n_A) \sum_{i=1}^{L} \hat{\theta}_i$$

$$\sum_{i=1}^{L} S_i^B = C(n_B)\hat{f} \sum_{i=1}^{L} \hat{\theta}_i$$

$$\sum_{i=1}^{L} D_i = (\hat{T} + (1+\hat{f})/2) \sum_{i=1}^{L} \hat{\theta}_i$$

and for $1 \le i \le L - 1$

$$S_i^A + S_i^B + D_i = \hat{\theta}_i \left\{ \hat{T} + (1+\hat{f})/2 + C(n_A) + C(n_B) \right\}$$

These equations may look complicated, but they are simple to solve. The first can be used to compute $\sum_{i=1}^{L} \hat{\theta}_i$, the second can then be used to find $\hat{f}$, the third to compute $\hat{T}$, and then the individual $\hat{\theta}_i$ can be computed from the remaining $L - 1$. We do not need the equation with $i = L$ since we have already computed the sum of the $\hat{\theta}_i$.

To measure the goodness of fit of these parameters, we can use

$$X^2 = \sum_{i=1}^{L}(S_i^A - \hat{E}(S_i^A))^2/\widehat{\operatorname{var}}(S_i^A)$$

$$+ \sum_{i=1}^{L}(S_i^B - \hat{E}(S_i^B))^2/\widehat{\operatorname{var}}(S_i^B)$$

$$+ \sum_{i=1}^{L}(D_i - \hat{E}(D_i))^2/\widehat{\operatorname{var}}(D_i)$$

If the quantities S_i^A, S_i^B, and D_i were stochastically independent of each other and normally distributed, then the statistic X^2 should be approximately χ^2 with $3L - (L+2) = 2L - 2$ degrees of freedom. For n_A, n_B, and T sufficiently large, all of these quantities are approximately normally distributed. Since the loci are unlinked, S_i^A is independent of S_j^A and S_j^B when $j \ne i$. Also, S_i^A is independent of S_i^B as long as T is large enough so that there are no shared polymorphisms. However, a small positive correlation is expected between S_i^A and D_i, and between S_i^B and D_i, because a positive fraction of the mutations that contribute to polymorphism also contribute to differences between species. The last observation, and the fact that the normality is only asymptotic, forces the test to be carried out by doing simulations with the estimated parameters.

Example 2.5. Adh. The first application of the HKA test was to the alcohol dehydrogenase locus in *Drosophila melanogaster*. The polymorphism data came from a four-cutter restriction enzyme survey of 81 isochromosomal lines of *D. melanogaster* studied by Kreitman and Aguadé (1986a,b). Nine polymorphic restriction sites were identified in the flanking region and eight in the *Adh* locus. They estimated the effective number of sites to be 414 in the flanking region and 79 in the *Adh* locus. Their interspecific data was based on a sequence comparison of one *D. melanogaster* sequence and one *D. sechelia* sequence. This comparison revealed 210 differences in 4052 bp of flanking sequence and 18 differences in 324 bp in the *Adh* locus. The next table summarizes the data:

	within *D. melanogaster*			between species		
	sites	variable	%	sites	variable	%
flanking region	414	9	0.022	4052	210	0.052
Adh locus	79	8	0.101	324	18	0.056

Note that the divergence between species is almost the same in the two regions, but there is a considerably higher rate of polymorphism in the *Adh* locus compared to the flanking sequence.

We have no data on the variability within *D. simulans*, so we will suppose that the ancestral population size is the same as the current population size, that is, $f = 1$. To take account of the differing number of sites in the comparisons within ($w_1 = 414$, $w_2 = 79$) and between ($b_1 = 4052$, $b_2 = 324$) species and to prepare for the fact that in the next example the two sample sizes will be different (here $n_1 = n_2 = 81$), we let μ_i be the per nucleotide mutation rate at the ith locus, let $\pi_i = 4N\mu_i$, and note that

$$ES_1^A = C(n_1) \cdot w_1 \pi_1$$
$$ES_2^A = C(n_2) \cdot w_2 \pi_2$$
$$ED_1 = b_1 \pi_1 (T + 1)$$
$$ED_2 = b_2 \pi_2 (T + 1)$$

Adding the equations as before, we arrive at

$$S_1^A + S_2^A = C(n_1) \cdot w_1 \hat{\pi}_1 + C(n_2) \cdot w_2 \hat{\pi}_2$$
$$D_1 + D_2 = (b_1 \hat{\pi}_1 + b_2 \hat{\pi}_2)(\hat{T} + 1)$$
$$S_1^A + D_1 = C(n_1) \cdot w_1 \hat{\pi}_1 + b_1 \hat{\pi}_1 (\hat{T} + 1)$$

These equations are not as easy to solve as the previous ones. Letting $x = \hat{\pi}_1$, $y = \hat{\pi}_2$, and $z = \hat{T} + 1$, they have the form

$$c = ax + by$$
$$f = dxz + eyz$$
$$i = gx + hxz$$

The three equations can be written as

$$z = \frac{f}{dx + ey} \qquad y = \frac{c - ax}{b} \qquad (1 - gx)\frac{f}{z} = fhx$$

Using the first two in the third equation leads to $\alpha x^2 + \beta x + \gamma = 0$, where

$$\alpha = g\left(d - \frac{ea}{b}\right)$$

$$\beta = \frac{gec}{b} + hf - i\left(d - \frac{ea}{b}\right)$$

$$\gamma = \frac{-iec}{b}$$

At this point, there are two cases to consider. If $n_1 = n_2$, $b_1 = w_1$, and $b_2 = w_2$, then

$$d - \frac{ea}{b} = b_1 - \frac{b_2 C(n_1)w_1}{C(n_2)w_2} = 0$$

In this case, $\alpha = 0$ so we solve the linear equation to get $x = -\gamma/\beta$. When $\alpha \neq 0$, the root of the quadratic equation that we want is

$$x = \frac{-\beta + \sqrt{\beta^2 - 4\alpha\gamma}}{2\alpha}$$

In either case, once x is found, we can compute y and z.

Carrying out the arithmetic in this example gives

$$\hat{\pi}_1 = 6.558 \times 10^{-3}, \qquad \hat{\pi}_2 = 8.971 \times 10^{-3}, \qquad \hat{T} = 6.734$$

Using the relationships

$$\text{var}\,(S_i^A) = ES_i^A + (w_i \pi_i)^2 C_2(n_i)$$
$$\text{var}\,(D_i) = ED_i + (b_i \pi_i)^2$$

we can compute $X^2 = 6.09$. Monte Carlo simulations with the parameters set equal to these estimates show that the probability of $X^2 > 6.09$ is approximately 0.016. As the reader may have noticed, the flanking sequence is not far enough from the gene region to make it reasonable to assume that the two are unlinked. However, the positive correlation that results from interlocus linkage will shift the distribution of X^2 toward smaller values and make rejections based on the model conservative. Likewise, the intralocus recombination we are ignoring will reduce the variance of the quantities estimated and tend to decrease the value of X^2.

Having identified a significant departure from neutrality, the next step is to seek an explanation. The fact that there is more polymorphism in the coding region than in the adjacent flanking sequence suggests that something is acting there to make the genealogies larger than they would be under the neutral model. In Section 6.2, we will see that one possible explanation for this is balancing selection acting on the fast/slow polymorphism.

Example 2.6. Drosophila fourth chromosome. Berry, Ajioka, and Kreitman (1991) studied a 1.1kb fragment of the *cubitus interruptus Dominant* (ci^D) locus on the small nonrecombining fourth chromosome for 10 lines of *Drosophila melanogaster* and 9 of *Drosophila simulans*. They found no polymorphism within *Drosophila melanogaster* and a single polymorphism within *Drosophila simulans*. To perform the HKA test, they used data on the 5' region of *Adh* from 11 sequences of Kreitman and Hudson (1991) as their comparison neutral locus. This yielded the following data:

	ci^D	5' *Adh*
nucleotides	1106	3326
polymorphism	0	30
divergence	54	78

Calculating as in the previous example we find

$$\hat{\pi}_1 = 3.136 \times 10^{-3} \quad \hat{\pi}_2 = 2.072 \times 10^{-2} \quad \hat{T} = 11.74$$

and $X^2 = 6.85$. Using the result of Hudson, Kreitman, and Aguadé (1987) that in this case the statistic has approximately a chi-square distribution with 1 degree of freedom, Berry, Ajioka, and Kreitman (1991) concluded that the probability of an X^2 value this large is < 0.01. (Note that the value of 1 here contrasts with the $2L - 2 = 2$ degrees of freedom that the statistic would have if S_i^A and D_i were independent.)

One explanation for these data is purifying selection. The original population sizes in both species were small, permitting effectively neutral drift of mildly deleterious alleles and causing the accumulation of fixed differences between the two species. Subsequent population expansion has increased the efficacy of selection against such mildly deleterious mutations, and what we see, within species, is the wholesale removal of variation by purifying selection. While this explanation is possible, it seems unlikely. Given the lack of variation at both silent and replacement sites, a slightly deleterious allele model would require that selection coefficients against both silent and replacement sites would fall between $1/2N_2$ and $1/2N_1$, where N_1 and N_2 are the pre- and post-expansion population sizes. It is unlikely that these two types of mutations, which have entirely different functional consequences, would have similar selection coefficients.

A second explanation is that a selective sweep eliminated variation in this region for both species. In order to estimate the time of occurrence of such a sweep, we note that if T_{tot} is the total time in the genealogy of our sample, μ is the mutation rate per nucleotide per generation, and k is the number of silent sites, then the expected number of segregating sites

$$ES = T_{tot}\mu k$$

To simplify calculations, we will suppose that the sweep was recent enough so that the resulting genealogy is star-shaped. In this case, $T_{tot} = nt$, where n is the sample size and t is the time of the sweep. For the *Drosophila melanogaster* sample, $S = 0$, so we are left with an estimate of $t = 0$. For *D. simulans* substituting 1 for ES, and taking $n = 9$, $k = 331$, and $\mu = 1 \times 10^{-9}$, we arrive at

$$t = \frac{ES}{nk\mu} = \frac{1}{9 \cdot 331 \cdot 10^{-9}} = 3.35 \times 10^5 \text{ generations ago}$$

Assuming 10 generations per year, this translates into 33,500 years.

Having assumed a star-shaped phylogeny and calculated a time, we should go back and check to see if our assumption is justified. The probability of no coalescence in a sample of size n during t generations in a population of size N is

$$\approx \exp\left(-\binom{n}{2}\frac{t}{2N}\right)$$

If we take $2N = 5 \times 10^6$, $n = 9$, and $t = 3.35 \times 10^5$, then the above

$$= \exp\left(-36\frac{3.35}{50}\right) = e^{-2.412} = 0.0896$$

i.e., it is very likely that there has been at least one coalescence. Once one abandons the assumption of a star-shaped phylogeny, calculations become difficult and it is natural to turn to simulation. Using $4N\mu = 3$ for *D. simulans*, Berry, Ajioka, and Kreitman (1991) computed that there was a 50% probability of sweep in the last $0.36N$ generations, or 72,000 years.

2.6 McDonald-Kreitman test

To describe the test of McDonald and Kreitman (1991), we need some notation. Of M possible mutations in a coding region, let M_r be the number of possible neutral replacement mutations (i.e., ones that change the amino acid but not the effectiveness of the protein) and let M_s be the number of possible neutral synonymous mutations. By definition, all of the $M - M_r - M_s$ remaining mutations are deleterious.

Let μ be the mutation rate per nucleotide, so that the mutation rate for any one of the three possible changes at a site is $\mu/3$. Under the neutral theory, the expected number of fixed replacement substitutions in a set of alleles is $T_b(\mu/3)M_r$, where T_b is the total time on between-species branches. The expected number of fixed synonymous substitutions in a set of alleles is $T_b(\mu/3)M_s$. For a particular phylogeny and mutation rate, the number of replacement substitutions is independent of the number of synonymous substitutions. Therefore, the ratio of expected replacement to expected synonymous fixed mutations is

$$\frac{T_b(\mu/3)M_r}{T_b(\mu/3)M_s} = \frac{M_r}{M_s}$$

If T_w is the total time on within-species branches, then the ratio of expected replacement to expected synonymous polymorphic mutations is

$$\frac{T_w(\mu/3)M_r}{T_w(\mu/3)M_s} = \frac{M_r}{M_s}$$

Thus, if protein evolution occurs by neutral processes, the two ratios are the same and we can use standard statistical tests for 2×2 contingency tables to test this null hypothesis.

Under the alternative model of adaptive protein evolution, there should be relatively more replacement substitution between species than replacement polymorphism within a species, so a deviation in this direction is interpreted as evidence for positive selection.

Example 2.7. To explain the workings of the test, we will begin with the original data set of McDonald and Kreitman (1991). They compared DNA sequences of the *Adh* locus in *Drosophila melanogaster*, *D. simulans*, and *D. yakuba*. The DNA sequence data can be found on page 653 of their paper. To carry out the test, the following summary is sufficient:

	Fixed	Polymorphic
Replacement	7	2
Synonymous	17	42

To analyze the table, we first compute the number of observations we expect to find in each cell (given in parentheses in the next table):

	Fixed	Polymorphic	Total
Replacement	7 (3.176)	2 (5.824)	9
Synonymous	17 (20.824)	42 (38.176)	59
Total	24	44	68

Then we compute the χ^2 statistic:

$$\frac{(7-3.176)^2}{3.176} + \frac{(2-5.824)^2}{5.824} + \frac{(17-20.824)^2}{20.824} + \frac{(42-38.176)^2}{38.176} = 8.198$$

The number of degrees of freedom in this case is 1, so the χ^2 distribution is just the square of a standard normal, χ, and we can use a table of the standard normal to conclude that the probability of a deviation this large by chance is $2P(\chi > \sqrt{8.198}) = 0.0042$. McDonald and Kreitman analyzed the contingency table with a G test of independence (with the Williams correction for continuity), finding $G = 7.43$ and $p = 0.006$.

Geneticists have embraced the McDonald-Kreitman test as a useful tool for looking for positive selection. However, the initial paper did not get such a warm reception. Graur and Li, and Whittam and Nei, each wrote letters that appeared in the November 14, 1991 issue of *Nature* suggesting that the test had serious problems. Both pairs of authors objected to some of the bookkeeping involved in the three-species comparison. For this reason, we will now consider only pairwise comparisons. The data for *D. melanogaster* and *D. simulans*, is

```
11111111111111111111111111
78889900122222334444455555555556
81375768902378012345901255566991
16400489939513465132048715700064

GGTCGGCCCCTCACCCCTTCACCCCCCGCGC
TT.T..T......A..AC...T.T..A.T..
TT.T..T.T....A..AC...T.T..A.T..
TT.T......C.........T.T..A.TA.
TT.T......C.........T.T..A.TA.
T..T......C.........T.T..A.T..
T..T......C........T.TTT....TA.
T..T......C.......GTCTT.....T..
T..T......C.......GTCTT.....T..
T..T......C.......GTCTT.....T..
T..T......C.......GTCTT.....T..
T..T......CA......GTCTT.....T..
TT.T......C.......GTCTT..T..TA.
```

The contingency table is now much different, with the 24 fixed differences having been reduced to just 4.

	Fixed	Polymorphic	Total
Replacement	2	2	4
Synonymous	2	26	28
Total	4	28	32

Since the cell counts are small, we analyze the results with *Fisher's exact test*. To derive this test, we note that if we condition on the number of replacement substitutions n_r, then the number of fixed replacements, n_{fr}, is binomial(n_r, p), where $p = T_b/(T_w + T_b)$. Likewise, if we condition on the number of synonymous substitutions, n_s, then the number of fixed synonymous substitutions, n_{fs}, is binomial(n_s, p). Let n_f and n_p be the number of fixed and polymorphic substitutions. The probability of a given table conditioned on the marginal values n_r, n_s, n_f, n_p is

$$\frac{n_r!}{n_{fr}! n_{pr}!} p^{n_{fr}} (1-p)^{n_{pr}} \cdot \frac{n_s!}{n_{fs}! n_{ps}!} p^{n_{fs}} (1-p)^{n_{ps}} = \frac{C}{n_{fr}! n_{pr}! n_{fs}! n_{ps}!}$$

where C is a constant independent of $(n_{fr}, n_{pr}, n_{fs}, n_{ps})$.

There are only five 2×2 tables with the indicated row and column sums: n_{fr} can be 0, 1, 2, 3, or 4 and this determines the rest of the entries. Of these, the ones with $n_{fr} = 2, 3, 4$ are more extreme than the indicated table. Using the preceding formula, it is easy to compute the conditional probability that $n_{fr} = k$ given the row and column sums:

k	0	1	2	3	4
prob.	0.569383	0.364405	0.063070	0.003115	0.000028

From this we see that the probability of a table this extreme is $0.066212 > 0.05$.

In contrast, if we compare *D. simulans* with *D. yakuba* using the data in McDonald and Kreitman (1991),

```
                  11111111111111111111111111
         788899001222223344445555555556
         813757689023780123459012555699 1
         16400489939513465132048715700 64

         GGTCGGCCCCTCACCCCTTCACCCCCCGCGC
         .TC..T.A....................T..
         .TC.A..A....T...............T.G
         .T...T.A.T.....T............T..
         .T...T.A....T..T........T..A...
         .T...T.A......T...............
         .TC..T.A......................
         RSSSSSSRSSSSSSSSSSSSSSRSSSSRSSSSS
         PPFPPPFPPPPPPPPPPPPPPFPPPPPPPPP
```

The contingency table is:

	Fixed	Polymorphic	Total
Replacement	6	0	6
Synonymous	17	29	46
Total	23	29	52

Fisher's exact test gives that the probability $n_{fr} = 6$ given the row and column sums is 0.00493, so there is a clear departure from neutral evolution.

Example 2.8. Eanes, Kirchner, and Yoon (1993) sequenced 32 and 12 copies of the gene (*G6pd*) in *Drosophila melanogaster* and *D. simulans* respectively. This revealed the following results (the number of observations we expect to find in each cell is given in parentheses):

	Fixed	Polymorphic	Total
Replacement	21 (12.718)	2 (10.282)	23
Synonymous	26 (34.282)	36 (27.717)	62
Total	47	38	85

The χ^2 statistic is 16.541. The probability of a χ^2 value this large by chance is < 0.0001. Thus, there is a very strong signal of departure from neutral evolution. The most likely explanation is that replacement substitutions are not neutral but have been periodically selected through the populations of one or both species as advantageous amino acid mutations.

Example 2.9. Accessory gland proteins are specialized proteins in the seminal fluid of *Drosophila*. They have been suggested to be involved in egg-laying stimulation, remating inhibition, and sperm competition, so there is reason to suspect that they are under positive selection. Tsaur, Ting, and Wu (1998) studied the evolution of *Acp26Aa*. They sequenced 39 *D. melanogaster* chromosomes, which they combined with 10 published *D. melanogaster* sequences and 1 $D.$ *simulans* sequence in Aguadé, Miyashita, and Langley (1992). The reader's first reaction to the sample size of 1 for $D.$ *simulans* may be that this makes it impossible to determine whether sites are polymorphic in $D.$ *simulans*. This does not ruin the test, however. It just reduces T_w to the total time in the genealogy for the $D.$ *melanogaster* sample.

The next table gives the data as well as the number of observations we expect to find in each cell (given in parentheses):

	Fixed	Polymorphic	Total
Replacement	75 (69.493)	22 (27.507)	97
Synonymous	21 (26.507)	16 (10.492)	37
Total	96	38	134

The χ^2 statistic is 5.574. The probability of a χ^2 value this large by chance is $2P(\chi \geq \sqrt{5.574}) = 0.0181$. It is interesting to note that while the McDonald-Kreitman test leads to a rejection of the neutral model, Tajima's D, which is -0.875, and Fu and Li's D, which is -0.118, do not come close to rejection.

Example 2.10. The first three examples have all shown a larger ratio of replacement to silent changes between species. Mitochondrial DNA shows the opposite pattern. Nachman (1998) describes the results of 25 comparisons involving a wide variety of organisms. Seventeen of the contingency tables deviate from the neutral expectation, and most of the deviations (15 of 17) are in the direction of greater ratio of replacement to silent variation within species. A typical example is the comparison of the ATPase gene 6 from *Drosophila melanogaster* and $D.$ *simulans* from Kaneko, Satta, Matsura, and Chigusa (1993). As before, the number of observations we expect to find in each cell is given in parentheses:

	Fixed	Polymorphic	Total
Replacement	4 (1.482)	4 (6.518)	8
Synonymous	1 (3.518)	18 (15.482)	19
Total	5	22	27

The χ^2 statistic is 7.467. The probability of a χ^2 value this large by chance is $2P(\chi \geq \sqrt{7.467}) = 0.0064$. One explanation for a larger ratio of replacement to silent changes within populations is that many of the replacement polymorphisms are mildly deleterious.

3

Recombination

"In mathematics you don't understand things. You get used to them."
John von Neumann

In this chapter, we will begin to investigate the effects of recombination on the patterns of genetic variability.

3.1 Two loci

Going forward in time, the dynamics of a Wright-Fisher model with two loci may be described as follows. To generate an individual in the next generation, with probability $1 - r$ we copy both loci from one randomly chosen individual, while with probability r a recombination occurs and we copy the two loci from two randomly chosen individuals. Reversing our perspective, suppose we sample m individuals from the population. Each locus will have its own genetic history leading to a coalescent, but the two coalescent processes will be correlated, due to the fact that in the absence of recombination the two loci will be copied from the same parent.

3.1.1 Sample of size 2

We call the two loci the a locus and the b locus. Ignoring the numbering of the copies, using parentheses to indicate the loci at which the individual has genetic material ancestral to the sample, the possible states of the system are given in the first column of the next table. The second column describes the state by giving the number of (a)'s, (b)'s, and (ab)'s. The third gives the number of a's, n_a, and the number of b's, n_b,

R. Durrett, *Probability Models for DNA Sequence Evolution*,
DOI: 10.1007/978-0-387-78168-6_3, © Springer Science+Business Media, LLC 2008

	$(a), (b), (ab)$	n_a, n_b	
$(a\,b)\,(a\,b)$	$(0,0,2)$	$2,2$	
$(a\,b)\,(a)\,(b)$	$(1,1,1)$	$2,2$	
$(a)\,(a)\,(b)\,(b)$	$(2,2,0)$	$2,2$	
$(a)\,(b)\,(b)$	$(1,2,0)$	$1,2$	Δ_1
$(b)\,(a)\,(a)$	$(2,1,0)$	$2,1$	Δ_1
$(a\,b)\,(b)$	$(0,1,1)$	$1,2$	Δ_1
$(a\,b)\,(a)$	$(1,0,1)$	$2,1$	Δ_1
$(a\,b)$	$(0,0,1)$	$1,1$	Δ_2
$(a)\,(b)$	$(1,1,0)$	$1,1$	Δ_2

To explain the notation, the initial state is $(ab)(ab)$, i.e., we have sampled two individuals and examined the states of their two loci. If the first event going backwards in time is a coalescence, the chain enters (ab) and the process stops before it gets interesting. If the first event is a recombination, the two copies in one individual become separated and the state is $(ab)(a)(b)$. At this point, a second recombination might produce $(a)(b)(a)(b)$, or a coalescence might produce $(ab)(b)$, $(ab)(a)$ or return us to the original $(ab)(ab)$.

The next figure shows a possible realization. The a locus is denoted by a solid dot and the b locus by an open circle. At the bottom, the first individual is on the left and the second on the right. On the right edge of the figure, we have indicated the sequence of states. The parts are sometimes listed in a different order from the table of states to make it easier to connect the state with the picture.

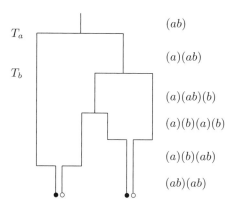

Fig. 3.1. Realization of the two-locus coalescent with recombination.

Let T_a be the coalescence time for the a locus and T_b be the coalescence time for the b locus. Our next task is to compute $E_x(T_a T_b)$ for the possible initial states x. We will use the notation in the table above. If $n_a = 1$ or $n_b = 1$, then $T_a T_b = 0$, so we will use Δ_1 to denote the set of configurations where one coalescence has occurred, and Δ_2 the configurations where both have. Considering the Wright-Fisher model, speeding up time by a factor of $2N$, and following the standard practice of letting $\rho = 4Nr$, the rates become

from/to	$(0,0,2)$	$(1,1,1)$	$(2,2,0)$	Δ_1	Δ_2	total
$(0,0,2)$	.	ρ	.	.	1	$\rho+1$
$(1,1,1)$	1	.	$\rho/2$	2	.	$(\rho/2)+3$
$(2,2,0)$	.	4	.	2	.	6

In the first row, $(0,0,2) \rightarrow (1,1,1)$, i.e., $(ab),(ab) \rightarrow (a),(b),(ab)$ by two possible recombinations, or $(0,0,2) \rightarrow \Delta_2$ by coalescence. In the second row, if the (a) and (b) coalesce, $(1,1,1) \rightarrow (0,0,2)$, while the other two possible coalescence events lead to Δ_1, and if recombination happens to (ab), $(1,1,1) \rightarrow (2,2,0)$. When the state is $(2,0,0) = (a),(a),(b),(b)$, four possible coalescence events lead back to $(1,1,1)$, while two lead to Δ_1.

Theorem 3.1. *Let $v(x)$ be $cov(T_a, T_b)$ when the initial state is x.*

$$v(0,0,2) = \frac{\rho + 18}{\rho^2 + 13\rho + 18} \quad v(1,1,1) = \frac{6}{\rho^2 + 13\rho + 18}$$

$$v(2,2,0) = \frac{4}{\rho^2 + 13\rho + 18} \tag{3.1}$$

These are sometimes referred to as $C_{ij,ij}$, $C_{ij,ik}$, $C_{ij,k\ell}$ with i, j, k, ℓ distinct numbers that indicate the individuals sampled at the two locations.

Proof. Since $ET_a = ET_b = 1$, we have

$$\mathrm{cov}\,(T_a, T_b) = ET_a T_b - ET_a ET_b = ET_a T_b - 1$$

Let $u(x) = E_x(T_a T_b)$. To get an equation for $u(x)$, let J be the time of the first jump, and let X_J be the state at time J:

$$E(T_a T_b | J, X_J) = E((T_a - J + J)(T_b - J + J)|J, X_J)$$
$$= E((T_a - J)(T_b - J)|J, X_J) + JE(T_a - J|J, X_J)$$
$$+ JE(T_b - J|J, X_J) + J^2$$

Let $v_a(X_J)$ and $v_b(X_J)$ be 1 if coalescence has not occurred at the indicated locus and 0 if it has. Noting that J is independent of X_J and taking the expected value gives

$$u(x) = E_x u(X_J) + E_x J \cdot (E_x v_a(X_J) + E_x v_b(X_J)) + E_x J^2$$

If J is exponential with rate λ, then $EJ = 1/\lambda$ and $EJ^2 = 2/\lambda^2$. Using this with our table of rates gives

$$u(0,0,2) = \frac{\rho}{\rho+1}\left(u(1,1,1) + 2\cdot\frac{1}{\rho+1}\right)$$
$$+ \frac{1}{\rho+1}\left(u(\Delta_2) + 0\cdot\frac{1}{\rho+1}\right) + \frac{2}{(\rho+1)^2}$$

since the probabilities of the two transitions are their rate over the total. Here and in the next two equations $E_x J^2$ is last and does not depend on the state X_J the chain jumps to. Similar reasoning gives

$$u(1,1,1) = \frac{(\rho/2)}{(\rho/2)+3}\left(u(2,2,0) + 2\cdot\frac{1}{(\rho/2)+3}\right)$$
$$+ \frac{1}{(\rho/2)+3}\left(u(0,0,2) + 2\cdot\frac{1}{(\rho/2)+3}\right)$$
$$+ \frac{2}{(\rho/2)+3}\left(u(\Delta_1) + 1\cdot\frac{1}{(\rho/2)+3}\right) + \frac{2}{((\rho/2)+3)^2}$$
$$u(2,2,0) = \frac{4}{6}\left(u(1,1,1) + 2\cdot\frac{1}{6}\right) + \frac{2}{6}\left(u(\Delta_1) + 1\cdot\frac{1}{6}\right) + \frac{2}{6^2}$$

Simplifying and using $u(\Delta_i) = 0$ gives

$$u(0,0,2) = \frac{\rho}{\rho+1}u(1,1,1) + \frac{2}{\rho+1}$$
$$u(1,1,1) = \frac{1}{(\rho/2)+3}u(0,0,2) + \frac{(\rho/2)}{(\rho/2)+3}u(2,2,0) + \frac{2}{(\rho/2)+3}$$
$$u(2,2,0) = \frac{2}{3}u(1,1,1) + \frac{1}{3}$$

$u(x) = E_x(T_a T_b)$. What we want to compute is the covariance $v(x) = u(x) - 1$, which satisfies the following equations (substitute $u(x) = v(x) + 1$):

$$v(0,0,2) = \frac{\rho}{\rho+1}v(1,1,1) + \frac{1}{\rho+1}$$
$$v(1,1,1) = \frac{1}{(\rho/2)+3}v(0,0,2) + \frac{(\rho/2)}{(\rho/2)+3}v(2,2,0) \qquad (3.2)$$
$$v(2,2,0) = 2v(1,1,1)/3$$

Rearranging the second equation, and then using the third equation,

$$v(0,0,2) = \left(\frac{\rho}{2}+3\right)v(1,1,1) - \frac{\rho}{2}v(2,2,0)$$
$$= \left(\frac{\rho}{2}+3-\frac{\rho}{3}\right)v(1,1,1) = \frac{\rho+18}{6}v(1,1,1)$$

Using this in the first equation, we have

$$v(0,0,2) = \frac{\rho}{\rho+1}\cdot\frac{6}{\rho+18}v(0,0,2) + \frac{1}{\rho+1}$$

Multiplying each side by $\rho + 1$, we have

$$1 = \left(\rho + 1 - \frac{6\rho}{\rho + 18}\right) v(0,0,2) = \frac{\rho^2 + 13\rho + 18}{\rho + 18} v(0,0,2)$$

which gives the first formula. The first equation in (3.2) implies

$$v(1,1,1) = \frac{(\rho + 1)v(0,0,2) - 1}{\rho}$$

$$= \frac{1}{\rho}\left(\frac{\rho^2 + 19\rho + 18}{\rho^2 + 13\rho + 18} - 1\right) = \frac{6}{\rho^2 + 13\rho + 18}$$

The final result follows from $v(2,2,0) = 2v(1,1,1)/3$. □

3.1.2 Sample of size n

To generalize the previous argument to larger sample sizes, we give up on trying to solve the equations to get formulas and content ourselves with a recursion that can be solved numerically. Let τ_a and τ_b be the total size of the genealogical trees for the two loci. Kaplan and Hudson (1985, see pages 386–387), wrote equations for $E(\tau_a \tau_b)$. However, Pluzhnikov and Donnelly (1996, see page 1260) noticed that the recursion for the covariance is much nicer. Suppose that $n_a = i + k$ chromosomes are sampled at locus a and $n_b = j + k$ chromosomes are sampled at locus b, in such a way that k chromosomes are common to the two samples, and let $F(i, j, k) = \text{cov}(\tau_a, \tau_b)$ when the initial configuration is (i, j, k). Let $\ell = i + j + k$. The configuration $x = (i, j, k)$ changes at rate

$$\beta_x = \frac{\ell(\ell - 1) + \rho k}{2}$$

due to coalescence of two of the ℓ chromosomes or recombination separating the a and b loci on one of the k chromosomes with both.

Theorem 3.2. *If $x = (i, j, k)$ and X is the state after the first jump, then*

$$F(x) = E_x F(X) + \frac{2k(k-1)}{\beta_x(n_a - 1)(n_b - 1)}$$

Proof. This clever proof is from Tavaré's St. Flour notes. We start by deriving the formula for conditional covariances. Since $E\{E(Y|X)\} = EY$, we have

$$\begin{aligned}
\text{cov}(\tau_a, \tau_b) &= E(\tau_a \tau_b) - E(\tau_a)E(\tau_b) \\
&= E\{E(\tau_a \tau_b | X)\} - E\{E(\tau_a | X)E(\tau_b | X)\} \\
&\quad + E\{E(\tau_a | X)E(\tau_b | X)\} - E(\tau_a)E(\tau_b) \\
&= E\{\text{cov}(\tau_a, \tau_b | X)\} + \text{cov}(E(\tau_a | X), E(\tau_b | X)) \quad (3.3)
\end{aligned}$$

Letting J be the time of the first jump, we can write $\tau_a = n_a J + \tau'_a$ and $\tau_b = n_b J + \tau'_b$, where the waiting times after J, τ'_a and τ'_b, and the state X are independent of J. From this we get

$$E\{\operatorname{cov}(\tau_a, \tau_b | X)\} = n_a n_b \operatorname{var}(J) + E\{\operatorname{cov}(\tau'_a, \tau'_b | X)\}$$
$$= (n_a n_b)/\beta^2 + E_x F(X) \tag{3.4}$$

where for simplicity we have dropped the subscript x on β. To evaluate the second term in (3.3), we recall that $E(E(\tau_a | X)) = E\tau_a$, so

$$\operatorname{cov}(E(\tau_a | X), E(\tau_b | X)) = E\{(E(\tau_a | X) - E\tau_a)) \cdot (E(\tau_b | X) - E\tau_b))\}$$

Let N_a and N_b be the number of a and b lineages after the jump. If $h(m) = 2 \sum_{j=2}^{m} 1/j$, then

$$E(\tau_c | X) - E\tau_c = \frac{n_c}{\beta} + h(N_c) - h(n_c) \tag{3.5}$$

To compute the expected value of the product, we have to look in detail at the transition rates:

(i, j, k) to	at rate	N_a	N_b
$(i+1, j+1, k-1)$	$r_1 = \rho k/2$	n_a	n_b
$(i-1, j-1, k+1)$	$r_2 = ij$	n_a	n_b
$(i-1, j, k)$	$r_3 = ik + i(i-1)/2$	$n_a - 1$	n_b
$(i, j-1, k)$	$r_4 = jk + j(j-1)/2$	n_a	$n_b - 1$
$(i, j, k-1)$	$r_5 = k(k-1)/2$	$n_a - 1$	$n_b - 1$

From this it follows that

$$Eh(N_a) - h(n_a) = -\frac{2}{n_a - 1} \cdot \frac{r_3 + r_5}{\beta} \qquad Eh(N_b) - h(n_b) = -\frac{2}{n_b - 1} \cdot \frac{r_4 + r_5}{\beta}$$

Hence, using (3.5),

$$E\{(E(\tau_a | X) - E\tau_a)) \cdot (E(\tau_b | X) - E\tau_b))\}$$
$$= \frac{n_a n_b}{\beta^2} - \frac{n_b}{\beta} \cdot \frac{2}{n_a - 1} \cdot \frac{r_3 + r_5}{\beta} - \frac{n_a}{\beta} \cdot \frac{2}{n_b - 1} \cdot \frac{r_4 + r_5}{\beta}$$
$$+ \frac{4}{(n_a - 1)(n_b - 1)} \cdot \frac{r_5}{\beta}$$

To simplify, notice that $E(E(\tau_c | X) - E\tau_c) = 0$ and (3.5) imply

$$\frac{n_a}{\beta} = \frac{2}{n_a - 1} \frac{r_3 + r_5}{\beta} \qquad \frac{n_b}{\beta} = \frac{2}{n_b - 1} \frac{r_4 + r_5}{\beta}$$

so the second and third terms from the previous formula are equal to -1 times the first and we have

$$\operatorname{cov}(E(\tau_a | X), E(\tau_b | X)) = -\frac{n_a n_b}{\beta^2} + \frac{2k(k-1)}{\beta(n_a - 1)(n_b - 1)}$$

Adding this to (3.4) gives the desired result. $\square$

Samples of size 2

To make the connection between the new formula and the old one, recall that the jump rate in state $x = (i, j, k)$ is $\beta_x = [\ell(\ell - 1) + \rho k]/2$, where $\ell = i + j + k$, so we have

$$
\begin{array}{ccc}
\text{state} & \ell & \beta_x \\
(0, 0, 2) & 2 & 1 + \rho \\
(1, 1, 1) & 3 & 3 + (\rho/2) \\
(2, 2, 0) & 4 & 6
\end{array}
$$

Since $F(i, j, k) = 0$ for all the other states, and $2k(k - 1) = 0$ unless $k = 2$, consulting the table of rates before Theorem 3.1, and using Theorem 3.2 gives

$$
F(0, 0, 2) = \frac{\rho}{\rho + 1} F(1, 1, 1) + \frac{4}{\rho + 1}
$$
$$
F(1, 1, 1) = \frac{1}{(\rho/2) + 3} F(0, 0, 2) + \frac{(\rho/2)}{(\rho/2) + 3} F(2, 2, 0)
$$
$$
F(2, 2, 0) = 2F(1, 1, 1)/3
$$

and we have $F(i, j, k) = 4v(i, j, k)$. The factor of 4 comes from the fact that we are considering the total size of the tree, which for a sample of size 2 is 2 times the coalescence time, so the covariance is four times as large.

Samples of size 3

To explain how the equations in Theorem 3.2 can be solved, consider the problem of computing $F(0, 0, 3)$. To do this we have to consider the other configurations with $n_a = 3$ and $n_b = 3$: (1,1,2), (2,2,1), and (3,3,0); the configurations with $n_a = 2$ and $n_b = 3$: (0,1,2), (0,2,1), and (2,3,0); and of course the configurations with $n_a = 3$ and $n_b = 2$. We have already computed the values for $n_a = 2$, $n_b = 2$. Those with $n_a = 1$ or $n_b = 1$ are 0.

To compute the values for the states with $n_a = 2$ and $n_b = 3$, we begin by identifying the transition rates:

n_a, n_b	to/from	(0,1,2)	(1,2,1)	(2,3,0)
2,3	(0,1,2)		2	
2,3	(1,2,1)	ρ		6
2,3	(2,3,0)		$\rho/2$	
1,3	(0,2,1)		1+0	
1,3	(1,3,0)			0+1
2,2	(0,0,2)	2+0		
2,2	(1,1,1)		2+1	
2,2	(2,2,0)			0+3
1,2	(0,1,1)	1		
	total $= \beta_x$	$\rho + 3$	$(\rho/2) + 6$	10
	$g(x)$	$2/\beta_x$	0	0

The final row gives the value of $g(x) = 2k(k-1)/[\beta_x(n_a-1)(n_a-2)]$. It is comforting to note that the total rate in each case $\beta_x = [\ell(\ell-1) + \rho k]/2$, where $\ell = i + j + k$. Using the notation from the proof of Theorem 3.2, the entries in the table with ρ's come from r_1. The numbers on the first two rows come from r_2. The entries with plus signs come from r_3 or r_4. The remaining rate $(0,1,2) \to (0,1,1)$ in the lower left is an r_5.

Using the values we have computed for $n_a = 2$, $n_b = 2$, and recalling that when $n_a = 1$, $F = 0$, we get three equations in three unknowns. To make the equations easier to write, we let $b = \rho+3$, $c = (\rho/2)+6$, and $d = \rho^2+13\rho+18$.

$$F(0,1,2) = \frac{\rho}{b}F(1,2,1) + \frac{2}{b} \cdot \frac{\rho+18}{d} + \frac{2}{b}$$

$$F(1,2,1) = \frac{2}{c}F(0,1,2) + \frac{\rho/2}{c}F(2,3,0) + \frac{3}{c} \cdot \frac{6}{d}$$

$$F(2,3,0) = \frac{6}{10}F(1,2,1) + \frac{3}{10} \cdot \frac{4}{d}$$

In matrix form, they become

$$\begin{pmatrix} 1 & -\frac{\rho}{b} & 0 \\ -\frac{2}{c} & 1 & -\frac{\rho}{2c} \\ 0 & -\frac{6}{10} & 1 \end{pmatrix} \begin{pmatrix} F(0,1,2) \\ F(1,2,1) \\ F(2,3,0) \end{pmatrix} = \begin{pmatrix} 2(\rho+18+d)/bd \\ 18/cd \\ 12/10d \end{pmatrix}$$

The equations can be solved by row reducing the matrix to upper triangular form, but the answer is not very pretty.

For a sample of size n there are $(n-1)^2$ values of $2 \le n_a, n_b \le n$, which one must tackle by considering all the $n_a + n_b = m$ for $m = 4, 5, \ldots 2n$. For a given value of n_a, n_b we have a system of $\min\{n_a, n_b\} + 1$ equations to solve. Each system can be solved in at most $O(n^2)$ operations, so of order n^4 computations are needed, which is feasible for samples of size 100. One could, as Ethier and Griffiths (1990) suggested, write $O(n^3)$ equations for $F(i,j,k)$ with $2 \le n_a, n_b \le n$, but then solving the matrix equations would take n^6 steps.

3.2 m loci

Our task in this section is to generalize the two-locus results to m linearly arranged loci, each of which follows the infinite sites model, and to let $m \to \infty$ to get a model of a segment of DNA where recombination can occur between any two adjacent nucleotides. The number of mutations per generation per locus is assumed to have a Poisson distribution with mean u/m. Recombination does not occur within subloci, but occurs between adjacent subloci at rate $r/(m-1)$ per generation. With this assumption, the recombination rate between the most distant subloci is r.

3.2.1 Samples of size 2

Let S_2 be the number of segregating sites in a sample of size n and let $f_2(x) = (x + 18)/(x^2 + 13x + 18)$ be the covariance of the coalescence time for samples of size 2 at two loci with scaled recombination rate x between them.

Theorem 3.3. *For the infinite sites model with recombination*

$$\mathrm{var}\,(S_2) = \theta + \theta^2 \int_0^1 2(1 - y) f_2(y\rho)\, dy \tag{3.6}$$

If $\rho = 0$, $f_2(0) = 1$ and this reduces to (1.22): $\mathrm{var}\,(S_2) = \theta + \theta^2$.

Proof. Let S_2^j be the number of segregating sites in the jth locus in a sample of two alleles.

$$\mathrm{var}\,(S_2) = \sum_{i=1}^m \mathrm{var}\,(S_2^j) + \sum_{1 \le i \ne j \le m} \mathrm{cov}\,(S_2^i, S_2^j)$$

If we let $\theta = 4Nu$, then it follows from (1.22) that

$$\mathrm{var}\,(S_2^j) = \frac{\theta}{m} + \left(\frac{\theta}{m}\right)^2$$

Let T_2^i be the coalescence time of the two copies of locus i. This distribution of S_2^i given T_2^i is Poisson with mean $(\theta/m)T_2^i$, so $E(S_2^i | T_2^i) = (\theta/m)T_2^i$. The numbers of segregating sites S_2^i, S_2^j are conditionally independent given T_2^i and T_2^j, so

$$E(S_2^i S_2^j | T_2^i, T_2^j) = \left(\frac{\theta}{m}\right)^2 T_2^i T_2^j$$

and $\mathrm{cov}\,(S_2^i, S_2^j) = (\theta/m)^2 \,\mathrm{cov}\,(T_2^i, T_2^j)$. Using (3.1), we see that the variance of the total number of segregating sites is

$$\mathrm{var}\,(S_2) = \theta + \frac{\theta^2}{m} + \frac{\theta^2}{m^2} \sum_{k=1}^{m-1} 2(m - k) f_2\left(\frac{k\rho}{m - 1}\right)$$

where $\rho = 4Nr$, since there are $2(m - k)$ pairs $1 \le i, j \le n$ with $|i - j| = k$. Letting $m \to \infty$, setting $y = k/m$, and noting that the sum approximates an integral gives the indicated result. □

3.2.2 Samples of size n

Let S_n be the number of segregating sites in a sample of size n, recall $h_n = \sum_{i=1}^{n-1} 1/i$, and let $f_n(x)$ be the covariance between the total time in the genealogical trees for two loci with scaled recombination rate x between them. This can be computed numerically using Theorem 3.2.

Theorem 3.4. *For the infinite sites model with recombination,*

$$var\left(S_n\right) = \theta h_n + \frac{\theta^2}{4}\int_0^1 2(1-y)f_n(y\rho)\,dy \tag{3.7}$$

Here, in contrast to (3.6), θ^2 is divided by 4. This is due to the fact that we consider the total size of the tree, which for a sample of size 2 is two times the coalescence time, and hence has a variance four times as large. Note that the mutational variance θh_n is the same as the case of no recombination, but the genealogical variance is reduced by recombination. Comparing with (1.23)

$$var\left(S_n\right) = \frac{\theta}{2}E(T_{tot}) + \left(\frac{\theta}{2}\right)^2 var\left(T_{tot}\right)$$

we see that the integral gives $var\left(T_{tot}\right)$, a fact that can be seen from the derivation.

Proof. To compute the variance of the number of segregating sites for a sample of size $n > 2$, we again begin with

$$var\left(S_n\right) = \sum_{i=1}^m var\left(S_n^j\right) + \sum_{1\le i\neq j\le m} cov\left(S_n^i, S_n^j\right)$$

(1.22) implies that

$$var\left(S_n^i\right) = \frac{\theta}{m}\sum_{j=1}^{n-1}\frac{1}{j} + \left(\frac{\theta}{m}\right)^2\sum_{j=1}^{n-1}\frac{1}{j^2}$$

Let τ_n^i be the total time in the tree for the ith locus. This distribution of S_n^i given τ_n^i is Poisson with mean $(\theta/2m)\tau_n^i$, so $E(S_n^i|\tau_n^i) = (\theta/2m)\tau_n^i$,

$$E(S_n^i S_n^j|\tau_n^i, \tau_n^j) = \left(\frac{\theta}{2m}\right)^2 \tau_n^i \tau_n^j$$

and $cov\left(S_n^i, S_n^j\right) = (\theta/2m)^2 cov\left(\tau_n^i, \tau_n^j\right)$. The scaled recombination rate between i and j is $4Nr(j-i)/(m-1)$. Combining our results gives

$$var\left(S_n\right) = \theta\sum_{j=1}^{n-1}\frac{1}{j} + \frac{\theta^2}{m}\sum_{j=1}^{n-1}\frac{1}{j^2}$$
$$+ \frac{\theta^2}{4m^2}\sum_{k=1}^{m-1} 2(m-k)f_n\left(\frac{k\rho}{(m-1)}\right)$$

Letting $m \to \infty$, setting $y = j/m$, and noting that the sum approximates an integral gives the indicated result. □

3.2.3 Pairwise differences

The computations above can also be used to study the variance of the number of pairwise differences, Δ_n, but the formula is nicer since it is explicit rather than in terms of $f_n(x)$, which must be computed numerically for $n > 2$.

Theorem 3.5. *For the infinite sites model with recombination,*

$$var(\Delta_n) = \frac{\theta(n+1)}{3(n-1)} + \frac{2\theta^2}{n(n-1)} \int_0^1 2(1-x) \frac{\rho x + (2n^2 + 2n + 6)}{(\rho x)^2 + 13(\rho x) + 18} \, dx \quad (3.8)$$

When $\rho = 0$, this reduces to Tajima's result in (1.30).

$$var(\Delta_n) = \theta \frac{n+1}{3(n-1)} + \theta^2 \frac{2(n^2 + n + 3)}{9n(n-1)}$$

Proof. Following the appendix of Pluzhnikov and Donnelly (1996), we suppose the m loci are nucleotides and write

$$\Delta_n = \sum_{a=1}^{m} \binom{n}{2}^{-1} \sum_{i<j} \delta_{i,j}^a$$

where $\delta_{i,j}^a = 1$ if nucleotide a is different in sequences i and j.

$$var(\Delta_n) = \sum_{a=1}^{m} var\left(\binom{n}{2}^{-1} \sum_{i<j} \delta_{i,j}^a \right)$$
$$+ \binom{n}{2}^{-2} \sum_{a \neq b} \sum_{i<j} \sum_{k<\ell} cov\left(\delta_{i,j}^a, \delta_{k,\ell}^b \right)$$

The summand in the first term is the variance of the heterozygosity, which was computed in (1.15). Plugging in the per locus mutation rate θ/m and noting that when m is large $2 + \theta/m \approx 2$, (1.15) simplifies to

$$\frac{\theta}{m} \cdot \frac{2}{n(n-1)} \left[1 + \frac{2(n-2)}{2} + \frac{(n-2)(n-3)}{6} \right]$$
$$= \frac{\theta}{m} \cdot \frac{6 + (6n - 12) + (n^2 - 5n + 6)}{3n(n-1)} = \frac{\theta}{m} \cdot \frac{n+1}{3(n-1)}$$

As in the previous calculation,

$$cov\left(\delta_{i,j}^a, \delta_{k,\ell}^b \right) = \left(\frac{\theta}{2m} \right)^2 cov\left(\tau_{i,j}^a, \tau_{k,\ell}^b \right)$$

where the τ's are the tree lengths for the indicated samples of size 2. If we let $z = (b-a)\rho/(m-1)$ be the scaled recombination rate between loci a and b

then by the calculation for (1.16) and the covariance for coalescence times in Theorem 3.1 (multiplied by 4),

$$\binom{n}{2}^{-2} \sum_{i<j} \sum_{k<\ell} \text{cov}\left(\tau_{i,j}^a, \tau_{k,\ell}^b\right)$$

$$= \binom{n}{2}^{-1} \frac{4}{z^2 + 13z + 18} \left[(z+18)\cdot 1 + 6\cdot 2(n-2) + 4\cdot \binom{n-2}{2} \right]$$

since for each of the $\binom{n}{2}$ values of $i < j$ there is one $k < \ell$ with $i = k$ and $j = \ell$, $2(n-2)$ values with $|\{i,j\} \cap \{k,\ell\}| = 1$, and $\binom{n-2}{2}$ with $|\{i,j\} \cap \{k,\ell\}| = 0$. A little algebra now gives

$$= \binom{n}{2}^{-1} \frac{4[z + (2n^2 + 2n + 6)]}{z^2 + 13z + 18}$$

Using the fact that there are $2(m-k)$ pairs a, b with $|b-a| = k$, we have

$$\binom{n}{2}^{-2} \sum_{a\neq b} \sum_{i<j} \sum_{k<\ell} \text{cov}\left(\delta_{i,j}^a, \delta_{k,\ell}^b\right)$$

$$= \left(\frac{\theta}{2}\right)^2 \frac{2}{n(n-1)} \frac{1}{m} \sum_{k=1}^{m} \frac{2(m-k)}{m} \frac{4[\frac{\rho k}{m-1} + (2n^2 + 2n + 6)]}{\left(\frac{\rho k}{m-1}\right)^2 + 13\frac{\rho k}{m-1} + 18}$$

Writing $x = k/(m-1)$ and letting $m \to \infty$ gives the indicated result. □

Although the calculus is somewhat unpleasant, one can evaluate the integral in (3.8) to get a formula first derived by Wakeley (1997).

Theorem 3.6. *For the infinite sites model with recombination,*

$$var(\Delta_n) = \theta \frac{(n+1)}{3(n-1)} + \theta^2 f(\rho, n) \tag{3.9}$$

where $a_n = \rho - 2n(n+1) + 7$, $b_n = 2n(n+1)(13+2\rho) - \rho - 55$, *and*

$$f(\rho, n) = \frac{2}{n(n-1)\rho^2}[-2\rho + a_n L_1 + b_n L_2]$$

with $L_1 = \log\left(\frac{\rho^2 + 13\rho + 18}{18}\right)$ *and* $L_2 = \log\left(\frac{(2\rho + 13 - \sqrt{97})(13 + \sqrt{97})}{(2\rho + 13 + \sqrt{97})(13 - \sqrt{97})}\right)$.

Proof. Changing variables $y = \rho x$, and letting $u_n = 2n^2 + 2n + 6$, the integral in (3.8) becomes

$$\frac{2}{\rho^2} \int_0^\rho (\rho - y) \frac{y + u_n}{y^2 + 13y + 18} \, dy$$

The quadratic in the denominator has roots

$$r_1 > r_2 \quad \text{where} \quad r_i = \frac{-13 \pm \sqrt{97}}{2}$$

so we write the integrand as

$$\frac{-y^2 + (\rho - u_n)y + u_n\rho}{y^2 + 13y + 18} = -1 + \frac{(\rho - u_n + 13)y + (u_n\rho + 18)}{(y - r_1)(y - r_2)}$$

To evaluate the integral, we note that

$$\int_0^\rho \frac{1}{y^2 + 13y + 18} \, dy = \frac{1}{r_1 - r_2} \int_0^\rho \frac{1}{y - r_1} - \frac{1}{y - r_2} \, dy$$

$$= \frac{1}{r_1 - r_2} \log \left(\frac{\rho - r_1}{-r_1} \cdot \frac{-r_2}{\rho - r_2} \right) = L_2$$

$$\int_0^\rho \frac{y}{y^2 + 13y + 18} \, dy = \frac{1}{r_1 - r_2} \int_0^\rho \frac{r_1}{y - r_1} - \frac{r_2}{y - r_2} \, dy$$

Recalling $r_i = (-13 \pm \sqrt{97})/2$ and $r_1 - r_2 = \sqrt{97}$, we have

$$= -\frac{13}{2} L_2 + \frac{1}{2} \int_0^\rho \frac{1}{y - r_1} + \frac{1}{y - r_2} \, dy$$

$$= -\frac{13}{2} L_2 + \frac{1}{2} L_1$$

since $L_1 = \log \left(\frac{\rho - r_1}{-r_1} \cdot \frac{\rho - r_2}{-r_2} \right)$. Using these formulas and recalling that $u_n = 2n(n+1) + 6$, the integral becomes

$$J_n = -\rho + (\rho - 2n(n+1) + 7) \left(-\frac{13}{2} L_2 + \frac{1}{2} L_1 \right)$$

$$+ ([2n(n+1) + 6]\rho + 18) \cdot \frac{2}{2} L_2$$

$$= -\rho + \frac{\rho - 2n(n+1) + 7}{2} L_1 + \frac{2n(n+1)(13 + 2\rho) - \rho - 55}{2} L_2$$

$$= -\rho + \frac{a_n}{2} L_1 + \frac{b_n}{2} L_2$$

Remembering the factors of $2/n(n-1)$ and $2/\rho^2$ we have left behind, the desired result follows. □

The reason for interest in (3.9) is that it allows us to construct an estimator for ρ. Hudson (1987) was the first to do this, but we will follow Wakeley's (1997) improvement.

Theorem 3.7. *Let $k_{\ell,m}$ be the number of differences between sequences ℓ and m, and let*

$$S_\pi^2 = \frac{2}{n(n-1)} \sum_{\ell < m} (k_{\ell m} - \Delta_n)^2$$

be the variance of the $\binom{n}{2}$ pairwise differences.

$$E(S_\pi^2) = \theta\frac{2(n-2)}{3(n-1)} + \theta^2 g_\pi(\rho, n) \tag{3.10}$$

where $\alpha_n = (n+1)\rho - (n-7)$, $\beta_n = (15n-1)\rho + (49n-55)$, and

$$g_\pi(\rho, n) = \frac{(n-2)}{n(n-1)\rho^2}\{-2\rho(n+1) + \alpha_n L_1 + \beta_n L_2\}$$

and L_1 and L_2 are as in (3.9).

Proof. S_π^2 can be rewritten as the second moment of the sample minus the square of the mean of the sample:

$$S_\pi^2 = \left[\frac{2}{n(n-1)}\sum_{\ell<m} k_{\ell m}^2\right] - \Delta_n^2$$

Since $Ek_{\ell m} = \theta = E\Delta_n$, we have

$$E(S_\pi^2) = E(k_{\ell m}^2) - E(\Delta_n^2) = \text{var}\,(\Delta_2) - \text{var}\,(\Delta_n)$$

since $k_{\ell m}$ is the number of pairwise differences between two sequences. From this and (3.9) it follows that

$$E(S_\pi^2) = \theta - \theta\frac{n+1}{3(n-1)} + \theta^2[f(\rho, 2) - f(\rho, n)]$$

This gives the θ term in (3.10). To compute $n(n-1)\rho^2[f(\rho, 2) - f(\rho, n)]$, we write

$$-n(n-1)\rho^2 f(\rho, n) = -2\{-2\rho + (\rho - 2n(n+1) + 7)L_1$$
$$+([4n^2 + 4n - 1]\rho + [26n(n+1) - 55])L_2\}$$
$$n(n-1)\rho^2 f(\rho, 2) = (n^2 - n)\{-2\rho + (\rho - 5)L_1 + (23\rho + 101)L_2\}$$

in order to prepare for the miracle that every term in the difference has $(n-2)$ as a factor

$$-2\rho(n^2 - n - 2) = -2\rho(n-2)(n+1)$$
$$\rho L_1(n^2 - n - 2) = \rho L_1(n-2)(n+1)$$
$$L_1(-5n^2 + 5n + 4n^2 + 4n - 14) = -L_1(n-2)(n-7)$$
$$\rho L_2(23n^2 - 23n - 8n^2 - 8n + 2) = \rho L_2(n-2)(15n-1)$$
$$L_2(101n^2 - 101n - 52n^2 - 52n + 110) = L_2(n-2)(49n-55)$$

Combining our calculations, we see that the coefficient of θ^2 in $E(S_\pi^2)$ is given by the indicated formula. $\qquad\square$

To introduce the estimator now, we follow Wakeley (1997) and write π for Δ_n. Since $E\pi^2 = \text{var}(\pi) + (E\pi)^2$, (3.9) implies

$$E\pi^2 = \theta \frac{(n+1)}{3(n-1)} + \theta^2[f(\rho, n) + 1]$$

so the following is an unbiased estimator of θ^2:

$$\frac{\pi^2 - [(n+1)/3(n-1)]\pi}{f(\rho, n) + 1}$$

and one can estimate ρ by solving

$$S_\pi^2 = \pi \frac{2(n-2)}{3(n-1)} + g_\pi(\rho, n) \frac{\pi^2 - [(n+1)/3(n-1)]\pi}{f(\rho, n) + 1}$$

Example 3.1. Wakeley (1997) applied his estimator to a data set of Schaeffer and Miller (1993), who sequenced $n = 99$ individuals in a 3.5 kb region containing the alcohol dehydrogenase genes *Adh* and *Adh-dup* of *Drosophila psuedoobscura*. The data set had 359 polymorphic sites with 27 having 3 nucleotides segregating, for a total of 386 mutations. Since

$$\frac{386 \cdot 385}{2} \cdot 3500 = 21.23$$

the large number of double hits is only a few more than we expect. Wakeley discarded the sites that were hit twice, computing $\pi = 31.7$ and a moment estimator of $\rho = 282$. Simulations suggested a 95% confidence interval of [172, 453]. Note that the estimate of ρ is about nine times the estimate of θ, i.e., in this region of the genome the recombination rate is about nine times the mutation rate.

3.3 Linkage disequilibrium

Linkage disequilibrium (LD) refers to the nonindependence of alleles at different loci. For example, suppose that allele A at locus 1 and allele B at locus 2 are at frequencies π_A and π_B, respectively. If the two loci were independent, then the AB haplotype would have frequency $\pi_{AB} = \pi_A \pi_B$. If this is not the case, then the two loci are in linkage disequilibrium and we let

$$D_{AB} = \pi_{AB} - \pi_A \pi_B$$

If r is the recombination probability between the two loci, then adding a superscript to indicate the generation number

$$\pi_{AB}^t = (1-r)\pi_{AB}^{t-1} + r\pi_A^{t-1}\pi_B^{t-1}$$

so if we ignore fluctuations in gene frequencies,

$$D_{AB}^t = (1-r)D_{AB}^{t-1} \cdots = (1-r)^t D_{AB}^0 \tag{3.11}$$

A wide variety of statistics have been proposed to measure LD. For probabilists and statisticians, the most natural is the square of the correlation coefficient

$$r^2 = \frac{D_{AB}^2}{\pi_A \pi_a \pi_B \pi_b} \tag{3.12}$$

A second commonly used measure, introduced by Lewontin (1964), is

$$D' = \begin{cases} \dfrac{D_{AB}}{\min(\pi_A \pi_b, \pi_B \pi_a)} & \text{if } D_{AB} > 0 \\[2mm] \dfrac{D_{AB}}{\min(\pi_A \pi_B, \pi_a \pi_b)} & \text{if } D_{AB} < 0 \end{cases}$$

To explain this formula, we will prove

Theorem 3.8. $D' \in [-1, 1]$ *with the extremes achieved when one of the four combinations is absent from the population.*

In contrast, $r^2 = 1$ when there are only two combinations in the population: AB and ab, or Ab and aB.

Proof. Clearly, $\pi_{AB} \leq \min\{\pi_A, \pi_B\}$. This implies

$$\pi_{AB} - \pi_A \pi_B \leq \min\{\pi_A(1-\pi_B), \pi_B(1-\pi_A)\} = \min(\pi_A \pi_b, \pi_B \pi_a)$$

and $D' \leq 1$. To get the bound $D' \geq -1$, note that $D_{AB} = -D_{Ab}$ and use the first result with the roles of B and b interchanged to conclude

$$D_{Ab} \leq \min(\pi_A \pi_B, \pi_b \pi_a)$$

At this point, we have shown $D' \in [-1, 1]$. If $D' = 1$, then $\pi_{AB} = \min\{\pi_A, \pi_B\}$. If $\pi_{AB} = \pi_A$, then $\pi_{Ab} = 0$. If $\pi_{AB} = \pi_B$, then $\pi_{aB} = 0$. If $D' = -1$, then we note $D_{AB} = -D_{Ab}$ and use the previous argument. $\qquad\square$

The computation of Er^2 is made difficult by the correlation between the numerator and denominator. Many people believe, see e.g., Hartl and Clark (2007, page 532), the following:

Mythical result. If $\rho = 4Nr$ then $Er^2 = 1/(1+\rho)$.

An Internet article on LD I found attributes this result to Hill and Robertson (1968). However, all they have to say about this is that the limiting value of Er^2 appears to approach $1/\rho$ as Nr increases, see on page 229. Song and Song (2007) have developed numerical methods to compute Er^2 which allow them to prove that $Er^2 \sim 1/\rho$ as $\rho \to \infty$.

Ohta and Kimura (1971) argued that unless one or both of the allele frequencies took values near 0 or 1,

$$Er^2 \approx \frac{ED_{AB}^2}{E(\pi_A \pi_a \pi_B \pi_b)} \equiv \sigma_d^2 \tag{3.13}$$

(where the $\equiv$ indicates the second equality is the definition of σ_d^2), and they used diffusion theory (see Theorem 8.13 at the end of Section 8.2) to show

Theorem 3.9.
$$\sigma_d^2 = \frac{10 + \rho}{22 + 13\rho + \rho^2}$$

Note that when ρ is large, $\sigma_d^2 \approx 1/\rho$. Here we will give McVean's (2002) derivation based on coalescent theory. Note that here the genotype frequencies are those of the population. One can compute the expected value for a sample, but the result is quite messy; see (10) in McVean (2002).

Proof. To simplify the first calculation, it is useful to consider the general case where there may be more than two alleles. In this case,

$$D_{\alpha,\beta} = f_{\alpha,\beta} - p_\alpha q_\beta$$

where p_α and q_β are the frequencies of α and β at the two loci, $f_{\alpha,\beta}$ is the frequency of the α, β haplotype, and one defines the square of the disequilibrium by

$$D^2 = \sum_{\alpha,\beta} (f_{\alpha,\beta} - p_\alpha q_\beta)^2$$

In the case of two alleles, $D_{11} = -D_{12} = -D_{21} = D_{22}$, so $D^2 = 4D_{\alpha,\beta}^2$. Note that $\sum_{\alpha,\beta} D_{\alpha,\beta} = 1 - 1 = 0$, so for two alleles we have

$$ED_{\alpha,\beta} = 0 \tag{3.14}$$

without any assumption other than a symmetric role for the two alleles.

To compute the second moment, we note that

$$\begin{aligned} ED^2 &= \sum_{\alpha,\beta} (f_{\alpha,\beta} - p_\alpha q_\beta)^2 \\ &= \sum_{\alpha,\beta} f_{\alpha,\beta}^2 - 2f_{\alpha,\beta} p_\alpha q_\beta + p_\alpha^2 q_\beta^2 \\ &= F_{ij,ij} - 2F_{ij,ik} + F_{ij,k\ell} \end{aligned} \tag{3.15}$$

where the $F_{ij,k\ell}$ is the probability that sequences i and j are equal at the first locus and sequences k and ℓ are equal at the second locus, and the sequences i, j, k, and ℓ are chosen at random. This is (A6) from Hudson (1985), who attributed the result to Strobeck and Morgan (1978). Since we are doing our calculation for the entire population, we can suppose that the indices i, j, k, and ℓ are distinct.

When there are only two alleles at each locus, the square of the disequilibrium coefficient is independent of how the alleles are defined, so we can consider $F^*_{ij,k\ell}$ the probability that the derived mutation occurs in sequences i and j at the first locus and in sequences k and ℓ at the second locus. It follows from (3.15) that we have

$$ED^2_{AB} = F^*_{ij,ij} - 2F^*_{ij,ik} + F^*_{ij,k\ell}$$

Let I^h_{ij} be the branch length leading from the most recent common ancestor of i and j to the most recent common ancestor of the sample at locus $h = 1, 2$, and τ^h the total size of the tree for the population at locus h. We assume that the units for these times are $2N$ generations. Assuming the mutation rate μ is the same at the two sites and letting $u = 2N\mu$, we have

$$F^*_{ij,k\ell} = \frac{E(uI^1_{ij}e^{-u\tau^1} uI^2_{k\ell}e^{-u\tau^2})}{E(u\tau^1 e^{-u\tau^1} u\tau^2 e^{-u\tau^2})}$$

Taking the limit as $u \to 0$ to eliminate the mutation rate gives

$$F^*_{ij,k\ell} = \frac{E(I^1_{ij}I^2_{k\ell})}{E(\tau^1\tau^2)}$$

Let t^h_{ij} be the time of the most recent common ancestor of i and j at locus h, and let T^h be the time of the most recent common ancestor of the population at locus h. Writing $I^h_{ij} = T^h - t^h_{i,j}$ and using symmetry to conclude that $E(T^1 t^2_{ij})$ does not depend on i, j (assuming they are distinct), we have

$$E(I^1_{ij}I^2_{ij}) - 2E(I^1_{ij}I^2_{ik}) + E(I^1_{ij}I^2_{k\ell})$$
$$= (1 - 2 + 1)E(T^1T^2) - (1 - 2 + 1)E(T^1 t^2_{ij} + T^2 t^1_{ij})$$
$$+ E(t^1_{ij}, t^2_{ij}) - 2E(t^1_{ij}, t^2_{ik}) + E(t^1_{ij}, t^2_{k\ell})$$

The means $Et^h_{ij} = 1$, so we have

$$ED^2_{AB} = \frac{\operatorname{cov}(t^1_{ij}, t^2_{ij}) - 2\operatorname{cov}(t^1_{ij}, t^2_{ik}) + \operatorname{cov}(t^1_{ij}, t^2_{k\ell})}{E(\tau^1\tau^2)}$$

A similar approach can be used on the denominator of σ^2_d, which represents the probability that when two alleles i and j are drawn with replacement at the first locus and another two k and ℓ are drawn with replacement at the second locus, then i and k have the mutant alleles, and j and ℓ do not.

$$E(\pi_A \pi_a \pi_B \pi_b) = \lim_{\mu \to 0} \frac{E(ut^1_{ij}e^{-u\tau^1} ut^2_{k\ell}e^{-u\tau^2})}{E(u\tau^1 e^{-u\tau^1} u\tau^2 e^{-u\tau^2})}$$
$$= \frac{E(t^1_{ij}t^2_{k\ell})}{E(\tau^1\tau^2)} = \frac{\operatorname{cov}(t^1_{ij}, t^2_{k\ell}) + 1}{E(\tau^1\tau^2)}$$

Combining this with the formula for ED^2_{AB}, we have

$$\sigma_d^2 = \frac{\mathrm{cov}\,(t_{ij}^1, t_{ij}^2) - 2\,\mathrm{cov}\,(t_{ij}^1, t_{ik}^2) + \mathrm{cov}\,(t_{ij}^1, t_{k\ell}^2)}{\mathrm{cov}\,(t_{ij}^1, t_{k\ell}^2) + 1}$$

Using (3.1) now, we have

$$\sigma_d^2 = \frac{18 + \rho - 2(6) + 4}{4 + 18 + 13\rho + \rho^2} = \frac{10 + \rho}{22 + 13\rho + \rho^2}$$

which proves the desired result.   □

LD in humans

Understanding the extent and distribution of linkage disequilibrium in humans is an important question because LD plays a fundamental role in the fine scale mapping of human disease loci, see e.g., Risch and Merikangas (1996) and Cardon and Bell (2001). The technique known as association mapping differs from traditional pedigree studies in that marker-disease associations are sought in populations of unrelated individuals. To explain the idea behind this approach, imagine that a disease causing mutation has just occurred in a population. The chromosome on which this mutation occurred contains specific alleles in neighboring polymorphic loci. At first, the mutation will only be observed in conjuction with these alleles, so the association (or LD) with these alleles will be high. Through time these associations will dissipate because of recombination, but the closest loci will experience the fewest recombinations and hence retain the highest levels of LD. Thus, by looking for significant correlations between disease state and alleles, we can hope to identify the region in which the disease causing genetic mutation lies.

One of the surprising patterns revealed by the construction of a dense genome-wide map of single nucleotide polymorphisms (SNPs) was the slow decay of LD. Kruglyak's (1999) simulation study suggested that useful levels of LD were unlikely to extend beyond an average distance of about 3 kilobases (kb) in the general population. In contrast, Reich et al. (2001) found in a study of 19 randomly selected regions in the human genome that LD in a United States population of northern European descent typically extends 60 kb from common alleles. These findings were confirmed and further quantified by Dawson et al. (2002), who measured LD along the complete sequence of human chromosome 22, and Ke et al. (2004), who studied a contiguous 10 megabase segment of chromosome 20. In both cases, see Figure 1 of Dawson et al. (2002) and Figure 3 of Ke et al. (2004), LD as measured by the square of the correlation coefficient r^2 is about 0.1 at 100 kb. If, as Ardlie, Kruglyak, and Seielstad (2002) argued, $r^2 > 1/3$ is the limit of useful LD, this occurs at about 30 kb.

3.4 Ancestral recombination graph

As we have seen in the first two sections of this chapter, analytical results for genealogies with recombination are difficult and messy. In this section we will show that it is reasonably easy to simulate the process. Hudson (1983) was the first to do this. Here we follow the formulation of Griffiths and Marjoram (1997). Taking a continuous perspective, we consider a segment of DNA and rescale it to be the unit interval $[0, 1]$. If we suppose that the segment is small enough so that two recombinations in one generation can be ignored, then the dynamics of a Wright-Fisher model (going forward in time) may be described as follows. To generate a chromosome in the next generation, with probability $1 - r$ we copy all of the contents from one randomly chosen individual. With probability r a recombination occurs. We pick a point uniformly along the chromosome and two individuals at random from the population. We copy the genetic material to the left of that point from the first individual and copy the material to the right from the second.

Reversing our perspective leads to a genealogical process, which in the case of a sample of size 4 can be drawn as in the next figure. Lines merging indicate coalescence events and are marked with letters. Splits indicate a recombination at the indicated location, with the convention that at recombination events the left half comes from the individual on the left and the right half comes from the individual on the right.

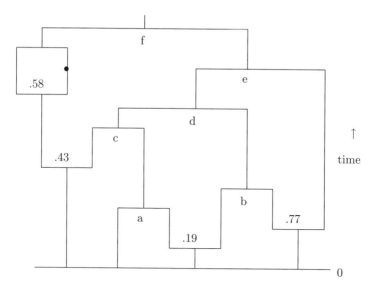

Fig. 3.2. Realization of the ancestral recombination graph for a sample of size 4.

The number of ancestors in the sample $2Nt$ units back in time Y_t is a birth and death process. When there are k ancestors, the death rate due to coalescence of lineages is $\mu_k = k(k-1)/2$,

the birth rate due to recombinations is $\lambda_k = k\rho/2$, where $\rho = 4Nr$.

The first rate is by now familiar. To check the second, note that the probability of no recombination in one generation is

$$(1-r)^k \approx 1 - kr = 1 - \frac{1}{2N} \cdot \frac{k\rho}{2}$$

Because of the quadratic birth rate compared to the linear birth rate, if we start with $m > 1$ sequences then with probability 1, there is a time τ_m where the number of ancestors is 1. Griffiths (1991) has calculated

Theorem 3.10.

$$E\tau_m = \frac{2}{\rho} \int_0^1 \left(\frac{1 - v^{m-1}}{1-v} \right) \left(e^{\rho(1-v)} - 1 \right) dv \qquad (3.16)$$

When $m = 2$, this becomes

$$E\tau_2 = \frac{2}{\rho} \int_0^1 e^{\rho(1-v)} - 1 \, dv = \frac{2}{\rho} \left(-\frac{1}{\rho} e^{\rho(1-v)} - v \right) \Big|_0^1$$

$$= \frac{2}{\rho} \left(-\frac{1}{\rho} - 1 + \frac{1}{\rho} e^\rho \right) = \frac{2}{\rho^2} \cdot (e^\rho - 1 - \rho)$$

Note that as $\rho \to 0$, $E\tau_2 \to 1$, the result for the ordinary coalescent, but $E\tau_2$ grows exponentially fast as ρ increases, which means that it will be very slow to simulate the ancestral recombination graph for large regions.

Proof. We use the methods of Section 1.5. We begin by computing hitting probabilities. The number of lineages in the genealogy, Y_t, changes

$$k \to k+1 \text{ at rate } \rho k/2$$
$$k \to k-1 \text{ at rate } k(k-1)/2$$

so the embedded jump chain, X_n,

$$k \to k+1 \text{ with prob. } p_k = \rho/(\rho + k - 1)$$
$$k \to k-1 \text{ with prob. } q_k = (k-1)/(\rho + k - 1)$$

We want to find a function ϕ so that $\phi(X_n)$ is a martingale. For this we need

$$\phi(k) = \phi(k+1)p_k + \phi(k-1)q_k$$

or, rearranging,

$$\phi(k+1) - \phi(k) = \frac{q_k}{p_k}(\phi(k) - \phi(k-1))$$

X_n has state space $\{1, 2, \ldots\}$. Setting $\phi(2) - \phi(1) = 1$ and iterating gives

$$\phi(k) - \phi(k-1) = \prod_{j=2}^{k-1} \frac{q_j}{p_j} = \frac{(k-2)!}{\rho^{k-2}}$$

(By convention, $\prod_{j=2}^{1} q_j/p_j = 1 = 0!/\rho^0$.) Taking $\phi(1) = 0$, we have

$$\phi(m) = \sum_{k=2}^{m} \frac{(k-2)!}{\rho^{k-2}}$$

Let $T_k = \min\{n \geq 0 : X_n = k\}$ be the time of the first visit to k. Since $\phi(X_n)$ is a martingale and the absorbing state 1 has $\phi(1) = 0$,

$$P_m(T_k < \infty) = \begin{cases} 1 & k \leq m \\ \frac{\phi(m)}{\phi(k)} & k > m \end{cases}$$

Let $T_k^+ = \min\{n \geq 1 : X_n = k\}$ be the time of the first return to k. If we start at k, the only way to avoid returning to k is to go from $k \rightarrow k-1$ on the first jump and then not come back, so

$$P_k(T_k^+ = \infty) = \frac{k-1}{\rho + k - 1} \left(1 - \frac{\phi(k-1)}{\phi(k)} \right)$$

If the process reaches k, then the number of visits to k, N_k, will have a geometric distribution with mean $1/P_k(T_k^+ = \infty)$, so

$$E_m N_k = \frac{P_m(T_k < \infty)}{P_k(T_k^+ = \infty)}$$

and it follows that

$$E_m N_k = \begin{cases} \frac{\rho+k-1}{k-1} \left(\frac{\phi(k)}{\phi(k)-\phi(k-1)} \right) & k \leq m \\ \frac{\phi(m)}{\phi(k)} \cdot \frac{\rho+k-1}{k-1} \left(\frac{\phi(k)}{\phi(k)-\phi(k-1)} \right) & k > m \end{cases}$$

Returning to Y_t, the continuous-time process that gives the size of the genealogy, we let S_k be the amount of time Y_t spends at k. Since the rate of jumps out of k is $k(\rho + k - 1)/2$, and $\phi(k) - \phi(k-1) = (k-2)!/\rho^{k-2}$,

$$E_m S_k = \frac{2}{k(\rho + k - 1)} E_m N_k = \begin{cases} \frac{2}{k(k-1)} \cdot \phi(k) \cdot \frac{\rho^{k-2}}{(k-2)!} & k \leq m \\ \phi(m) \cdot \frac{2}{k(k-1)} \cdot \frac{\rho^{k-2}}{(k-2)!} & k > m \end{cases}$$

Using $x \wedge y = \min\{x, y\}$, we can combine the two formulas into one:

$$E_m S_k = 2\phi(k \wedge m) \frac{\rho^{k-2}}{k!} = 2\frac{\rho^k}{k!} \sum_{j=2}^{k \wedge m} \frac{(j-2)!}{\rho^j}$$

where in the last expression we have multiplied the numerator and denominator by ρ^2.

Since $T_1 = \sum_{k=2}^{\infty} S_k$, we have

$$E_m T_1 = 2 \sum_{k=2}^{\infty} \frac{\rho^k}{k!} \sum_{j=2}^{k \wedge m} \frac{(j-2)!}{\rho^j}$$

Interchanging the order of the summation and letting $i = k - j$, the above

$$= 2 \sum_{j=2}^{m} \sum_{k=j}^{\infty} (j-2)! \frac{\rho^{k-j}}{k!} = 2 \sum_{j=2}^{m} \sum_{i=0}^{\infty} (j-2)! \frac{\rho^i}{(i+j)!}$$

To see the first step note that the sum is over $k \geq 2$, $2 \leq j \leq m$, $j \leq k$.

To relate the last formula to the answer given above, we note that

$$\frac{2}{\rho} \int_0^1 \frac{1 - v^{m-1}}{1 - v} \left(e^{\rho(1-v)} - 1 \right) dv$$

$$= \frac{2}{\rho} \int_0^1 \sum_{k=0}^{m-2} v^k \sum_{\ell=1}^{\infty} \frac{\rho^\ell (1-v)^\ell}{\ell!} dv$$

Letting $\ell = i + 1$ and $k = j - 2$, this becomes

$$2 \sum_{j=2}^{m} \sum_{i=0}^{\infty} (j-2)! \rho^i \int_0^1 \frac{v^{j-2}}{(j-2)!} \frac{(1-v)^{i+1}}{(i+1)!} dv$$

Integrating by parts $j - 2$ times gives

$$\int_0^1 \frac{v^{j-2}}{(j-2)!} \frac{(1-v)^{i+1}}{(i+1)!} dv = \int_0^1 \frac{(1-v)^{i+j-1}}{(i+j-1)!} dv = \frac{1}{(i+j)!}$$

and we have proved the desired result.   □

The time τ_m is only an upper bound on the MRCAs for all nucleotides because (i) some of the ancestors in the genealogy, such as the one marked by the black dot in the example in Figure 3.2 above, may have no genetic material that is ancestral to that in the sample, and (ii) different segments of the chromosome will have MRCAs at different times $\leq \tau_m$. In the example, the four chromosome segments have the following genealogies. Here the letters correspond to the coalescence events in the ancestral recombination graph, while × indicates where the recombination occurred to change the tree to the next one.

3.4.1 Simulation

In simulating the ancestral recombination graph, we do not want to create ancestors that have no genetic material in common with the sample. To explain

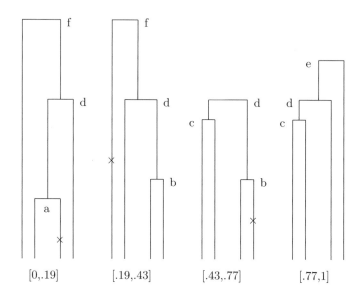

Fig. 3.3. Coalescent trees for the four segments in our example.

how to avoid this, we will, for simplicity, consider the two-locus case. Generation t signifies the population t generations before the present. We consider a sample of size n from the current population (generation 0). Let $g(t)$ be the number of chromosomes of generation t that contain genetic material at either locus that is directly ancestral to genetic material of the sample. Let $d(t)$ be the number of ancestral chromosomes that contain copies of both loci.

Let t_k be the time of the kth event (i.e., recombination or coalescence). If time is written in units of $2N$ generations, then $t_{k+1} - t_k$ has an exponential distribution with rate

$$\lambda(t_k) = d(t_k)\rho/2 + g(t_k)(g(t_k) - 1)/2$$

The next event is a recombination with probability

$$\frac{d(t_k)\rho/2}{\lambda(t_k)}$$

In this case, we pick one of the $d(t_k)$ chromosomes that contain copies of both loci to split. The event is a coalescence with probability

$$\frac{g(t_k)(g(t_k) - 1)/2}{\lambda(t_k)}$$

In this case, we pick one pair of ancestral chromosomes to coalesce. Of course, after either of these events, we must compute the new values of $g(t_{k+1})$ and $d(t_{k+1})$. When we finally reach the point at which there is only one chromosome, then we can move forward in time assigning mutations to the branches as in the ordinary coalescent. See Hudson (1991) for more details. The extension to a chromosomal segment is fairly straightforward. However, we are all fortunate that Hudson has created and made publicly available a program *ms* that will do these simulations.

$$\text{http://home.uchicago.edu/~rhudson1/source.html}$$

Coalescence trees along the chromosome

Wiuf and Hein (1999) developed a different way of constructing the ancestral recombination graph for a sequence. To start, one generates a genealogy for the endpoint $x_0 = 0$ in the usual way. Let r be the recombination rate per unit distance and $\rho = 4Nr$. If τ^0 is the total size of the tree (measured in units of $2N$ generations), then the distance x_1 until we encounter the next recombination is exponential with rate $\tau^0 \rho/2$. To generate the genealogy at x_1, pick a point U^1 uniformly from the tree. At this point, the portion of the chromosome to the right of the recombination has a different path, so we simulate to see where it should rejoin the tree.

The simplest, but unfortunately incorrect, way to do this is to erase the part of the branch from that point to the next coalescence event, creating a floating lineage, and then simulate to determine where the lineage should reattach to the rest of the tree. In the example drawn, from the time U^1 marked by the $\times$ to T_3^0 the coalescence rate is 3, from time T_3^0 to T_1^0 the rate is 2, and after T_1^0 at rate 1. Thus, the new time of the most recent common ancestor may be larger than the old one. The coalescence times in the new tree are labeled T_4^1, T_3^1, T_2^1, and T_1^1, i.e., the superscript indicates tree numbers along the chromosome, and the subscript is the number of lineages just before the coalescence event.

This recipe is simple, but, unfortunately, it is wrong. If the new lineage coalesces with the erased one, then until there is another recombination the new lineage must follow the choices made by the erased lineage until it reattaches to the tree. This event will happen with significant probability, since from the $\times$ until time T_3^0 the erased lineage is one of four possible targets for coalescence, and from T_3^0 to T_2^0 it is one of three. The unfortunate consequence of this is that one cannot erase the lineage but must keep adding paths to the diagram, increasing the complexity and the computer storage requirements of the computation. We refer the reader to Wiuf and Hein (1999) for details about how to correctly implement the algorithm.

3.4.2 Two approximate algorithms

Given the problems that arise from the ghosts of previous branches, it is natural to forget them to produce a Markovian process that approximates

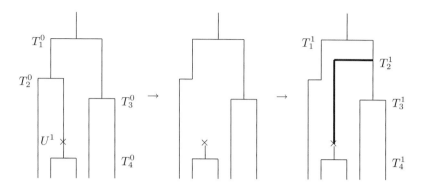

Fig. 3.4. The wrong way of moving from one tree to the next.

how the genealogical tree of a sample changes as we move along a chromosome. McVean and Cardin (2005) implemented the approximation as it was described above, and named it the spatial Markovian coalescent (SMC). Marjoram and Wall (2006) modified the approximation so the old lineage is not erased until after the coalescence point of the new one has been determined, which allows for the possibility that the new lineage coalesces with the one that was to be erased and no change occurs in the genealogy. In both papers, simulations show that the improved algorithm produces simulated data that for many statistics is close to that resulting from simulation of the full coalescent. Marjoram and Wall developed software (*FastCoal*) for implementing these approximations, which is available from the authors. The most important reason for being interested in the approximation is that the amount of computation needed for the SMC and the MW algorithms increases linearly with the length of chromosome simulated, while the ancestral recombination graph requires an exponentially increasing amount of work.

To obtain some insight into the workings of these approximations, we will consider $n = 2$ and for simplicity only the SMC, which is less accurate than the MW chain, but nicer to compute with. When $n = 2$, there is only one coalescence time, so we let H_k be the height of the kth tree. The place at which the recombination occurs, $U^1 = \xi_1 H_0$, where ξ_1 is uniform on $(0, 1)$. The new height $H_1 = U^1 + \eta_1$ and η_1 is exponential with mean 1. Since $H_1 = \xi_1 H_0 + \eta_1$, it is easy to see that if

$$H_0 =_d \sum_{n=1}^{\infty} \eta_n \prod_{m=1}^{n-1} \xi_m$$

then H_1 also has this distribution, so this is the stationary distribution π for the discrete-time chain. Since trees of height y stay around for a mean time

$1/2y$, the continuous-time stationary distribution has

$$\mu(dy) = \frac{1}{c} \cdot \frac{1}{2y} \pi(dy) \quad \text{where} \quad c = \int \frac{1}{2y} \pi(dy)$$

The first thing to prove is that

Theorem 3.11. *The stationary distribution for the height of the SMC tree for a sample of size 2, μ, is exponential with mean 1, which is the correct distribution for the coalescent.*

Proof 1. To prove this, we begin by taking the expected value

$$E(H_1|H_0) = \frac{H_0}{2} + 1$$

to conclude that in equilibrium $EH = EH/2 + 1$ and $EH = 2$. We will now prove by induction that $EH^k = (k+1)!$. To do this we begin with the observations that (i) the uniform distribution on $(0,1)$, ξ, has

$$E\xi^m = \int_0^1 x^m \, dx = \frac{1}{m+1}$$

and (ii) integration by parts and induction shows that the mean 1 exponential distribution, η, has

$$E\eta^m = \int_0^\infty x^m e^{-x} \, dx = m \int_0^\infty x^{m-1} e^{-x} \, dx = m!$$

Now, since H and $H\xi + \eta$ have the same distribution, if the formula $EH^m = (m+1)!$ is correct for all powers $m < k$, then

$$EH^k = \sum_{m=0}^{k} \binom{k}{m} E(H^m) E(\xi^m) E(\eta^{k-m})$$

$$= EH^k \frac{1}{k+1} + \sum_{m=0}^{k-1} \frac{k!}{m!(k-m)!} (m+1)! \frac{1}{m+1} (k-m)!$$

Each term in the sum is $k!$, so we have

$$\frac{k}{k+1} EH^k = k \cdot k! \quad \text{and hence} \quad EH_k = (k+1)!$$

From the moments we see that $\pi(dx) = xe^{-x} \, dx$, so $c = 1/2$ and $\mu(dy) = e^{-y} \, dy$. $\qquad\square$

Proof 2. We begin by computing the transition kernel for the discrete-time Markov chain $H_0, H_1, \ldots$. Suppose $H_n = x$. Breaking things down according to the value of $z = H_n \xi$, we have

$$\text{if } y > x \quad K(x,y) = \int_0^x \frac{dz}{x} e^{-(y-z)} = \frac{e^{x-y} - e^{-y}}{x}$$

$$\text{if } y < x \quad K(x,y) = \int_0^y \frac{dz}{x} e^{-(y-z)} = \frac{1 - e^{-y}}{x}$$

We will now show that $\pi(x) = xe^{-x}$ satisfies the *detailed balance condition*: $\pi(x)p(x,y) = \pi(y)p(y,x)$, which implies

$$\int \pi(x)p(x,y)\, dx = \pi(y) \int p(y,x)\, dx = \pi(y)$$

To check the detailed balance condition, we can assume that $y > x$. In this case, our formulas imply

$$\begin{aligned}
\pi(x)K(x,y) &= xe^{-x} \cdot (e^{x-y} - e^{-y})/x \\
&= e^{-y} - e^{-(x+y)} \\
&= ye^{-y} \cdot (1 - e^{-x})/y = \pi(y)K(y,x)
\end{aligned}$$

The detailed balance condition implies not only that the π is a stationary distribution, but also that the chain is *reversible*; in equilibrium the chain looks the same going forwards or backwards. This is natural since a chromosome consists of two complementary strands of DNA and has no inherent orientation. □

As McVean and Cardin (2005) explained, a nice way of thinking about the SMC is that it is a modification of the ancestral recombination graph in which coalescence is forbidden if the ancestors do not have genetic material in common. In the two-locus case, this means that (a)'s may coalesce with (ab)'s and (b)'s with (ab)'s but (a)'s and (b)'s are forbidden from coalescing. From this we see that, in a large population, the covariance of the coalescence times T_a and T_b is ≈ 0 if the initial sampling configuration is $(a)(ab)(b)$ or $(a)(a)(b)(b)$. Thus, the covariance for the state $(ab)(ab)$ can only be nonzero if coalescence occurs before recombination, and in this case it is 1, so for the SMC

$$E_\pi(H_0 H_t - 1) = \frac{1}{1+t}$$

since $\rho/2 = t/2$ is the scaled recombination rate for the interval $[0, t]$, and there are two individuals in the sample subject to recombination. In comparison to the exact answer given in (3.1),

$$\frac{1}{1+t} < \frac{t+18}{t^2 + 13t + 18} < 1.2797 \frac{1}{1+t}$$

the maximum relative error of 28% occurring when $t = 4.24$. This is a pretty large error. However, if we consider the variance of the number of segregating sites, $\mathrm{var}(S_2)$, computed exactly in (3.6), the error will be less since we will integrate $E_\pi(H_0 H_t - 1)$ over a range of t values.

Using the forbidden coalescence reasoning, one can compute the joint distribution $P_\pi(H_0 = x, H_t = y)$. If coalescence occurs before recombination, an event of probability $1/(1+t)$, then the joint distribution (H_0, H_t) is $(t+1)e^{-(1+t)x}1_{\{x=y\}}$. When recombination occurs before coalescence, an event of probability $t/(1+t)$, the additional time for the two coalescences are independent mean 1 exponentials, so if $x < y$, the joint density is

$$\frac{t}{t+1} \int_0^x (1+t)e^{-(1+t)z}e^{-(x-z)}e^{-(y-z)} \, dz$$

$$= t \int_0^x e^{(1-t)z}e^{-x}e^{-y} \, dz = t\frac{e^{(1-t)x} - 1}{1 - t}e^{-(x+y)}$$

By symmetry, the formula for $y < x$ is

$$t\frac{e^{(1-t)y} - 1}{1 - t}e^{-(x+y)}$$

Dividing by $P_\pi(H_0 = x) = e^{-x}$, we have the transition probability

$$P(H_t = y | H_0 = x) = \begin{cases} e^{-tx}\delta_x & x = y \\ t\frac{e^{(1-t)x}-1}{1-t}e^{-y} & x < y \\ t\frac{e^{(1-t)y}-1}{1-t}e^{-y} & x > y \end{cases}$$

3.5 Counting recombinations

Suppose that there are two alleles, A and a, at site i and two alleles, B and b, at site j. As we observed in the analysis of the data of Ward et al. (1991) in Section 1.4, if there is no recombination and each site has been hit only once by mutation, then at most three of the four gametic types AB, Ab, aB, and ab can be present. Thus, if we assume that each site has been hit only once by mutation, then there must have been a recombination in between i and j. To get a lower bound, R_M, on the number of recombinations that have occurred, Hudson and Kaplan (1985) set $d(i, j) = 1$ when all four gametes are present and $d(i, j) = 0$ otherwise. To compute R_M, we represent the (i, j) with $d(i, j) = 1$ as an open interval and apply the following algorithm:

- Delete all (m, n) that contain another interval (i, j).
- Let (i_1, j_1) be the first interval not disjoint from all the others. If (m, n) has $i_1 < m < j_1$, then delete (m, n). Repeat until done.

Following Hudson and Kaplan (1985), we will analyze Kreitman's (1983) data on the alcohol dehydrogenase locus of *Drosophila melanogaster*. In the data set, F and S indicate the fast and slow alleles that are caused by the substitution at position 32. The first sequence is a reference sequence, the other 11 are the data. Here we have ignored the six sites in Kreitman's data at which there have been insertions and/or deletions.

```
                        1111111111222222222233333333334444
             12345678901234567890123456789012345678901234
ref  CCGCAATATGGGCGCTACCCCCGGAATCTCCACTAGACAGCCT
 1S  ........AT........TT.ACA.TAAC.............
 2S  ..C.............TT.ACA.TAAC..............
 3S  ...............................A....T.A
 4S  ...............GT................A..TA...
 5S  ...AG...A.TC..AGGT.................C......
 6S  ..C...........G..............T.T.CAC....T.
 1F  ..C...........G.............GTCTCC.C......
 2F  TGCAG...A.TCG..G.............GTCTCC.CG.....
 3F  TGCAG...A.TCG..G.............GTCTCC.CG.....
 4F  TGCAG...A.TCG..G.............GTCTCC.CG.....
 5F  TGCAGGGGA....T.G....A...G....GTCTCC.C......
```

It should be clear from the first step in the algorithm that for each i we only have to locate $j_i = \min\{j > i : d(i,j) = 1\}$, because the other intervals with left endpoint i will contain (i, j_i). An example should help clarify the procedure. The first stage is to find (i, j_i) for $1 \le i \le 43$. In doing this, we can ignore sites 6, 7, 8, 10, 14, 15, 21, 25, and 39–43, which have only one mutation since these cannot be part of a pair with $d(i,j) = 1$. Note that 13, 19–30, 32, 34, 37, and 38 are consistent with all of the rows that followed, so they do not produce intervals. The remaining (i, j_i) pairs are as follows:

$$
\begin{array}{ccc}
(1,11) & (2,11) & \mathbf{(3,4)} \\
(4,17) & (5,17) & \mathbf{(9,16)} \\
(11,17) & (12,17) & \mathbf{(16,17)} \\
(17,36) \quad (18,36) \quad (31,36) \quad (33,36) & & \mathbf{(35,36)} \\
& & \mathbf{(36,37)}
\end{array}
$$

On each row, the interval at the end in bold-faced type is a subset of the previous intervals. The five intervals in boldface are disjoint, so $R_M = 5$. This conflicts with the value $R_M = 4$ reported by Hudson and Kaplan (1985); however, one can verify the five comparisons in the table above.

By looking at the data more closely, one can infer a larger number of recombinations. To make this easier to see, we delete sequences 3F and 4F, which are identical to 2F, and then delete the columns with singletons:

```
            |    11|1|11122222222333333|3|3
            123|45912|6|78902346789012345|6|7
    1:1S  ...|..A..|.|.|..TTACATAAC......|.|.
    2:2S  ..C|.....|.|.|..TTACATAAC......|.|.
    3:3S  ...|.....|.|.|..............|A|.
    4:4S  ...|.....|.|.|GT..............|A|.
    5:5S  ...|AGATC|G|.|GT..............|.|C
    6:6S  ..C|.....|.|G|..........T.T.C|A|C
    7:1F  ..C|.....|.|G|..........GTCTCC|.|C
    8:2F  TGC|AGATC|G|.|..........GTCTCC|.|C
    9:5F  TGC|AGA..|.|G|..........GTCTCC|.|C
```

Myers and Griffiths (2003) improved Hudson's bound by observing that if in a set of S columns there are K haplotypes, then there must be at least $K - S - 1$ recombinations. ($S = 2$ and $K = 4$ gives the four-gamete test.) In the example above, if we look at columns 3, 9, and 12, we see six haplotypes:

.A.   C..   ...   .AT   CAT   CA.

so there must have been at least $6 - 3 - 1 = 2$ recombinations in the interval $(3, 12)$.

Given some method of computing lower bounds $b_{i,j}$ on the number of recombinations in the interval (i, j), we can compute a better bound by

$$B_{i,j} = \max_{i < k < j} b_{i,k} + B_{k,j}$$

and a lower bound on the number in whole region by B_{1,S_n}, where S_n is the number of segregating sites. If $b_{i,k} = 1$ when all four gametes are present, this reduces to Hudson and Kaplan's R_M. Myers and Griffiths (2003) developed an estimator R_h where $b_{i,k}$ is computed from the haplotype bound applied to various subsets of $[i, j]$. For this data set it turns out that $R_h = 6$. Bafna and Bansal (2006) further improved the method and showed that there must be at least seven recombinations in this data set.

Song and Hein (2005) have a different approach, which in this case shows that the lower bound of 7 is optimal. The next figure gives trees that are consistent with the mutation patterns in the six regions of the sequence separated by vertical lines. Black dots indicate where recombinations can be introduced to move to the next tree in the seqeunce. Circles indicate the locations of mutations needed in the regions.

While the lower bounds are ingenious, they do not get close to the number of recombinations. Bafna and Bansal (2006) investigated the mean values of Hudsons' R_M, Myers and Griffiths' (2003) program Recmin, and their own bound R_g in comparison with the actual number of recombinations R for various values of the scaled recombination rate $\rho = 4Nr$ in a sample of size 100, when the scaled mutation rate $\theta = 10$.

ρ	1	5	20	50
R_M	1.02	3.03	6.29	9.39
Recmin	1.23	4.88	13.58	24.86
R_g	1.23	5.09	15.80	31.54
R	5.21	27.19	126.45	388.76

An assessment of R_M was done much earlier by Hudson and Kaplan (1985), who used simulations to obtain an estimate from their lower bound. In Kreitman's Adh data, the number of segregating sites is $S = 43$, so $\theta = 4Nu$ can be estimated by

$$S \bigg/ \sum_{i=1}^{10} 1/i = \frac{43}{2.928968} = 14.68$$

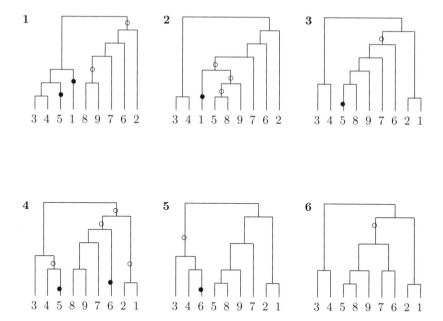

Fig. 3.5. A seven recombination scenario for the Adh data.

If $\theta = 15$, the average values of R_M were 3.7 for $R = 10$, 5.5 for $R = 20$, and 8.8 for $r = 50$. Interpolating suggests $R \approx 18$. To get a confidence interval for R, Hudson and Kaplan (1985) did further simulations with a fixed number of segregating sites to argue that R should be between 5 and 150, which is not very informative.

3.6 Estimating recombination rates

Earlier we saw two not very satisfactory methods of estimating the scaled recombination rate $\rho = 4Nr$. Hudson and Kaplan's (1985) R_M based on the four-gamete test, and its more sophisticated relatives, typically detect only a small fraction of the recombination events that have occurred. Wakeley's (1997) estimator based on the variance of the pairwise differences is easy to calculate but is biased, and not very effective even for samples of size 99.

One of the problems with Wakeley's estimator is that it describes the data with a single summary statistic, ignoring most of the available information. At the other extreme, likelihood approaches estimate the probability of observing a given data set under an assumed population genetic model. Griffiths and Marjoram (1996), Kuhner, Yamata, and Felsenstein (2000), Nielsen (2000), and Fearnhead and Donnelly (2001) developed estimators of ρ based on likelihood methods. Each of these methods uses computationally intensive statistical methods to approximate the likelihood curve, and even the most efficient of these can accurately estimate this curve only for small data sets.

Given the difficulties of computing the full likelihood, various compromises have been introduced. Fearnhead and Donnelly (2002) split the region of interest into several subregions. They computed the full likelihood for the data in each of the subregions and then multiplied to obtain a composite likelihood. Wall (2000) used simulation to find the parameter values ρ_H and ρ_{HRM} that maximized the likelihood of $(\rho|H)$ and $(\rho|H, R_M)$, where H is the number of haplotypes, and R_M is Hudson and Kaplan's (1985) estimator. Hudson (2001) developed a composite likelihood method by examining pairs of segregating sites. In this section, we will explain his method, but first we consider

3.6.1 Equations for the two-locus sampling distribution

Suppose we sample n chromosomes and obtain information about two loci with alleles A_0 and A_1 and B_0 and B_1. Let n_{ij} be the number of sampled chromosomes that carry allele A_i at locus a and B_j at locus b. Let $q(\mathbf{n}; \theta, \rho)$ be the probability of observing $\mathbf{n} = (n_{00}, n_{10}, n_{01}, n_{11})$ for the indicated parameters. The first sign of trouble is the large number of probabilities we have to compute. The number of vectors $\mathbf{n}$ with $n_{00}+n_{10}+n_{01}+n_{11} = n$ and $n_{ij} \geq 0$ is the same as the number of ways of writing $n + 4 = m_{00} + m_{10} + m_{01} + m_{11}$ with $m_{ij} = n_{ij} + 1 \geq 1$, which, by an earlier calculation, is $\binom{n+3}{3}$. For $n = 10$ this is

$$\frac{13 \cdot 12 \cdot 11}{3!} = 286$$

For $n = 40$ it is 12,341.

Golding (1984) was the first to develop a recursive equation for these probabilities. Here, we will follow the approach of Ethier and Griffiths (1990). As in the case of the recursion of Pluzhnikov and Donnelly (1996) given in Theorem 3.2, in order to obtain a closed set of equations we must consider a larger class of probabilities that allow some members of the sample to be specified only at one location. Let c_{ij} be the number of (A_i, B_j) chromosomes (sampled at both loci with allele A_i at locus a and B_j at locus b), let a_i be the number of $(A_i, \cdot)$ chromosomes (sampled only at the a locus with allele A_i), and let b_j be the number of $(\cdot, B_j)$ chromosomes (sampled only at the b locus with allele B_j). In what follows, we will assume that it is not known which of the alleles is ancestral.

Despite its enormous size, the recursion is straightforward to write. We use a subscript $\cdot$ to indicate a variable that has been summed over. Let $a. = \sum_i a_i$, $b. = \sum_j b_j$, $c_{.j} = \sum_i c_{ij}$, $c_{i.} = \sum_j c_{ij}$ and $c_{..} = \sum_{ij} c_{ij}$. $n_a = a. + c_{..}$ is the sample size at the a locus, $n_b = b. + c_{..}$ is the sample size at the b locus, and

$n = a. + b. + c..$ is the total sample size. If $n_a = 1$ or $n_b = 1$, there is only one locus and probabilities can be computed from the folded site frequency spectrum.

Jumps happen at total rate

$$\lambda_{a,b,c} = [n(n-1) + \rho c.. + \theta_A(a. + c..) + \theta_B(b. + c..)]/2$$

due to coalescence, recombination, and mutation. Letting $e_i \in R^2$ and $e_{ij} \in R^4$ be vectors with 1's in the indicated position and 0 otherwise, the jumps due to coalescence and recombination occur as follows:

$$
\begin{array}{ll}
(a - e_i, b, c) & a_i(a_i - 1 + 2c_{i.})/2 \\
(a, b - e_j, c) & b_j(b_j - 1 + 2c_{.j})/2 \\
(a, b, c - e_{ij}) & c_{ij}(c_{ij} - 1)/2 \\
(a - e_i, b - e_j, c + e_{ij}) & a_i b_j \\
(a + e_i, b + e_j, c - e_{ij}) & \rho c_{ij}/2
\end{array}
$$

To check these, note that $(A_i, \cdot)$ chromosomes can only coalesce with $(A_i, \cdot)$ or (A_i, B_j) chromosomes. Finally, as in the dual of the Wright-Fisher model with mutation, mutations can kill lineages. However, since we assume that all mutants were created by one mutation, jumps can only occur when there is one of the type left:

$$
\begin{array}{ll}
(a, b + e_j, c - e_{ij}) & \theta_A/2 \text{ if } a_i = 0, c_{ij} = 1, c_{i.} = 1 \\
(a - e_i, b, c) & \theta_A/2 \text{ if } a_i = 1, c_{i.} = 0 \\
(a + e_i, b, c - e_{ij}) & \theta_B/2 \text{ if } a_j = 0, c_{ij} = 1, c_{.j} = 1 \\
(a, b - e_j, c) & \theta_B/2 \text{ if } a_i = 1, c_{i.} = 0
\end{array}
$$

Other mutations kill the chain because they create a configuration not consistent with the pattern we are looking for.

To illustrate the use of these equations, we will consider the simplest possible situation, $n = 2$.

Theorem 3.12. *Let x_1, x_2 and y_1, y_2 be the alleles present in our sample at the two loci. If we let $\theta = \theta_A + \theta_B$,*

$$q(2) = \frac{2(3 + \theta)(6 + \theta) + \rho[(2 + \theta) + 2(6 + \theta)\phi] + \rho^2\phi}{2(1 + \theta)(3 + \theta)(6 + \theta) + \rho(2 + \theta)(13 + 3\theta) + \rho^2(2 + \theta)}$$

and $\phi = 1/(1 + \theta_A) + 1/(1 + \theta_B)$, then

$$
\begin{aligned}
P(x_1 = x_2, y_1 = y_2) &= q(2) \\
P(x_1 = x_2, y_1 \neq y_2) &= (1 + \theta_A)^{-1} - q(2) \\
P(x_1 \neq x_2, y_1 = y_2) &= (1 + \theta_B)^{-1} - q(2) \\
P(x_1 \neq x_2, y_1 \neq y_2) &= 1 - (1 + \theta_A)^{-1} - (1 + \theta_B)^{-1} - q(2)
\end{aligned}
$$

This agrees with (2.9) in Ethier and Griffiths (1990). As a check, we note that when $\rho = 0$ this is $1/(1 + \theta)$, while if we let $\rho \to \infty$, the limit is

$$\frac{\phi}{2+\theta} = \frac{1}{1+\theta_A+1+\theta_B} \cdot \left(\frac{1}{1+\theta_A} + \frac{1}{1+\theta_B}\right) = \frac{1}{1+\theta_A} \cdot \frac{1}{1+\theta_B}$$

the answer for independent loci.

Proof. One locus results imply $P(x_1 = x_2) = (1+\theta_A)^{-1}$ and $P(y_1 = y_2) = (1+\theta_B)^{-1}$, so the second and third equations follow from the first one. The fourth one then follows from the fact that the four probabilities add to 1. Thus, it is enough to show that $P(x_1 = x_2, y_1 = y_2) = q(2)$.

The event $x_1 = x_2$, $y_1 = y_2$ occurs if and only if there is no mutation before coalescence, so we use the three states from calculations with Section 3.1: (0,0,2), (1,1,1), and (2,2,0), which we will abbreviate by giving their third coordinates. When we do this, the total transition rates are

$$\lambda_2 = 1+\theta+\rho \qquad \lambda_1 = 3+\theta+\frac{\rho}{2} \qquad \lambda_0 = 6+\theta$$

For states with $(n_a, n_b) = (2,1)$, $(1,2)$, or $(1,1)$, the two-locus homozygosities are $1/(1+\theta_A)$, $1/(1+\theta_B)$, and 1, respectively. Using this with the transition rates, and letting $\phi = 1/(1+\theta_A) + 1/(1+\theta_B)$, we have

$$q(2) = \frac{\rho}{\lambda_2}q(1) + \frac{1}{\lambda_2}$$

$$q(1) = \frac{1}{\lambda_1}q(2) + \frac{\rho/2}{\lambda_1}q(0) + \frac{1}{\lambda_1}\phi \qquad (3.17)$$

$$q(0) = \frac{4}{\lambda_0}q(1) + \frac{1}{\lambda_0}\phi$$

Inserting the third equation into the second, we have

$$q(1) = \frac{1}{\lambda_1}q(2) + \frac{2\rho}{\lambda_0\lambda_1}q(1) + \frac{\rho+2\lambda_0}{2\lambda_0\lambda_1}\phi$$

Rearranging, we have

$$q(2) = \frac{\lambda_0\lambda_1 - 2\rho}{\lambda_0}q(1) - \frac{\rho+2\lambda_0}{2\lambda_0}\phi$$

The first equation in (3.17) implies $q(1) = (\lambda_2/\rho)q(2) - 1/\rho$. Inserting this in the previous equation, we have

$$q(2) = \frac{\lambda_0\lambda_1\lambda_2 - 2\rho\lambda_2}{\rho\lambda_0}q(2) - \frac{\lambda_0\lambda_1 - 2\rho}{\rho\lambda_0} - \frac{\rho+2\lambda_0}{2\lambda_0}\phi$$

Rearranging gives

$$\frac{\lambda_0\lambda_1\lambda_2 - \rho\lambda_0 - 2\rho\lambda_2}{\rho\lambda_0}q(2) = \frac{\lambda_0\lambda_1 - 2\rho}{\rho\lambda_0} + \frac{\rho+2\lambda_0}{2\lambda_0}\phi$$

and hence we have

$$q(2) = \frac{2\lambda_0\lambda_1 - 4\rho + (\rho^2 + 2\lambda_0\rho)\phi}{2[\lambda_0\lambda_1\lambda_2 - \rho\lambda_0 - 2\rho\lambda_2]} \tag{3.18}$$

Recalling the definitions of the λ_k, we see that the denominator is

$$\begin{aligned}
= {}& 2(1+\theta)(3+\theta)(6+\theta) + \rho(1+\theta)(6+\theta) + 2\rho(3+\theta)(6+\theta) + \rho^2(6+\theta) \\
& - 2\rho(6+\theta) - 4\rho(1+\theta+\rho)
\end{aligned}$$

The coefficient of ρ^2 in the denominator is $(6+\theta) - 4 = 2+\theta$, while that of ρ is

$$\begin{aligned}
(1+\theta+6+2\theta-2)(6+\theta) - 4(1+\theta) &= (3\theta+5)(\theta+6) - 4 - 4\theta \\
&= 3\theta^2 + 19\theta + 26 = (3\theta+13)(\theta+2)
\end{aligned}$$

The numerator of the formula for $q(2)$ in (3.18) is

$$2(3+\theta)(6+\theta) + \rho(6+\theta) - 4\rho + \rho^2\phi + 2(6+\theta)\rho\phi$$

Combining our computations gives

$$q(2) = \frac{2(3+\theta)(6+\theta) + \rho[(2+\theta) + 2(6+\theta)\phi] + \rho^2\phi}{2(1+\theta)(3+\theta)(6+\theta) + \rho(2+\theta)(13+3\theta) + \rho^2(2+\theta)}$$

which proves the desired result.   □

3.6.2 Simulation methods

Given the analytical complications of Golding's (1984) equation, Hudson (1985) used a clever simulation technique to estimate the probabilities. He generated a two-locus genealogy of a sample of n chromosomes, but then instead of generating a single sample from a pair of trees, he calculated the distribution of $\mathbf{n} = (n_{00}, n_{01}, n_{10}, n_{11})$ conditional on the pair of trees. To explain this we need some notation. Let $\mathcal{E}$ be the sequence of coalescence and recombination events that define the tree. Given the sequence of events E_i, the time T_i between E_{i-1} and E_i has an exponential distribution with mean determined by the configuration of ancestral lineages during the interval.

Let τ_A and τ_B be the total length of the trees, measured in units of $4N$ generations so that the scaled mutation rate is θ. Let $I(\mathcal{E}, \mathbf{n}, j, k) = 1$ if mutations on the jth branch of the a locus tree and on the kth branch of the b locus tree would produce the sample configuration $\mathbf{n}$, and 0 otherwise. As the notation indicates, this depends on the event sequence $\mathcal{E}$ and not on the time sequence $\mathcal{T}$. Letting a_j and b_k be the length of the two branches, the probability of the sample configuration $\mathbf{n}$ being produced is

$$I(\mathcal{E}, \mathbf{n}, j, k)(1 - e^{-\theta a_j})(1 - e^{-\theta b_k})e^{-\theta(\tau_A - a_j)}e^{-\theta(\tau_B - b_k)}$$

since we need at least one mutation on each of the selected branches and none on the rest of the trees. Note that the identification of the branches that produce the desired mutation pattern depends on whether or not we consider A_0 and B_0 to be the ancestral alleles and A_1 and B_1 to be derived, which Hudson calls $a - d$ specified samples, but this is otherwise irrelevant to the details of the procedure.

We will apply this formula when the a and b loci are single nucleotides, so θ is small and the above is

$$\approx \theta^2 I(\mathcal{E}, \mathbf{n}, j, k) a_j b_k$$

Summing over j and k and then taking expected value with respect to the joint distribution of the $\mathcal{E}$ and $\mathcal{T}$ sequences, we have

$$q_u(\mathbf{n}; \theta, \rho) \approx \theta^2 h_u(\mathbf{n}; \rho)$$

where $h_u(\mathbf{n}; \rho) = E\left(\sum_{j,k} I(\mathcal{E}, \mathbf{n}, j, k) a_j b_k\right)$. The subscript u indicates that this is the unconditioned version of the probability, i.e., we are not conditioning on the event that the two sites are variable.

To estimate $h_u(\mathbf{n}; \rho)$, we generate a sequence of genealogies with events $\mathcal{E}_i$, $1 \le i \le m$, and let

$$\hat{h}_u(\mathbf{n}; \rho) = \frac{1}{m} \sum_{i=1}^{m} \sum_{j,k} I(\mathcal{E}_i, \mathbf{n}, j, k) a_j(i) b_k(i)$$

In the case of a constant-size population model, this can be done more efficiently by using

$$\tilde{h}_u(\mathbf{n}; \rho) = \frac{1}{m} \sum_{i=1}^{m} \sum_{j,k} I(\mathcal{E}_i, \mathbf{n}, j, k) E(a_j b_k | \mathcal{E}_i)$$

In words, we replace the observed times by their conditional expectation given the event sequence $\mathcal{E}_i$. This reduces the variance of the estimator and eliminates the need to simulate the time sequence.

Most applications of the two-locus sampling distribution will focus on pairs of sites in which both are polymorphic in the sample. That is, we must consider the probability

$$q(\mathbf{n}, \theta, \rho | \text{two alleles at each locus}) = \frac{q_u(\mathbf{n}; \theta, \rho)}{\sum_{\mathbf{m}} q_u(\mathbf{m}; \theta, \rho)}$$

Again, we are interested in

$$q_c(\mathbf{n}, \rho) = \lim_{\theta \to 0} q(\mathbf{n}, \theta, \rho | \text{two alleles at each locus}) = \frac{h_u(\mathbf{n}; \rho)}{\sum_{\mathbf{m}} h_u(\mathbf{m}; \rho)}$$

which we can estimate without specifying θ. Hudson has done this for sample sizes 20, 30, 40, 50, and 100 and a range of ρ values between 0 and 100, and these are available on his web page.

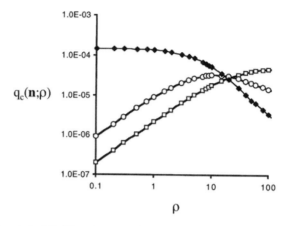

$q_c(\mathbf{n};\rho)$

ρ

Fig. 3.6. Likelihood curves for three sample configurations.

3.6.3 Composite likelihood estimation of ρ

The graph above, which is Figure 2 in Hudson (2001), shows that if we try
to use one pair of sites to estimate ρ, then the likelihood curves are often
monotonic, resulting in estimates of 0 or ∞. Consider, for the moment, the
mythical situation that we have k independent pairs of loci, each with scaled
recombination rate ρ. The overall likelihood is

$$L(\mathbf{n}_1, \mathbf{n}_2, \ldots \mathbf{n}_k; \rho) = \prod_{i=1}^{k} q_c(\mathbf{n}_i; \rho)$$

and this is maximized to produce the estimate $\hat{\rho}$. To characterize statistical
properties of $\hat{\rho}$, we consider

$$E_{\rho_0}(\log q_c(\mathbf{n}; z)) = \sum_{\mathbf{n}} q_c(\mathbf{n}; \rho_0) \log q_c(\mathbf{n}; z)$$

which is the expected log-likelihood given that the true parameter is ρ_0. The
second derivative of this function with respect to z, evaluated at ρ_0 is inversely
proportional to the asymptotic variance of the maximum likelihood estimate.
More precisely, for a sample of size k

$$\mathrm{var}_{\rho_0, k}(\hat{\rho}) \approx \frac{-1}{k(\partial^2/\partial z^2)E_{\rho_0}(\log q_c(\mathbf{n}; z))|_{z=\rho_0}}$$

The next graph, which is Figure 5 in Hudson (2001), shows $E_{\rho_0}(\log q_c(\mathbf{n}; z))$
and a quadratic function fitted to several points near $\rho_0 = 5$, which suggests
that this approximation may be accurate when k is not large.

Since it is trivial that smaller values of ρ can be estimated with less abso-
lute error than larger values, it is interesting to look instead at the coefficient
of variation:

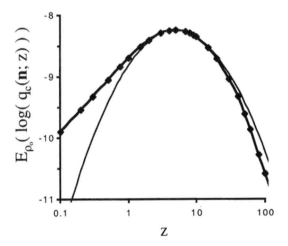

Fig. 3.7. Expected log-likelihood curve and its quadratic approximation.

$$\frac{\operatorname{var}_{\rho_0,k}(\hat{\rho})}{\rho_0^2} \approx \operatorname{var}_{\rho_0,k}(\log(\hat{\rho})) \approx \frac{-1}{k(\partial^2/\partial(\log z)^2)E_{\rho_0}(\log q_c(\mathbf{n};z))|_{z=\rho_0}}$$

Figure 6 of Hudson (2001), which is given here as Figure 3.8, shows this quantity as a function of ρ_0. It achieves a minimum at $\rho_0 = 5$ and to quote Hudson, "shows that pairs separated by ρ in the range $2 - 15$ are best for estimating ρ."

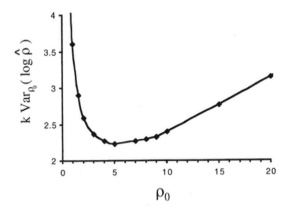

Fig. 3.8. Asymptotic variance of the log of the MLE

In practice, different pairs of polymorphic sites will be different distances apart and hence will have different recombination rates. Letting ρ_b be the recombination probability per base pair and d_i the distance between the ith pair, we can write the likelihood as

$$L(\mathbf{n}_1, \mathbf{n}_2, \ldots \mathbf{n}_k; \rho_b) = \prod_{i=1}^{k} q_c(\mathbf{n}_i; \rho_b d_i)$$

To define his composite likelihood estimator ρ_{CL}, Hudson uses the log-likelihood:

$$\sum_{i<j} \log q_c(\mathbf{n}_{ij}; \rho_b d_{ij})$$

where $\mathbf{n}_{ij}$ is the observed sampling distribution at sites i and j and d_{ij} is the distance between the two sites. Hudson investigated ρ_{CL} by simulation and showed that it and Wall's estimator ρ_{HRM} performed much better than Wakeley's moment estimator ρ_{wak} and Hey and Wakeley's (1997) γ.

Example 3.2. Hudson (2001) applied his method to estimating ρ from a survey of human variation on the X chromosome by Taillon-Miller et al. (2000). In this study, 39 SNPs were surveyed in three population samples, but Hudson only considered the sample of 92 CEPH males. The parameter ρ_b was estimated by maximizing the composite likelihood for (i) all 39 SNPs, (ii) the 14 SNPs in Xq25, and (iii) the 10 SNPs in or near Xq28. Here the q refers to the q arm of the X chromosome, and 25 and 28 refer to chromosome bands, which can be observed under microscopes, and in the era before whole genome sequencing was used to describe the locations of genes. For loci on the X chromosome, the methods described above will result in an estimate of $2Nr$ since these loci do not experience recombination in males. Hudson multiplied his computer output by 2 to give an estimate of $4Nr$ and reported estimates of (i) 9×10^{-5}, (ii) 8.8×10^{-5}, and (iii) 9×10^{-5}. These conclusions contrast with those of Taillon-Miller et al. (2000), who observed that linkage disequilibrium was high in Xq25 but almost nonexistent in Xq28.

Fearnhead (2003) investigated the consistency of the composite likelihood estimators of Fearnhead and Donnelly (2002), and the pairwise likelihood of Hudson (2001), when one examines an increasing number of segregating sites for a fixed sample size. He proved that the composite likelihood is consistent and proved the consistency of a truncated pairwise likelihood, which is based on the product of the likelihoods for all pairs of sites that are less than some distance R apart. Smith and Fearnhead (2005) compared the accuracy of these two methods and the pseudolikelihood method of Li and Stephens (2003) using simulated sequence data. They found that the performance was similar but that the pairwise likelihood method could be improved by including contributions to the log-likelihood only for pairs of sites that are separated by some prespecified distance.

3.7 Haplotypes and hot spots

Sequencing of the human genome revealed that there are sizable regions over which there is little evidence of recombination and a small number of SNPs is sufficient to describe most of the genetic variation. For example, Patil et al. (2001) found that 80% of human chromosome 21 variation can be described by only three SNPs per block. The average observed block length was 7.8 kb, but the longest block stretched more than 115 kb and contained 114 SNPs. Results of Daly et al. (2001) on a 500 kb region on chromosome 5q31, implicated as containing a genetic risk factor for Crohn's disease, showed large haplotype blocks with limited diversity punctuated by apparent sites of recombination. Gabriel et al. (2002) studied haplotype patterns across 51 autosomal regions and showed that the human genome could be parsed objectively into haplotype blocks.

Reich et al. (2002) argued that the simplest explanation for LD is that the population under study experienced an extreme founder effect or bottleneck, and that a severe bottleneck occurring 800-1600 generations ago could have generated the LD they observed. Frisee et al. (2001) argued that the higher levels of LD in European populations, when compared to African populations, support a contribution from the bottleneck in the founding of non-African populations. However, Ardlie, Kruglyak, and Seielstad (2002) carried out simulations of several models of demographic history, including bottlenecks and expansion, and found that any model with nucleotide diversity in the empirically observed range had useful levels of LD limited to approximately 10 kb. In the other direction, Anderson and Slatkin (2004) argued that rapid population growth could account for the patterns observed in the data from 5q31. For more on the patterns of LD found in various population genetics models, see Pritchard and Przeworski (2001).

One possible explanation for the observed haplotype blocks is that recombinations are not uniformly spread along a chromosome but preferentially occur in hot spots. Analysis of recombination breakpoints and crossover events in sperm typing experiments have demonstrated the presence of recombination hot spots in several genomic locations. The first work was done by Jeffreys, Ritchie, Neumann (2000) and Jeffreys, Kauppi, Nuemann (2001). For a survey, see Kauppi, Jeffreys, and Keeney (2004).

Since experimental confirmation of hot spots is technically challenging and time consuming, Li and Stephens (2003) introduced statistical procedures based on a heuristic formula for haplotype probabilities to detect fine scale variation in recombination rates from population data. Using a method based on Hudson's (2001) two-sample distributions, McVean et al. (2004) found evidence of rate variation spanning four orders of magnitude and suggested that 50% of all recombinations take place in less than 10% of the sequence.

Myers et al. (2005) followed up on this work applying the method to 1.6 million SNPs genotype in samples from three samples: 24 European Americans, 23 African Americans, and 24 Han Chinese from Los Angeles by Perlegen, see Hinds et al. (2005). They estimated that there is a hotspot every 50 kb or so in the human genome, with approximately 80% of recombinations occuring in 10 to 20% of the sequence. With more than 25,000 hotspots at their disposal, Myers et al. (2005) also found the first set of sequence features substantially overrepresented in hot spots relative to cold spots. Although the motifs are neither necessary nor sufficient for hot spot activity, their top scoring candidate ($CCCTCCCT$) played a role in 11% of them.

While recombination hot spots exist and will produce haplotype blocks, it remains unclear how much rate variation, if any, is needed to account for observed haplotype blocks. Phillips et al. (2003) found that only about one third of chromosome 19 was covered in haplotype blocks and that there was no reason to invoke recombination hot spots in order to explain the observed blocks. Wang et al. (2002) and Zhang et al. (2003) found through extensive coalescent simulations that haplotype blocks were observed in models where recombination crossovers were randomly and uniformly distributed.

A second mystery is that despite 99% identity between human and chimpanzee DNA sequences, ther is virtually no overlap between these two species in the locations of their hot spots. Ptak et al. (2005) showed that the well studied TAP2 hotspot in humans was absent in chimpanzees. Winckler et al. (2005) found that 18 recombination hot spots covering 1.5 megabases in humans were absent in chimpanzees and observed no correlation between estimates of fine scale recombination rates.

4

Population Complications

"Mathematics is not a careful march down a well-cleared highway, but a journey into a strange wilderness, where explorers get lost. Rigour should be a signal to the historian that maps have been made, and the real explorers have gone elsewhere." W. S. Anglin

4.1 Large family sizes

In the Wright-Fisher model the ith individual in the population in generation t has $\nu_i = m$ descendants in generation $t+1$ with a probability given by the binomial distribution

$$\binom{2N}{m} \left(\frac{1}{2N}\right)^m \left(1 - \frac{1}{2N}\right)^{2N-m}$$

Expanding out the binomial coefficient, and letting $N \to \infty$

$$\frac{2N(2N-1)\cdots(2N-m+1)}{m!} \left(\frac{1}{2N}\right)^m \left(1 - \frac{1}{2N}\right)^{2N-m}$$

$$= \frac{2N(2N-1)\cdots(2N-m+1)}{(2N)^m} \frac{1^m}{m!} \left(1 - \frac{1}{2N}\right)^{2N-m} \to 1 \cdot \frac{1^m}{m!} e^{-1}$$

since $(1-1/2N)^{2N-m} \to e^{-1}$. In words, the binomial$(2N, 1/2N)$ distribution is approximately Poisson with mean 1 when N is large.

If one is concerned that family sizes in a population don't follow a Poisson distribution then one can instead use *Cannings' model*. Suppose that the $2N$ members of the generation t have $\nu_1, \nu_2, \ldots \nu_{2N}$ offspring. By symmetry the ν_i are *exchangeable*, i.e., the joint distribution of any subset of size k, $(\nu_{i(1)}, \ldots \nu_{i(k)})$ where the $i(j)$ are distinct, does not depend on the indices, and $\nu_1 + \nu_2 + \cdots \nu_{2N} = 2N$. The distribution of the ν_i will depend on N but we will not record the dependence in the notation.

R. Durrett, *Probability Models for DNA Sequence Evolution*,
DOI: 10.1007/978-0-387-78168-6_4, © Springer Science+Business Media, LLC 2008

Note that by exchangeability all of the ν_i have the same mean, which must be 1 since the sum is $2N$. Given the family sizes, the probability that two individuals chosen at random in generation $t+1$ have a common ancestor in generation t is

$$\pi_2 = \sum_{i=1}^{2N} \frac{\nu_i(\nu_i - 1)}{2N(2N - 1)}$$

$E\nu_i(\nu_i - 1) = E\nu_i^2 - E\nu_i = E\nu_i^2 - 1^2 = \text{var}(\nu_i)$. For a sample of size 2, the probability of coalescence per generation is constant. This suggests that if we let

$$c_N = \frac{\text{var}(\nu_i)}{2N - 1}$$

then as $N \to \infty$, $c_N T_2$ will have an exponential distribution.

Kingman (1982b) showed that if $\sup_N E\nu_1^k < \infty$ for all k then the genealogy on time scale t/c_N converges to the coalescent. If $\text{var}(\nu_i) \to \sigma^2$ then the time scale is $2Nt/\sigma^2$. Möhle (2000) has shown, see (16) on page 989,

Theorem 4.1. *Convergence to Kingman's coalescent occurs if and only if*

$$\frac{E[\nu_1(\nu_1 - 1)(\nu_1 - 2)]/N^2}{E[\nu_1(\nu_1 - 1)]/N} \to 0$$

The condition says that mergers of three lineages occur at a slower rate than mergers of two.

Seeing the result above, it is natural to ask if there are limits when the variances tend to ∞. Sagitov (1999) and Möhle and Sagitov (2001) obtained a definitive solution to this problem. We begin by describing the limit for the corresponding coalescent process. To do this we assume, as we did in Section 1.2, that the state of the coalescent at time t is described by a partition $\xi = \{A_1, \ldots, A_k\}$ of $\{1, 2, \ldots n\}$ where n is the sample size, and i and j are in the same set if the ith and jth individuals have a common ancestor t generations back in time.

A pair (ξ, η) of partitions is said to be a k *merger* if all but one of the sets in η are inherited from ξ without change, and the exceptional set in η is the union of k sets of ξ. Writing $|\xi|$ for the number of sets in the partition, $|\eta| = |\xi| - k + 1$. The limit process is characterized by an arbitrary finite measure Λ on $[0, 1]$, and is called the Λ-*coalescent*. If (ξ, η) is a k-merger then the rate of jumps from ξ to η is

$$q_{\xi,\eta} = \int_0^1 x^{k-2}(1 - x)^{|\xi|-k} \Lambda(dx) \tag{4.1}$$

Note that Kingman's coalescent corresponds to the special case in which Λ is a point mass at 0, i.e., each 2-merger occurs at rate 1.

The transition rates may look odd, but Pitman (1999) proved that they are natural. Let $\lambda_{b,k}$ be rate at which a specific k-merger occurs when the partition ξ has b blocks. Pitman (1999) showed that if the transition rates satisfy the consistency condition:

$$\lambda_{b,k} = \lambda_{b+1,k} + \lambda_{b+1,k+1}$$

then we must have

$$\lambda_{b,k} = \int_0^1 x^{k-2}(1-x)^{b-k} \Lambda(dx)$$

for some finite measure Λ on $[0,1]$.

Example 4.1. p-merger. At times of a rate λ Poisson process we toss coins with probability p of heads and coalesce all of the lineages for which a heads occurs. In this case each k-merger occurs at rate

$$\lambda p^k(1-p)^{b-k} \qquad \text{for } k \geq 2$$

so $\Lambda(\{p\}) = \lambda p^2$. A Λ-coalescent with $\lambda = \int x^{-2}\Lambda(dx) < \infty$ can be thought of as a p-merger with a randomly distributed p, chosen according to $x^{-2}\Lambda(dx)/\lambda$.

The total rate at which mergers occur when there are b blocks is

$$\lambda_b = \sum_{k=2}^b \binom{b}{k}\lambda_{b,k} = \int_0^1 \sum_{k=2}^b \binom{b}{k} x^{k-2}(1-x)^{b-k}\Lambda(dx)$$
$$= \int_0^1 \left(1 - (1-x)^k - bx(1-x)^{k-1}\right) x^{-2}\Lambda(dx) \qquad (4.2)$$

When $b = 2$, $1 - (1-x)^2 - 2x(1-x) = x^2$, so $\lambda_2 = \Lambda([0,1])$.

Sagitov (1999) found necessary and sufficient conditions for the genealogies of a sequence of Cannings' models to converge to a Λ-coalescent. For convenience we formulate the result using his notation, which in particular replaces $2N$ by N

Theorem 4.2. *Suppose that the family sizes $\nu_1, \ldots \nu_N$ are exchangeable with fixed sum $\nu_1 + \cdots + \nu_N = N$. Let $\sigma^2(N) = E(\nu_1 - 1)^2$, which we suppose is $o(N)$ and let $T_N = N\sigma^{-2}(N)$. If $NT_N P(\nu_1 > x) \to \int_x^1 y^{-2}\Lambda(dy)$ at all points where the limit function is continuous and*

$$N^{-a}T_N E(\nu_1 - 1)^2(\nu_2 - 1)^2 \to 0$$

then, when time is scaled by T_N, the genealogical process converges to the Λ-coalescent.

Sketch of proof. The key to the proof is the observation that if we condition on the family sizes $(\nu_1, \ldots \nu_N)$ and write $(N)_c = N(N-1)\cdots(N-c+1)$ then

$$p_{\xi,\eta}(N) = \frac{(N)_a}{(N)_b} E[(\nu_1)_{b(1)} \cdots (\nu_a)_{b(a)}]$$

when $|\eta| = a$ and the ith set of η is the union of $b(i)$ sets of η. From this we see that the second condition in the theorem excludes two mergers from happening simultaneously. The first condition allows us to conclude that in the limit $q_{\xi,\eta}$ satisfies (4.1). We refer the reader to Sagitov's paper for further details.   □

Eldon and Wakeley (2006) have argued that Λ-coalescents are appropriate for modeling genealogies of some marine species. They considered a model in which at each discrete time step, exactly one individual reproduces and is the parent of $U-1$ new individuals. The parent persists while its offspring replace $U-1$ individuals who die. Mutations can occur to any of the $U-1$ offspring but the other $N-U$ individuals persist without change. To have large family sizes they let $0 < \psi < 1$ be a fixed constant and suppose that

$$P(U = u) = \begin{cases} 1 - N^{-\gamma} & u = 2 \\ N^{-\gamma} & u = \psi N \\ 0 & \text{otherwise} \end{cases}$$

To use the Λ-coalescent for modeling, one needs to restrict attention to a subfamily of measures Λ. However, "beta coalescents," which we will now describe, seem to be a more attractive choice of model. Schweinsberg (2003) considered a population model in which the number of offspring of each individual X_i is chosen independently according to some distribution on $\{0, 1, \ldots\}$ with mean > 1, and then N of the offspring are chosen to make up the next generation. For a mental picture, consider annual plants. Each plant produces many seeds, but the total population size remains roughly constant since the environment can support a limited number of plants per acre.

Let $\nu_1, \ldots \nu_N$ be the family sizes after the N offspring have been chosen and define

$$c_N = \frac{E[\nu_1(\nu_1 - 1)]}{N - 1}$$

Part (c) of Theorem 4 in Schweinsberg (2003) shows

Theorem 4.3. *Suppose $EX_i = \mu > 1$ and $P(X_i \geq k) \sim Ck^{-\alpha}$ with $1 < \alpha < 2$. Then, when time is run at rate $1/c_N$, the genealogical process converges to a Λ-coalescent where Λ is the beta$(2 - \alpha, \alpha)$ distribution, i.e.,*

$$\Lambda(dx) = \frac{x^{1-\alpha}(1-x)^{\alpha-1}}{B(2-\alpha, \alpha)}$$

where $B(a,b) = \Gamma(a)\Gamma(b)/\Gamma(a+b)$, and $\Gamma(a) = \int_0^\infty x^{a-1}e^{-x}\, dx$ is the usual gamma function.

Parts (a) and (b) of his result show that when $\alpha \geq 2$ the limit is Kingman's coalescent, while part (d) shows that when $\alpha < 1$ the limit has simultaneous mergers. In the borderline case $\alpha = 1$ the conclusion of Theorem 4.3 also holds. We have excluded this case from our consideration so that we can use Lemma 13 in Schweinsberg (2003) to conclude:

$$\lim_{N \to \infty} N^{\alpha-1} c_N = C\alpha B(2 - \alpha, \alpha) \tag{4.3}$$

Note that time is run at rate $O(N^{\alpha-1})$ compared to time $O(N)$ in Kingman's coalescent or in Cannings' model with finite variance. To explain the rate we note that if the only effect of the subsampling was to truncate the distribution at N then

$$E\nu_1^2 \sim \int_0^N 2xP(X_i > x)\,dx = 2C \int_0^N x^{1-\alpha}\,dx = O(N^{2-\alpha})$$

so $c_N = O(N^{1-\alpha})$.

The limit processes that appear in Theorem 4.3 are commonly referred to as beta coalescents. They are convenient for modeling because they are a two parameter family described by (i) the index α determined by the size of the tail of the distribution and (ii) the constant A which appears in the time change: $AN^{\alpha-1}$. Of course, we cannot estimate A, only the composite parameter $\theta = AN^{\alpha-1}\mu$.

A second reason for using these processes is that many of their properties have been studied mathematically. Birkner et al. (2005) have shown that these coalescents arise as genealogical processes of continuous state branching processes that arise from stable branching mechanisms. However, for the developments here the work of J. Berestycki, N. Berestycki, and Schweinsberg (2006a,b), which we now describe, is more relevant.

Theorem 4.4. *Suppose we introduce mutations into the beta coalescent at rate θ, and let S_n be the number of segregating sites observed in a sample of size n. If $1 < \alpha < 2$ then*

$$\frac{S_n}{n^{2-\alpha}} \to \frac{\theta\alpha(\alpha - 1)\Gamma(\alpha)}{2 - \alpha}$$

in probability as $n \to \infty$.

In contrast, as Theorem 1.23 shows, in Kingman's coalescent

$$\frac{S_n}{\log n} \to \theta$$

However, in comparing the two results one must remember that $\theta = 2N\mu$ in Kingman's coalescent compared with $\theta = A\mu N^{\alpha-1}$ in the beta coalescent.

One can also compute the site frequency spectrum.

Theorem 4.5. *Suppose we introduce mutations into the beta coalescent at rate θ, and let $M_{n,k}$ be the number of mutations affecting k individuals in a sample of size n. Then*

$$\frac{M_{n,k}}{S_n} \to a_k = \frac{(2-\alpha)\Gamma(\alpha+k-2)}{\Gamma(\alpha-1)k!}$$

in probability as $n \to \infty$.

Using the fact that $\Gamma(n+a)/\Gamma(n) \to n^a$ as $n \to \infty$

$$a_k \sim C_\alpha k^{\alpha-3} \quad \text{as } k \to \infty$$

When $\alpha = 2$ this reduces to the $1/k$ behavior found in Kingman's coalescent.

When $k = 1$, $a_k = 2 - \alpha$ so one can, in principle, use the fraction of singletons in a large sample to estimate $2 - \alpha$.

Example 4.2. Boom, Boulding, and Beckenbach (1994) did a restriction enzyme digest of mtDNA on a sample of 159 Pacific Oysters from British Columbia and Washington. They found 51 segregating sites and 30 singleton mutations, resulting in an estimate of

$$\alpha = 2 - \frac{30}{51} = 1.41$$

However, this estimate is biased. If the underlying data was generated by Kingman's coalescent, we would expect a fraction $1/\ln(159) = 0.2$ of singletons, resulting in an estimate of $\alpha = 1.8$. Since there are no formulas for the expected site frequency spectrum for fixed n, this quantity needs to be computed by simulation in order to correct the estimate.

A second data set to which these methods can be applied is Arnason's (2004) sequencing of 250 bp of the mitochondrial cytochrome b gene in 1278 Atlantic cod *Gadus morhua*. However, the analysis of this data is made problematic by the fact that 39 mutations define 59 haplotypes, so there have been at least 19 repeat mutations. Birkner and Blath (2007a,b) have analyzed this data set by extending methods of Ethier and Griffiths (1987) and Griffiths and Tavaré (1994a) to compute likelihoods for data from nonrecombining regions.

4.2 Population growth

In some cases, e.g., humans, it is clear that the population size has not stayed constant in time. This motivates the consideration of a version of the Wright-Fisher model in which the number of diploid individuals in generation t is given by a function $N(t)$, but the rest of the details of the model stay the same. That is, in the case of no mutation, generation t is built up by choosing with replacement from generation $t - 1$. Reversing time leads as before to a coalescent.

4.2.1 Exponential growth

Let $N(s) = N_0 e^{\rho s}$ for $0 \geq s \geq s_0$, where $s_0 = -\rho^{-1}\ln(N_0)$ is defined so that $N(s_0) = 1$. A concrete example of this can be found in the work of Harding et al. (1997), who used $N_0 = 18,807$ and $\rho = 0.7/2N_0 = 1.861 \times 10^{-5}$ as a best fit model for data from the β-globin locus in the human population.

The probability that the lineages of two chromosomes sampled at time 0 do not coalesce before generation $t < 0$ is (using $1 - x \approx e^{-x}$)

$$\prod_{s=t}^{-1} \left(1 - \frac{1}{2N(s)}\right) \approx \exp\left(-\sum_{s=t}^{-1} \frac{1}{2N(s)}\right) \tag{4.4}$$

$$\approx \exp\left(-\int_{s=t}^{0} \frac{e^{-\rho s}}{2N_0}\,ds\right) = \exp\left(-\frac{e^{-\rho t} - 1}{2\rho N_0}\right)$$

The right-hand side is shown in the following diagram:

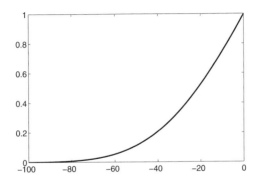

Fig. 4.1. Distribution of coalescence times in the example of Harding et al (1997).

To check the graph, we note that $2\rho N_0 = 0.7$ so the right-hand side is $e^{-1} = 0.3678$ when $t = -\rho^{-1}\ln(1 + 2\rho N_0) = -28,500$. Note that the time s_0 at which $N(s_0) = 1$ is $s_0 - \rho^{-1}\ln(N_0) = -528,850$, but as the graph shows, almost all of the coalescence has occurred by time $-80,000$.

The shape of the coalescence curve may surprise readers who have heard the phrase that genealogies in exponentially growing populations will tend to be *star-shaped*. That is, all of the coalescence tends to occur at the same time. Results of Slatkin and Hudson (1991) show that this is true provided $2N_0\rho$ is large, while in our first example $2N_0\rho = 0.7$.

Consider now a situation in which $N_0 = 100,000$ and $\rho = 5 \times 10^{-5}$ so $2N_0\rho = 10$. The distribution is now a little more concentrated, but there still is a considerable amount of variability in the distribution drawn in the next figure. To check the graph, note that the curve crosses $e^{-1/10} = 0.905$ when

$$\frac{e^{-\rho t} - 1}{10} = \frac{1}{10} \quad \text{or} \quad t = -\ln 2/\rho \approx -14,000$$

and crosses $e^{-2.3} = 0.1$ when $t = -\ln(24)/\rho \approx -63,500$.

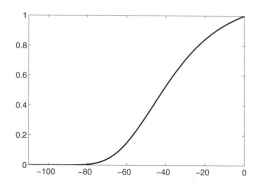

Fig. 4.2. Distribution of coalescence times when $N_0 = 100,000$ and $\rho = 5 \times 10^{-5}$.

To further understand the nature of the phylogenies in a population of variable size, we note that when there are k lineages at time $s+1$ a coalescence will occur at time s with probability

$$\frac{\binom{k}{2}}{2N(s)} + O\left(\frac{1}{N(s)^2}\right)$$

From this it follows that

Theorem 4.6. *If we represent the time interval $[s, s+1]$ as a segment of length $1/2N(s)$, then on the new time scale our process is almost the continuous time coalescent in which k lineages coalesce after an exponentially distributed amount of time with mean $1/\binom{k}{2}$.*

This idea, which is due to Kingman (1982b), allows us to reduce our computations for a population of variable size at times $t \leq s \leq 0$ to one for the ordinary coalescent run for an amount of time

$$\tau = \sum_{s=t}^{-1} \frac{1}{2N(s)} \tag{4.5}$$

Example 4.3. Kruglyak's (1999) model of the human population expansion. Humans expanded out of Africa about 100,000 years ago. Assuming a generation time of 20 years, this translates into $T = 5,000$ generations. If we assume that the population had size 10,000 before expansion and use a figure of 6 billion for the population size today, then the rate of expansion found by solving the equation

$$e^{5000\rho} = (6 \times 10^9)/(10,000)$$

is $\rho = (\ln(6 \times 10^5))/5000 = 0.00266$.

$$\underset{0}{\overset{N = 10,000}{\rule{6cm}{0.4pt}}} \quad \text{exp. growth} \quad T = 5000$$

Using (4.5), we see that genealogies in $[0, T]$ correspond to the coalescent run for time

$$\tau = \sum_{t=0}^{T-1} \frac{1}{20,000e^{\rho t}} \approx \frac{1}{20,000} \int_0^T e^{-\rho t}\, dt = \frac{1 - e^{-\rho T}}{20,000\rho}.$$

Recalling that $T = 5000$ and $e^{-\rho T} = 1/(6 \times 10^5) \approx 0$, we see that the above

$$\approx \frac{1}{20,000\rho} = \frac{1}{52.2} = 0.01916.$$

Let T_k be the number of generations we have to go back in time until there are only k lineages ancestral to the sample. If we consider two lineages, then the probability of no coalescence before time 0 as we work back from the present time T is

$$P(T_1 > 0.01916) = e^{-0.01916} = 0.9810.$$

When the two lineages do not coalesce before time 0, they are then two lineages in a population of constant size 10,000 and hence require an additional amount of time with mean 20,000 to coalesce. Ignoring the possibility of coalescence before time 0, which will introduce only a small error, we conclude that the expected coalescence time of two lineages is 25,000.

If we now consider three lineages, then there are three pairs potentially coalescing, so the probability of no coalescence before time 0 is

$$P(T_2 > 0.01916) = e^{-3(0.01916)} = 0.9441.$$

When none of the three lineages coalesces before time 0, the time to the first coalescence will require an average of $2N/3 = 13,000$ generations and then the final coalescence will require another 20,000 generations on the average. From this we can see that to a good first approximation, the effect of population expansion has been simply to add 5,000 generations to the end of each lineage in the ordinary coalescent.

Durrett and Limic (2001) investigated properties of Kruglyak's model in order to predict the number of SNPs in the human genome and to show that most of them date to the time before the population expansion. A key part of

their investigation is a result of Griffiths and Pakes (1988) which shows that
the genealogical structure of the population during the growth phase can be
well approximated by the lineages in a supercritical branching process that
have offspring alive at the end of the expansion.

The lengthening of the tips of the genealogical tree in Kruglyak's model
results in an excess of mutations that have one representative. This should re-
mind the reader of the Aquadro and Greenberg data set discussed in Example
1.5 in Section 1.4. Before we tackle the thorny question: "Do the patterns of
variation in human mitochondrial DNA show signs of the human population
expansion?", we will introduce another demographic scenario.

4.2.2 Sudden population expansion

Let $N(t) = N_1$ for $-t_1 < t \leq 0$, $N(t) = N_2$ for $t \leq -t_1$, where N_1 is much
larger than N_2.

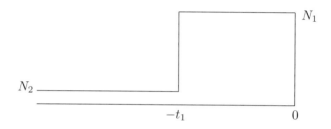

Example 4.4. DiRienzo et al. (1994) used a sudden expansion model with t_1
= 5,000 generations, $N_1 = 10^7$ and $N_2 = 10^3$, to model the Sardinian population. As the authors say "This model represents a drastic bottleneck in
the population's size that results in a starlike gene genealogy." To check this
for their sample of size 46, note that the initial probability of coalescence per
generation is

$$\frac{46 \cdot 45}{2} \cdot \frac{1}{2 \cdot 10^7} = 5.175 \times 10^{-5}$$

so the expected number of coalescence events in 5000 generations is ≤ 0.25875,
and the actual number will be roughly Poisson with this mean. Of course,
once the 46 (or 45) lineages reach time -5000 and the population size drops
to 1,000, the remainder of the coalescence will take an average of

$$2000 \cdot \left(2 - \frac{1}{46} \right) = 3956 \quad \text{generations}$$

Since this number is almost as large as the original 5000, we see that even
in this extreme scenario the genealogies are not quite star-shaped. There is,
of course, quite a bit of coalescence right after the population decrease. The

number of lineages will reduce to 10 in an average of $(2/10)1000 = 200$ generations.

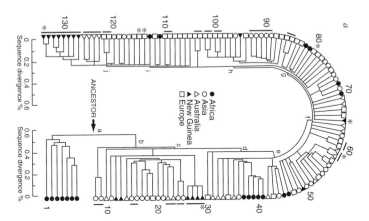

Fig. 4.3. Parsimony tree for mtDNA data of Cann et al. (1987).

Example 4.5. Human mitochondrial DNA. Cann, Stoneking, and Wilson (1987) collected mtDNA from 147 people drawn from five geographic populations and analyzed them using 12 restriction enyzmes, finding 398 mutations. Their analysis found 133 different mutation patterns. Adding in the pattern associated with the reference sequence of human mtDNA from Anderson (1981), they built a parsimony tree using the computer program PAUP designed by Swofford. Noting that one of the two primary branches leads exclusively to African mtDNA types, while the second primary branch also leads to African mtDNA types, they suggested that Africa was the likely source of the mtDNA gene pool.

To assign a time scale to the tree, they first used their data and previous sources to argue that the mean rate of mtDNA divergence for humans lies between 2 and 4 percent per million years; see Stoneking, Bhatia, and Wilson (1986). Using their tree, they concluded that the average divergence between the African group and the rest of the sample was 0.57%, giving an estimate of 140,000 to 290,000 years for the date of the common ancestor. Similar results were obtained in two other studies by Vigilant et al. (1989) and Horai and Hayasaka (1990).

The next figure shows the distribution of the number of differences when all possible pairs of the 147 data points are compared. The x-axis gives the number of differences, while the y-axis gives the number of pairwise comparisons with that number of differences. The circles show the best fitting Poisson

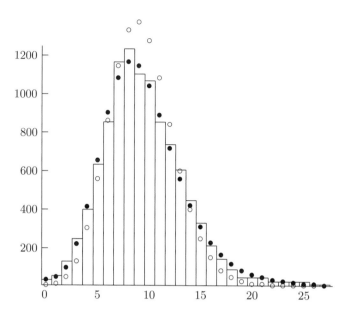

Fig. 4.4. Distribution of pairwise differences, compared to the best fitting Poisson (circles), and a more exact calculation of Rogers and Harpending (black dots).

distribution, which has mean 9.2. The idea behind fitting the Poisson distribution is that if there is rapid population growth then the genealogy tends to be star-shaped, and if the genealogy were exactly star-shaped then the number of pairwise differences between two individuals would have exactly a Poisson distribution.

Rogers and Harpending (1992) took a more sophisticated approach to fitting the data on pairwise differences, resulting in the fit given by the black dots. Their starting point was a result of Li (1977), which we will now derive. Shifting the time of the expansion to 0 for convenience we suppose $N(s) = N_0$ for $s < 0$, $N(s) = N_1$ for $s \geq 0$. Consider the situation $2N_1 t$ units of time after the population expansion, use the infinite sites model for the mutation process, and let $F_i(t)$ be the probability that two randomly chosen individuals differ at i sites. Breaking things down according to the time to coalescence of two lineages, which in a population of constant size N_1 has an exponential distribution with mean 1, and recalling that mutations occur on each lineage at rate $2N_1 u = \theta_1/2$, we have

$$F_i(t) = \int_0^t e^{-s} e^{-\theta_1 s} \frac{(\theta_1 s)^i}{i!} \, ds + e^{-t} \sum_{j=0}^{i} e^{-\theta_1 t} \frac{(\theta_1 t)^j}{j!} F_{i-j}(0),$$

where $F_k(0) = \theta_0^k/(1+\theta_0)^{k+1}$ is the probability that two individuals at time 0 differ at k sites.

To convert this into the form given in Li's paper, we note that the equilibrium probability of i differences is

$$\hat{F}_i = \int_0^\infty e^{-s} e^{-\theta_1 s} \frac{(\theta_1 s)^i}{i!} \, ds = \frac{\theta_1^i}{(1+\theta_1)^{i+1}},$$

where the second equality follows by recalling the formula for the ith moment of the exponential distribution or by integrating by parts i times. Noticing that in a population of constant size N_1 if the two lineages do not coalesce before time t then the number of mutations that occur after time t before coalescence has distribution $\hat{F}_k$, we have

$$\hat{F}_i = \int_0^t e^{-s} e^{-\theta_1 s} \frac{(\theta_1 s)^i}{i!} \, ds + e^{-t} \sum_{j=0}^i e^{-\theta_1 t} \frac{(\theta_1 t)^j}{j!} \hat{F}_{i-j}.$$

Subtracting this from the corresponding formula for $F_i(t)$ we have

$$F_i(t) = \hat{F}_i + e^{-t} \sum_{j=0}^i e^{-\theta_1 t} \frac{(\theta_1 t)^j}{j!} (F_{i-j}(0) - \hat{F}_{i-j})$$

Introducing the scaled time variable $\tau = \theta_1 t$, Rogers and Harpending found the best parameters to fit the data of Cann, Stoneking, and Wilson (1987) were as follows: $\theta_0 = 2.44$, $\theta_1 = 410.69$, $\tau = 7.18$. The predicted values given by the black dots in the figure give a much better fit to the data than the Poisson distribution. Of course, the Rogers and Harpending model has three parameters compared to one for the Poisson distribution.

To relate the fitted parameters to data, we must estimate u, which is the mutation rate in the region under study. Using the data of Cann, Stoneking, and Wilson (1987), Rogers and Harpending calculated that u is between 7.5×10^{-4} and 1.5×10^{-3}. The estimate $\hat{\tau} = 7.18$ implies that the population expansion began some $2N_1 t = \tau/2u$, or 2400 to 4800 generations ago. Assuming a generation time of 25 years, this translates into 60,000 to 120,000 years. The estimate of θ_0 puts the initial population at $\theta_0/4u$, or 400 to 800 females, and that of θ_1 puts the population after expansion at $\theta_1/4u$ or 68,500 to 137,000 females. The first prediction conflicts with the widely held belief, see Takahata (1995), that an effective population size of 10,000 individuals has been maintained for at least the last half million years.

Example 4.6. Harding et al. (1997) have studied a 3 kb region on chromosome 11 encompassing the β-globin gene in nine populations from Africa, Asia, and Europe. Eliminating sequences that showed evidence of recombination they used methods of Griffiths and Tavaré (1994a,b) to give a maximum-likelihood estimate of the genealogical tree, which has a most recent common ancestor

800,000 years ago. At first this may seem to disagree with the estimate of 200,000 years for mitochondrial eve, but mtDNA is haploid and maternally inherited, so the effective population size (and hence coalescence time) for nuclear genes is four times as large.

Maximum-likelihood estimates of the scaled mutation rate θ were made with a model of exponential growth for comparison with a constant-size model. The best fitting population model had $\theta = 4.5$ and a population size $18,087e^{-0.7t}$ at $2N_0t$ generations into the past. An improved fit compared with the constant population size model must occur because of the addition of a parameter to the model, but this improvement was not judged to be significant by a log-likelihood test ratio. Harding et al. (1997) concluded that any population size expansion has been too recent to be detectable in the surveyed patterns of β-globin diversity.

4.3 Founding effects and bottlenecks

Bottlenecks are a sudden reduction in the size of a population. For an extreme example, suppose that a single inseminated female from a large population migrates into an unoccupied geographical location and establishes a new colony followed by a rapid population growth in the new environment. This process is believed to have occurred repeatedly in the evolution of Hawaiian *Drosophila* species. We being by discussing work of Nei, Maruyama, and Chakraborty (1975), who did computations for a very general model.

Example 4.7. Founding effects. To study the effect of bottlenecks on genetic variability in populations, we will consider $J_t =$ the expected homozygosity in generation t.

Theorem 4.7. *Letting u denote the mutation rate per locus per generation and assuming that each mutation produces a new genetic type*

$$J_t \approx \int_0^t \frac{1}{2N_s} \exp\left(-2u(t-s) - \int_s^t \frac{1}{2N_r}\, dr\right)\, ds$$
$$+ J_0 \exp\left(-2ut - \int_0^t \frac{1}{2N_s}\, ds\right) \tag{4.6}$$

Proof. In order for two individuals to have the same genotype at time t, neither can be a new mutant, an event of probability $(1-u)^2$. If neither is a mutant and their parents in generation $t-1$ are the same, an event of probability $1/2N_{t-1}$ they will be the same with probability 1; if not they will be the same with probability J_{t-1}, so we have

$$J_t = (1-u)^2 \left[\frac{1}{2N_{t-1}} + \left(1 - \frac{1}{2N_{t-1}}\right) J_{t-1}\right] \tag{4.7}$$

Using the formula in (4.7) for J_{t-1} we have

$$J_t = (1-u)^2 \frac{1}{2N_{t-1}} + (1-u)^4 \left(1 - \frac{1}{2N_{t-1}}\right) \frac{1}{2N_{t-2}}$$
$$+ (1-u)^4 \left(1 - \frac{1}{2N_{t-1}}\right)\left(1 - \frac{1}{2N_{t-2}}\right) J_{t-2}$$

Iterating this relationship, we find

$$J_t = \sum_{j=1}^{t} \frac{(1-u)^{2j}}{2N_{t-j}} \prod_{i=1}^{j-1}\left(1 - \frac{1}{2N_{t-i}}\right)$$
$$+ J_0(1-u)^{2t} \prod_{i=1}^{t}\left(1 - \frac{1}{2N_{t-i}}\right)$$

where $\prod_{i=1}^{0} z_i = 1$ by convention. This formula can be derived directly by breaking things down according to the generation $t-j$ on which coalescence occurs, with the last term giving the probability of no coalescence. When v is small and the N_{t-i}'s are large, this can be approximated by

$$J_t = \sum_{j=1}^{t} \frac{1}{2N_{t-j}} \exp\left(-2uj - \sum_{i=1}^{j-1} \frac{1}{2N_{t-i}}\right)$$
$$+ J_0 \exp\left(-2ut - \sum_{i=1}^{t} \frac{1}{2N_{t-i}}\right)$$

Changing variables $k = t - j$ and $\ell = t - i$, which in the inner sum will range from $\ell = t - (j-1) = k+1$ up to $t-1$, we have

$$J_t = \sum_{k=0}^{t-1} \frac{1}{2N_k} \exp\left(-2u(t-k) - \sum_{\ell=k+1}^{t-1} \frac{1}{2N_\ell}\right)$$
$$+ J_0 \exp\left(-2ut - \sum_{\ell=0}^{t-1} \frac{1}{2N_\ell}\right)$$

Converting the sums into integrals gives the desired result. □

For a concrete example of the use of our formula (4.6), consider

Example 4.8. Logisitic population growth. In many situations it is natural to assume that the size of a population grows according to the logistic growth model

$$\frac{dN_t}{dt} = rN_t(K - N_t)$$

Here r is the intrinsic growth rate, which gives the exponential rate of growth when the population is small, and $K = \lim_{t\to\infty} N_t$ is the equilibrium population size, the maximum number of individuals that the habitat can support. The solution to this equation for initial condition N_0 is

$$N_t = K/(1 + Ce^{-rt}) \quad \text{where } C = (K - N_0)/N_0.$$

With the Hawaiian *Drosophila* example in mind, Nei, Maruyama, and Chakraborty took $N_0 = 2$ to indicate the founding of the population by one female that has been inseminated by one male. Thinking of the locus as a single nucleotide, they set the mutation rate $u = 10^{-8}$. The equilibrium population size $K = 4 \times 10^6$ was then chosen so that the heterozygosity

$$H_0 = \frac{4Ku}{4Ku + 1} = \frac{0.16}{1.16} = 0.138$$

matches that observed in nature. They did not have much information about the intrinsic growth rate r, so they considered several values of r ranging from 0.1 to 1.

$N_0 = 2 \ll K$ so $C = (K - N_0)/N_0 \approx K/2$ and $N_t \approx K/(1 + (K/2)e^{-rt})$. When $r = 0.1$ it takes $t = 10 \ln(K/2) = 145$ generations for the population to reach size $K/2$ and $t = 10 \ln(5K) = 161$ generations to reach size $K/1.1$. To begin to evaluate (4.6), we note that

$$\frac{1}{N_t} = \frac{1}{K} + \frac{1}{2}e^{-rt}$$

Inserting this into (4.6) leads to an ugly expression with two levels of exponentials. Because of this and the fact that the approximations that led to (4.6) are not valid when $N(s)$ is small, we use the recursion (4.7) to evaluate J_t for small t. Since $u = 10^{-8}$, we can write

$$J_t \approx \frac{1}{2N_{t-1}} + \left(1 - \frac{1}{2N_{t-1}}\right) J_{t-1}$$

or the heterozygosity $H_t = 1 - J_t$ satsifies

$$H_t \approx \left(1 - \frac{1}{2N_{t-1}}\right) J_{t-1}$$

The initial heterozygosity $H_0 = 0.138$, so using the recursion we can compute the values of J_t for $0 \le t \le 50$ and $r = 1.0, 0.5, 0.2, 0.1$.

If we continue to iterate up to time 300 then the values change very little.

	$r = 1.0$	$r = 0.5$	$r = 0.2$	$r = 0.1$
50	0.089022	0.069016	0.031176	0.008346
100	0.089022	0.069015	0.031174	0.008201
150	0.089021	0.069015	0.031174	0.008200
200	0.089021	0.069015	0.031174	0.008200
250	0.089020	0.069014	0.031174	0.008200
300	0.089020	0.069014	0.031173	0.008199

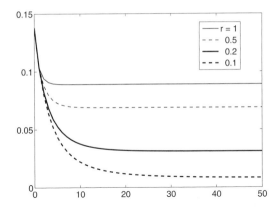

Fig. 4.5. Values of J_t for $0 \le t \le 50$.

At time 300, we have $N_t = 3,999,999.17$ when $r = 0.1$. Since N_t increases with r or with t we see that for $t \ge 300$, $N_t \approx K = 4,000,000$. Using this in (4.6), we see that for $t \ge 300$

$$J_t \approx \int_{300}^t \frac{1}{2K} \exp\left(-\left(2u + \frac{1}{2K}\right)(t-s)\right) ds$$
$$+ J_{300} \exp\left(-\left(2u + \frac{1}{2K}\right)(t-300)\right)$$

Evaluating the integral and letting $\gamma = 2u + 1/2K = 1.45 \times 10^{-7}$ we have

$$J_t \approx \frac{1}{4Ku+1}\left(1 - \exp^{-\gamma(t-300)}\right) + J_{300} \exp^{-\gamma(t-300)}$$

This shows that J_t eventually returns to the equilibrium value $1/(4Ku+1) = 0.862$. However, the time required is a multiple of $1/\gamma = 6,896,551$ generations. This is not quite as large as it sounds, since if we assume Drosophila have 10 generations a year this is about 700,000 years.

Bottlenecks

The use of the logisitic growth model in the previous example is somewhat unusual. In studying bottlenecks, it is much more common to suppose that the population sizes are piecewise constant. As drawn in the picture below, we assume that the current population size is N_1, at some point in the past it is reduced to fN_1 for an amount of time t_B, measured in units of $2N_1$ generations, and then returns to an ancestral population size of N_0.

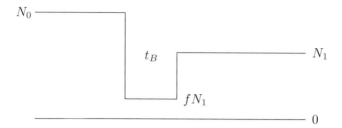

Suppose for the moment that $N_0 = N_1$. During the bottleneck the probability of coalescence is increased. The probability that a pair of chromosomes does not coalesce during the bottleneck is $1 - \exp(-t_B/f)$, so the strength of the bottleneck is mainly determined by the product of its length (t_B) and its severity $(1/f)$.

The effect of a bottleneck on a sample depends on the strength of the bottleneck and how long ago it occurred. During a recent severe bottleneck, all lineages coalesce leading to a star-shaped genealogy, and to an excess of rare alleles. In contrast, more moderate bottlenecks allow several lineages to survive the crash, leading to genealogies with long internal branches and resulting in an excess of alleles of intermediate frequency. Tajima's D is negative in the first case and positive in the second. Fay and Wu (1999) used simulations to show that Tajima's D becomes positive during the bottleneck and then decays to negative values after the bottleneck is over.

Example 4.9. One situation that leads to a bottleneck is the domestication of crop species. Eyre-Walker et al. (1998) studied the domestication of maize (*Zea mays* subspecies *mays*), which is thought to have occurred somewhere in southern or central Mexico about 7,500 years ago. To begin they determined the DNA sequence of a 1400 bp region of the *Adh1* locus from 24 individuals representing maize, its presumed progenitor (*Z. mays* subspecies *parviglumis*), and a more distant relative (*Z. luxurians*).

The data is given in the following table. There, n is the number of seqeunces, m the number of silent sites, S, the number of segregating silent sites (with the number of segregating replacement sites in parentheses), Wattersons' estimate of $\theta = 4N_e\mu$, and Tajima's D based on silent sites.

Taxa	n	m	S	θ	D
parviglumis	8	993	63(1)	24.3	- 0.241
maize	9	997	49(1)	18.03	0.785
luxurians	7	998	26(0)	10.61	0.258

The effective population size, N_A before the bottleneck can be estimated by using the value of $\theta = 0.0245$ for parviglumis. Using a mutation rate of $\mu = 6.5 \times 10^{-9}$ per generation for the approximately 1000 silent sites sequenced in this annual plant, we find

$$N_A = \frac{\theta}{4 \cdot 1000 \cdot \mu} = \frac{24.3}{26 \times 10^{-6}} \approx 940,000$$

Using coalescent simulations they estimated the number of plants N_B during the bottleneck. If the duration of bottleneck was $t_B = 10$ generations then they found $N_B = 23$, while for a bottleneck of $t_B = 500$ generations they found $N_B = 1,157$. Note that in each case the probability of coalescence during the bottleneck $t_B/N_B \approx 1/2.3 = 0.434$. This means that the nine sampled lineages are unlikely to coalesce to 1 during the bottleneck, which by remarks above is consistent with the observed positive value of Tajima's D.

4.4 Effective population size

In some cases when the population is subdivided or its size changes in time, measures of genetic variability are the same as in a homogeneously mixing population of effective size N_e. In Section 1.4, we mentioned the effective population size in our discussion of the nucleotide diversity. Specifically N_e was the population size needed to make $\pi = 4N\mu$ match the observed frequency of pairwise differences. Since π is 2μ times the expected coalescence time T_2 for two lineages, we can define

- the *nucleotide diversity effective population size*

$$N_e^\pi = ET_2/2$$

In general, one can define an effective population size by considering a convenient statistic. In a homogeneously mixing population, the probability, π_2, that two genes picked at random are descendants of the same parent gene $= 1/2N$, so we can let

- the *inbreeding effective population size*

$$N_e^i = (2\pi_2)^{-1}$$

The largest eigenvalue smaller than 1 for the Wright-Fisher model is $\lambda_{max} = 1 - 1/2N$ so

- the *eigenvalue effective population size*

$$N_e^e = \frac{1}{2}(1 - \lambda_{max})^{-1}$$

If $x_t =$ the fraction of A alleles then $\mathrm{var}\,(x_{t+1}|x_t) = x_t(1 - x_t)/2N$, so

- the *variance effective population size*

$$N_e^v = \frac{x_t(1 - x_t)}{\mathrm{var}\,(x_{t+1}|x_t)}$$

This notion is not easy to work with and does not make sense in some settings, e.g., subdivided populations, so we will ignore it here. The eigenvalue effective population size is difficult to compute and is covered in detail in Section 3.7 of Ewens (2004) book, so here we will concentrate on N_e^π, N_e^i, and a new notion introduced by Sjödin et al. (2005).

• If, when time is run at rate $2L$, genealogies follow Kingman's coalescent then we define the *coalescent effective population size* $N_e^c = L$.

This is the strongest and most satisfactory notion, since when in this case ALL of the statistics associated with the model are the same as in a homogeneously mixing population of size N_e^c.

To illustrate the use of these definitions, we will consider

Example 4.10. Periodically varying population.

Theorem 4.8. *Suppose that the population size $N(s)$ cyclically assumes the values $N_1, N_2, \ldots N_m$. The coalescent effective population size is*

$$N_e^c = m \left/ \sum_{s=1}^{m} \frac{1}{N(s)} \right.$$

which is the harmonic mean of the population sizes.

Since $1/x$ is convex, Jensen's inequality implies

$$\frac{1}{m} \sum_{s=1}^{m} \frac{1}{N(s)} \geq \frac{1}{(1/m) \sum_{s=1}^{m} N(s)}$$

That is, $N_e \leq (1/m) \sum_{s=1}^{m} N(s)$, the average population size.

Proof. The probability of no coalescence in one cycle is:

$$\prod_{s=1}^{m} \left(1 - \frac{1}{2N(s)}\right) \approx 1 - \sum_{s=1}^{k} \frac{1}{2N(s)}$$

Let $N_e = m / \sum_{s=1}^{m} 1/N(s)$. Assuming all of the $N(s)$ are large the probability that two lineages will not coalesce in $2N_e t$ generations is approximately

$$\left(1 - \sum_{s=1}^{m} \frac{1}{2N(s)}\right)^{2N_e t/m} \approx e^{-t}$$

A similar calculation shows that if there are k lineages and we scale time in units of $2N_e$ generations then coalescence occurs at rate $\binom{k}{2}$, i.e., the genealogy is given by Kingman's coalescent. $\qquad \square$

For a concrete example, if the population size alternates between 1,000 and 100,000 then it has effective population size

$$2 \bigg/ \left(\frac{1}{1,000} + \frac{1}{100,000} \right) = \frac{200,000}{101} = 1980$$

We would get the same result if, thinking of *Drosophila*, there were 10 generations a year, five when the population was small (1000) and five when it was large (100,000). Note that the size of N_e is dictated primarily by the small population size. If we replace 100,000 by 10,000,000 then $N_e \approx 2000$.

One has similar results when the population size experiences random fluctuations, e.g., the population sizes are independent or are given by a finite state Markov chain. In this case if $\pi_k = P(N(s) = k)$ then

$$N_e = 1 \bigg/ \sum_k \frac{\pi_k}{k}$$

See Kaj and Krone (2003) and Sano, Shimizu, and Iizuka (2004).

Example 4.11. Two-sex model. Suppose that in any generation there are N_1 diploid males and N_2 diploid females with $N_1 + N_2 = N$ and that each diploid offspring gets one chromosome chosen at random from the male population and one chosen at random from the female population.

Theorem 4.9. *For the two-sex model, the coalescent effective population size is*

$$N_e^c = \frac{4N_1 N_2}{N_1 + N_2} \tag{4.8}$$

Proof. To compute the inbreeding effective population size we note that two chromosomes taken at random from the population will have identical parent chromosomes if both are descended from the same male chromosome or both from the same female chromosome. Since there are $N(N-1)$ pairs of male chromosomes and $N(N-1)$ pairs of female chromosomes out of $2N(2N-1)$ pairs in the population, the probability of identical parentage is thus

$$\pi_2 = \frac{N(N-1)}{2N(2N-1)} \left\{ (2N_1)^{-1} + (2N_2)^{-1} \right\}$$

From this it follows that

$$N_e^i = (2\pi_2)^{-1} \approx \frac{2}{(2N_1)^{-1} + (2N_2)^{-1}} = \frac{4N_1 N_2}{N_1 + N_2}$$

Again, since the probability of coalescence per generation is constant, $T_2/2N_e^i$ will have approximately an exponential distribution and $N_e^c = N_e^i$. It is straightforward to extend the last conclusion to k lineages and show that rescaling time by $2N_e^i t$ produces Kingman's coalescent in the limit, so $N_e^c = N_e^i$. $\square$

For a concrete example, consider a herd of cattle where each generation has 10 bulls and 1000 cows. In this case

$$N_i^e = \frac{4 \cdot 10 \cdot 1000}{1010} = 39.6$$

In general, if $N_1 \ll N_2$ then $N_i^e \approx 4N_1$.

4.5 Matrix migration models

Bodmer and Cavalli-Sforza (1968) introduced a model in which there are a finite number of subpopulations with migrations between them described by a transition matrix. In this section, we will study the model in general. In the next, we will restrict our attention to the symmetric case in which all of populations have the same size and the migration rates are equal. To motivate the ideas we will consider the following data collected by Modiano et al. (1965) for four villages on the Lecce province. Let $a_{i,j}$ be the number of parents of individuals in village i who were from village j.

$$a_{ij} = \begin{array}{c|cccc|c} & 1 & 2 & 3 & 4 & \\ \hline 1 & 344 & 6 & 22 & 6 & 378 \\ 2 & 4 & 214 & 4 & 4 & 226 \\ 3 & 11 & 1 & 326 & 7 & 346 \\ 4 & 1 & 3 & 7 & 356 & 368 \\ \hline & 360 & 224 & 360 & 374 & 1318 \end{array}$$

If we normalize the columns to sum to 1, then we get the forward migration matrix $f_{i,j}$ which gives the fraction of parents from village j who sent their offspring to village i.

$$f_{ij} = \begin{array}{c|cccc} & 1 & 2 & 3 & 4 \\ \hline 1 & 0.955556 & 0.026786 & 0.061111 & 0.016043 \\ 2 & 0.011111 & 0.955357 & 0.011111 & 0.010695 \\ 3 & 0.030556 & 0.004464 & 0.905556 & 0.021390 \\ 4 & 0.002778 & 0.013393 & 0.022222 & 0.951872 \\ \hline & 1.0 & 1.0 & 1.0 & 1.0 \end{array}$$

If we normalize the rows to sum to 1, then we get the backward migration matrix $m_{i,j}$ which gives the fraction of offspring from village i whose parents are from village i.

$$m_{ij} = \begin{array}{c|ccccc} & 1 & 2 & 3 & 4 & \\ \hline 1 & 0.910053 & 0.015873 & 0.058201 & 0.015873 & 1.0 \\ 2 & 0.017699 & 0.946903 & 0.017699 & 0.017699 & 1.0 \\ 3 & 0.031792 & 0.002890 & 0.942197 & 0.023121 & 1.0 \\ 4 & 0.002717 & 0.008152 & 0.021739 & 0.967391 & 1.0 \end{array}$$

If we let $n_0 = (378, 226, 346, 368)$ be the column vector that gives the initial distribution of parents, then $n_1(i) = \sum_j f_{i,j} n_0(j)$ gives distribution of offspring, which can be found on the bottom row of the table that gives $a_{i,j}$. The evolution of the village sizes over time can be computed by

$$n_t(i) = \sum_i f_{ij} n_0(j)$$

The proportions of the individuals in the four villages in the 0th generation are $0.2867, 0.1714, 0.2625, 0.2792$. In the 1st generation they change to $0.2731, 0.1699, 0.2731$, and 0.2837. If we raise f to the 200th power we see

	1	2	3	4
1	0.442521	0.442505	0.442517	0.442507
$f_{i,j}^{200} = $ 2	0.198028	0.198038	0.198028	0.198027
3	0.190741	0.190735	0.190740	0.190739
4	0.168710	0.168722	0.168715	0.168727

so the current distribution is far from the equilibrium that will be attained through migration.

In order to predict the evolution of allele frequencies, the backwards matrix is used. If we let $p_t(j)$ be the frequencies in generation t then

$$p_{t+1}(i) = \sum_j m_{ij} p_t(j)$$

The next figure gives a concrete example of the use of this formula.

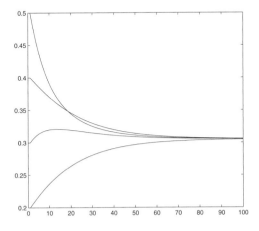

Fig. 4.6. Frequencies in our example when we start with $(0.5, 0.4, 0.3, 0.2)$.

Abstracting from the example, suppose that there are n subpopulations that have reached equilibrium sizes $N_1, \ldots N_n$ and let m_{ij} be the probability that an individual in subpopulation i is a descendant of an individual in subpopulation j. That is, to build up subpopulation i in generation $t + 1$, we sample N_i times with replacement from generation t, picking at random from subpopulation j with probability m_{ij}. Note that with the definition of the coalescent in mind, these are the backward migration probabilities. If we think of annual plants and a fraction f_{ij} of seeds of an individual in population i are sent to j then

$$m_{ij} = \frac{N_j f_{ji}}{\sum_k N_k f_{ki}} \tag{4.9}$$

For a concrete example consider the two sex model. In this case $m_{ij} = 1/2$ while $f_{ij} = N_j/(N_1 + N_2)$.

To define a genealogical process introduced by Notohara (1990) and called the *structured coalescent*, we choose let choose a real number N and run time at rate $2N$. The state of the process at time t is $\mathbf{k} = (k_1, \ldots, k_n)$ where k_i is the number of lineages in subpopulation i. Writing e_i for a vector with 1 in the ith place and 0 otherwise transitions occur

to	at rate
$\mathbf{k} - e_i + e_j$	$2N \cdot k_i m_{ij}$
$\mathbf{k} - e_i$	$2N \cdot \frac{k_i(k_i-1)}{2} \cdot \frac{1}{2N_i}$

The $2N$'s in front come from speeding up time. The first transition rate comes from one of the k_i lineages migrating from i to j with probability m_{ij}. The second transition comes from one of the $\binom{k_i}{2}$ pairs of lineages coalescing in subpopulation i, an event of probability $1/2N_i$. We assume that all of the m_{ij} are small so there is no factor of the form $1 - \sum_{j \neq i} m_{ij}$ for the lineages not migrating. As usual, we ignore the possibility that two events occur at one time. For more on the structured coalescent, see Wilkinson-Herbots (1998).

4.5.1 Strobeck's theorem

Let T_i be the coalescence time of two lineages sampled from colony i. Strobeck (1987) found general conditions that imply that $ET_i = 2Nn$ and hence do not depend on the migration rate. To state his result we need two assumptions:

- m_{ij} is *irreducible*, i.e., it is possible to get from any colony to any other by a sequence of migrations
- m_{ij} is *aperiodic*, i.e., for some i the greatest common divisor of $\{k : m_{ii}^k > 0\}$ is 1.

A sufficient condition for aperiodicity is $m_{ii} > 0$ for some i. A result from Markov chain theory implies that if m_{ij} is irreducible and aperiodic, then there is a stationary distribution π_i so that $\sum_i \pi_i m_{ij} = \pi_j$, and $m_{ij}^t \to \pi_j$ as $t \to \infty$.

Theorem 4.10. *Suppose that all subpopulations have the same size, $N_i = N$, and that the irreducible aperiodic migration probabilities are symmetric $m_{ij} = m_{ji}$ or, more generally, are doubly stochastic: $\sum_i m_{ij} = 1$. Let T_i be the coalescence time of two lineages sampled from subpopulation i.*

$$ET_i = 2Nn.$$

Proof. When both lineages are in the same colony, they have probability $1/2N$ per generation to coalesce, so $ET_i = 2NER_i$, where R_i is the first time $t \geq 1$ that the two lineages are in the same colony. The fact that the transition probability is doubly stochastic implies

$$\sum_i \frac{1}{n} m_{ij} = \frac{1}{n}$$

so stationary distribution for the migration process is uniform, $\pi_i = 1/n$. A well-known result in Markov chain theory implies $\pi_i = 1/ER_i$ and the desired result follows. $\square$

4.5.2 Fast migration limit

We will now describe results of Nagylaki (1980) and Notohara (1993), which show that in the presence of fast migration, the structured coalescent reduces to the ordinary coalescent.

Theorem 4.11. *Consider a matrix migration model with an irreducible aperiodic migration matrix m_{ij} having stationary distribution π_i. Let $N_T = N_1 + \cdots + N_n$ be the total population size and suppose that $N_T \to \infty$ with $N_i/N_T \to \kappa_i$. If N_T is large then the model has coalescent effective population size*

$$N_e^c = \beta N_T \quad \text{where} \quad \beta = \left(\sum_i \pi_i^2 \frac{1}{\kappa_i} \right)^{-1}$$

Sketch of proof. As noted earlier, since m_{ij} is irreducible and aperiodic, $m_{ij}^t \to \pi_j$ as $t \to \infty$. The amount of time we need for the Markov chain m_{ij} to reach equilibrium may be large, but it is a fixed number independent of the population size N_T. If we start the coalescent with two lineages and run it at rate $2N_T$ then the locations of the two lineages converge to equilibrium more rapidly than coalescence occurs. Thus, at all times their locations are independent and have the equilibrium distribution, so coalescence occurs at rate $\sum_i \pi_i^2/\kappa_i$. $\square$

In the special case $N_i = N$ for $1 \leq i \leq n$ and $\pi_i = 1/n$, we have $\kappa_i = 1/n$, $\beta = 1$, and $N_e^c = N_T$. This conclusion also holds if migration is conservative, i.e., $\sum_{k \neq j} N_k m_{k,j} = (1 - m_{jj})N_j$, or, in words, if the rate of migration into each colony is the same as the rate of migration out. Dividing each side by N_T

we have $\pi_i = \kappa_i$, and hence $\beta = (\sum_i \kappa_i)^{-1} = 1$. Expressing β as a harmonic mean, which cannot exceed the corresponding arithmetic mean, we see that in general

$$\beta = \left(\sum_i \pi_i (\kappa_1/\pi_i)^{-1} \right)^{-1} \leq \sum_i \pi_i (\kappa_1/\pi_i) = 1$$

so $N_e^c \leq N_T$. That is, with fast migration, population subdivision makes the effective population size smaller than the total population size. Nordborg and Krone (2002) generalized these results to migration rates of the form $m_{ij} = b_{ij}/N^\alpha$ where $0 \leq \alpha < 1$.

4.6 Symmetric island model

The simplest and the most commonly studied matrix migration model is the *symmetric island model*. In this case $N_i = N$ for all i, $m_{ii} = 1 - m$ and $m_{ij} = m/(n-1)$ for $j \neq i$. Let u be the per locus mutation rate and define the rescaled mutation and migration rates by $\theta = 4Nu$, and $M = 4Nm$. To explain the last two definitions, note that two lineages:

> coalesce at rate 1 when they are in the same population
>
> migrate at rate $2m \cdot 2N = M$
>
> experience mutations at rate $2u \cdot 2N = \theta$.

4.6.1 Identity by descent

Let $p_s(\theta)$ and $p_d(\theta)$ be the probabilities that two lineages are identical by descent when they are picked from the same or different populations.

Theorem 4.12. *If we let $\gamma = M/(n-1)$ and $D = \theta^2 + \theta(1 + n\gamma) + \gamma$ then in the symmetric island model we have*

$$p_s(\theta) = \frac{\theta + \gamma}{D} \qquad p_d(\theta) = \frac{\gamma}{D} \tag{4.10}$$

Note that in the limit of fast migration, $\gamma \to \infty$

$$p_s(\theta) = p_d(\theta) = \frac{1}{n\theta + 1}$$

This may look different from the usual result for a homogeneously mixing population but since $n\theta = (4Nn)u$ this is the, by now familiar, formula for a population with total size Nn.

Proof. By considering what happens at the first event, which is a coalescence with probability $1/(1 + \theta + M)$, migration with probability $M/(1 + \theta + M)$, and mutation with probability $\theta/(1 + \theta + M)$:

$$p_s(\theta) = \frac{1}{1 + \theta + M} \cdot 1 + \frac{M}{1 + \theta + M} \cdot p_d(\theta)$$

$$p_d(\theta) = \frac{M/(n-1)}{\theta + M} \cdot p_s(\theta) + \frac{M(n-2)/(n-1)}{\theta + M} \cdot p_d(\theta)$$

The second equation implies that

$$\frac{\theta + M/(n-1)}{\theta + M} \cdot p_d(\theta) = \frac{M/(n-1)}{\theta + M} \cdot p_s(\theta)$$

so cross multiplying gives:

$$p_d(\theta) = \frac{M/(n-1)}{\theta + M/(n-1)} \cdot p_s(\theta) = \frac{M}{(n-1)\theta + M} \cdot p_s(\theta) \qquad (4.11)$$

Using this in the first equation, we have

$$p_s(\theta) = \frac{1}{1 + \theta + M} + \frac{M}{1 + \theta + M} \cdot \frac{M}{(n-1)\theta + M} \cdot p_s(\theta)$$

A little algebra gives

$$(1 + \theta + M)((n-1)\theta + M) - M^2$$
$$= \theta(n-1) + \theta^2(n-1) + M\theta(n-1) + M + M\theta$$
$$= \theta^2(n-1) + \theta(n-1+Mn) + M$$

so we have

$$\frac{\theta^2(n-1) + \theta(n-1+Mn) + M}{(1 + \theta + M)((n-1)\theta + M)} \cdot p_s = \frac{1}{1 + \theta + M}$$

and it follows that

$$p_s(\theta) = \frac{\theta(n-1) + M}{\theta^2(n-1) + \theta(n-1+Mn) + M}$$

Using this in (4.11) we have

$$p_d(\theta) = \frac{M}{\theta^2(n-1) + \theta(n-1+Mn) + M}$$

Dividing the numerator and denominator of each fraction by $n-1$ and changing notation gives the formulas in the theorem. $\qquad \square$

4.6.2 Mean and variance of coalescence times

Let T_s be the coalescence time for two (distinct) individuals chosen from the same population and $t_s = T_s/2N$. Here and throughout this section, capital letter times are big times $O(N)$, while lower case letter times are small, i.e., have been divided by $2N$.

Theorem 4.13. *Writing* $\gamma = M/(n-1)$ *we have in the symmetric island model that*

$$Et_s = n \qquad Et_d = n + 1/\gamma \qquad (4.12)$$

Proof. The second result follows easily from the first: two lineages that start in different populations will require an average of $(n-1)/M = 1/\gamma$ units of time to enter the same population so $Et_d = (1/\gamma) + Et_s$. To compute Et_s we note that if $\theta = 4Nu$

$$p_s(\theta) = Ee^{-2uT_s} = Ee^{-\theta t_s}$$

In words, $p_s(\theta)$ is the Laplace transform of t_s. Differentiating the last formula, we see that $-p'_s(\theta) = E(t_s e^{-\theta t_s})$ so $-p'_s(0) = Et_s$. To use this formula, we note that $D' = 2\theta + (1 + n\gamma)$ so

$$-p'_s(\theta) = -\left\{ \frac{1}{D} - \frac{(\theta + \gamma)D'}{D^2} \right\} \qquad (4.13)$$

Setting $\theta = 0$, which implies $D = \gamma$ and $D' = 1 + n\gamma$, we have

$$Et_s = -p'_s(0) = -\left\{ \frac{1}{\gamma} - \frac{\gamma(1 + n\gamma)}{\gamma^2} \right\} = n$$

which completes the proof. $\qquad\qquad\qquad\qquad\qquad\qquad\qquad\qquad\square$

Continuing the analysis above we can compute the variances of t_s and t_d.

Theorem 4.14. *Writing* $\gamma = M/(n-1)$ *we have in the symmetric island model that*

$$var(t_s) = \frac{2(n-1)}{\gamma} + n^2 \qquad var(t_d) = var(t_s) + 1/\gamma^2 \qquad (4.14)$$

Proof. Again the second result follows from the first. Two lineages that start in different populations take an exponentially distributed amount of time with mean $1/\gamma$ (and hence variance $1/\gamma^2$) to come to the same population. The time to get to the same population is independent of the additional time needed to coalesce, so

$$var(t_d) = \frac{1}{\gamma^2} + var(t_s)$$

To begin to compute $var(t_s)$, we note that $p''_s(0) = Et_s^2$ and using (4.13)

$$p_s''(\theta) = \frac{-D'}{D^2} - \frac{D'}{D^2} - \frac{(\theta + \gamma)D''}{D^2} + 2\frac{(\theta + \gamma)(D')^2}{D^3}$$

Since $D = \theta^2 + \theta(1 + n\gamma) + \gamma$, $D' = 2\theta + 1 + n\gamma$, and $D'' = 2$ we have

$$E(t_s^2) = p_s''(0) = -\frac{2(1 + n\gamma)}{\gamma^2} - \frac{2\gamma}{\gamma^2} + \frac{2(1 + n\gamma)^2}{\gamma^2}$$

$$= \frac{2}{\gamma^2}((n-1)\gamma + (n\gamma)^2) = \frac{2(n-1)}{\gamma} + 2n^2$$

Since $Et_s = n$, and $\mathrm{var}\,(t_s) = Et_2^2 - (Et_s)^2$ the desired result follows.   □

4.6.3 Effective population sizes

Following Nei and Takahata (1993), we will now try to make sense of our collection of formulas by computing some of the effective population sizes introduced in Section 4.4. Let T_r be the (unscaled) coalescence time of two individuals sampled at random from the entire population. Using (4.12)

$$ET_r = 2N\left[\frac{1}{n} \cdot n + \frac{n-1}{n} \cdot \left(n + \frac{1}{\gamma}\right)\right] = 2Nn\left[1 + \frac{(n-1)}{n^2\gamma}\right]$$

Recalling $\gamma = M/(n-1)$, we have

$$ET_r = 2Nn\left[1 + \frac{(n-1)^2}{Mn^2}\right] \tag{4.15}$$

Since in a homogeneously mixing population of effective size N_e we have $ET_r = 2N_e$, this shows that

- a symmetric island model has *nucleotide diversity effective population size*

$$N_e^\pi = Nn\left[1 + \frac{(n-1)^2}{Mn^2}\right] \approx Nn\left[1 + \frac{1}{M}\right] \tag{4.16}$$

when n is large. Note that N_e^π is always larger than the actual population size Nn and can be much greater when $M = 4Nm$ is small.

To compute the inbreeding effective population size, N_e^i, we suppose that the migration rate is small enough so that we can ignore two migration events, and that the subpopulation size N is large enough so that sampling with or without replacement is the same. In this case chromosomes chosen at random have probability $1/n$ to be chosen from the same subpopulation and $(n-1)/n$ to be chosen from different subpopulations. If they are chosen from the same subpopulation and exactly one of them migrates coalescence is impossible. On the other hand if they are chosen from different subpopulations, coalescence is only possible if one migrates and lands in the subpopulation of the other. Thus the probability of coalescence in one generation is

$$\pi_2 = \frac{1}{n} \cdot (1 - 2m) \cdot \frac{1}{2N} + \frac{n-1}{n} \cdot \frac{2m}{n-1} \cdot \frac{1}{2N} = \frac{1}{2Nn}$$

and we have shown

- a symmetric island model has *inbreeding effective population size*

$$N_e^i = Nn \qquad (4.17)$$

To try to resolve the conflict between the two values for the effective population size, we will consider the homozygosity, F, which is the probability that two individuals chosen at random are identical by descent. Using (4.10) and $\gamma = M/(n-1)$, we have

$$F = \frac{1}{n} p_s(\theta) + \frac{n-1}{n} p_d(\theta) = \frac{M/(n-1) + \theta/n}{D}$$

Recalling that $D = \theta^2 + \theta(1 + n\gamma) + \gamma$, we have

$$\frac{1}{F} = \frac{\theta^2 + \theta \left(1 + \frac{nM}{n-1}\right) + \frac{M}{n-1}}{\frac{\theta}{n} + \frac{M}{n-1}}$$

Breaking the numerator up as $\theta^2 + \theta nM/(n-1)$ plus $M/(n-1) + \theta/n$ plus $\theta - \theta/n$, we can write the above as

$$\theta n + 1 + \frac{\frac{n-1}{n}\theta}{\frac{\theta}{n} + \frac{M}{n-1}} = 1 + 4Nun + \frac{(n-1)^2 4Nu}{\theta(n-1) + Mn}$$

Thus if we define an effective population size based on the heterozygosity by $1/F = 1 + 4N_e^h u$, we have

- a symmetric island model has *heterozygosity effective population size:*

$$N_e^h = \frac{\frac{1}{F} - 1}{4u} = Nn \left[1 + \frac{(n-1)^2}{\theta n(n-1) + Mn^2}\right] \qquad (4.18)$$

This definition has the unfortunate property that it depends on the mutation rate. However, when $M \gg \theta$, the formula for N_e^h in (4.18) reduces to the one for N_e^π given in (4.16).

4.6.4 Large n limit

Wakeley (1998) showed that the island model simplifies considerably when the number of subpopulations is large.

Theorem 4.15. *If we sample at most one individual from each subpopulation in a symmetric island model then in the limit as the number of subpopulations $n \to \infty$, the genealogy of the sample converges to that of a homogeneously mixing population of size*

$$N_e^c = Nn \left(1 + \frac{1}{M}\right) \quad \text{where } M = 4Nm.$$

Note that this is the same as the formula for N_e^π given in (4.16), except that in the large n limit, scaling by the nucleotide diversity effective population size gives Kingman's coalescent.

Proof. Let $T(k_1, k_2, \ldots k_\ell)$ be the total size of the tree (measured in units of $2Nn$ generations) for a sample in which k_i colonies are sampled i times, and the total sample size is $\ell = \sum_i ik_i$. When all ℓ lineages are in different colonies, migration occurs at rate $2Nn \cdot m\ell$ and a migration results in two individuals in the same colony with probability $(\ell - 1)/(n - 1)$. Thus the time until we see two lineages in the same population is exponential with rate $2Mmn\ell(\ell - 1)/(n - 1)$. Since there are ℓ lineages initially, the total time satisfies

$$ET(\ell, 0, \ldots, 0) = \ell \cdot \frac{n - 1}{2Nnm\ell(\ell - 1)} + ET(\ell - 2, 1, 0, \ldots 0) \qquad (4.19)$$

To shorten formulas we will drop the ending zeros, so that the variables above become $T(\ell)$ and $T(\ell - 2, 1)$ When the configuration is $(\ell - 2, 1)$, there are several possible transitions depending on which lineage migrates or if a coalescence occurs. Here R is the total rate of events that change the configuration. We will compute R once we have written down the individual rates but for the moment we get it out of the way by moving them to the left side of the equation.

$$R \cdot ET(\ell - 2, 1) = \ell + 2Nn \cdot (\ell - 2)m\frac{1}{n - 1}ET(\ell - 3, 0, 1)$$

$$+ 2Nn \cdot (\ell - 2)m \cdot \frac{\ell - 3}{n - 1}ET(\ell - 4, 2)$$

$$+ 2Nn \cdot 2m \cdot \left(1 - \frac{\ell - 1}{n - 1}\right) ET(\ell)$$

$$+ 2Nn \cdot \frac{1}{2N} \cdot ET(\ell - 1)$$

On the right-hand side, the first term in the four products, $2Nn$, is the rate at which time is run. In the first three cases, the second term is km where k is the number of lineages that can move to make the desired transition, and the third term $j/(n - 1)$ is the fraction of moves that result in the desired transition. The final term corresponds to coalescence of the two lineages in the same colony.

The third rate $\sim Mn$ where $M = 4Nm$. The fourth is n. The first and second are smaller, so we have

$$R \sim n(M + 1)$$

and the equation simplifies to

$$ET(\ell - 2, 1) = \frac{M}{M+1} ET(\ell) + \frac{1}{M+1} ET(\ell - 1) \qquad (4.20)$$

Using the new notation and $(n-1)/n \to 1$ we can write (4.19) as

$$ET(\ell) = \frac{2}{M(\ell - 1)} + ET(\ell - 2, 1)$$

Adding this to (4.20), and doing some cancellation, we have

$$\frac{1}{M+1} ET(\ell) = \frac{2}{M(\ell - 1)} + \frac{1}{M+1} ET(\ell - 1)$$

Solving gives

$$ET(\ell) = \frac{2}{(\ell - 1)} \cdot \frac{M+1}{M} + ET(\ell - 1)$$

and iterating we have

$$ET(\ell) = \left(1 + \frac{1}{M}\right) \sum_{i=1}^{\ell-1} \frac{2}{i}$$

Since our M is Wakeley's $2M$ this agrees with his formula (26) on page 169.
□

At the other extreme from sampling at most one individual from each subpopulation is sampling from only one subpopulation.

Theorem 4.16. *If we sample ℓ individuals from one subpopulation in a symmetric island model then in the limit as the number of subpopulations $n \to \infty$, the total time in the tree has the same distribution as when we sample one lineage from each of K_ℓ subpopulations, where K_ℓ is the distribution of the number of alleles in an infinite alleles model with $\theta = 4Nm$.*

Proof. If we run time at rate $2N$ and let $\theta = 4Nm$ then the initial rate of migration is $\theta/2$ and the initial rate of coalescence is $\ell(\ell - 1)/2$, exactly as in the coalescent with killing described in Section 1.3.1. In that model mutations create a new type. Here, migrations take a lineage to a new population, which, when n is large, will all be distinct. At the time when the number of lineages in the original population reaches 1, the *scattering phase*, as Wakeley calls it, is over and the lineages are all in the different populations. The number of surviving lineages K_ℓ has the distribution of the number of alleles in an infinite alleles model with $\theta = 4Nm$. To complete the proof we note that the scattering phase lasts for time $O(2N)$, while the subsequent coalescence of the surviving lineages, called the *collection phase*, will take time $O(2Nn)$, which is much larger. □

Wakeley has used the large n limit to simplify other population genetics problems. In Wakeley (1999), he explored the consequences of a sudden change in the population size. In Wakeley (2000), he investigated various two population models in which each population is a modeled as an island model with a large number of subpopulations. Wakeley and Aliacar (2001) studied metapopulation models, in which the populations are subject to extinction and recolonization. Wakeley and Lessard (2003) and Lessard and Wakeley (2004) investigated the limiting behavior of two loci with recombination when the number of subpopulations $n \to \infty$.

4.7 Fixation indices

To quantify the impact of population structure on genetic differentiation, Wright (1951) introduced a quantity called F_{ST} that can be defined as follows, see Nei (1975) page 151,

$$F_{ST} = \frac{f_0 - \bar{f}}{1 - \bar{f}} \tag{4.21}$$

where f_0 is the probability that two individuals sampled from the same subpopulation are identical by descent and $\bar{f}$ is that probability for two individuals sampled at random from the entire population.

The probability of identity by descent is the probability that no mutation occurred before their coalescence time T so

$$f = \sum_{t=1}^{\infty} (1 - \mu)^{2t} P(T = t) \approx \sum_{t=1}^{\infty} (1 - 2t\mu) P(T = t) = 1 - 2\mu ET$$

the middle approximation being valid if $\mu ET \ll 1$. Using this with the previous formula gives a simplification due to Slatkin (1991):

$$F_{ST} \approx \frac{\bar{T} - \bar{T}_0}{\bar{T}} \tag{4.22}$$

where $\bar{T}_0$ is the average coalescence time of two individuals drawn from the same population and $\bar{T}$ is the the average coalescence time for two genes sampled at random from the entire population. An advantage of (4.22) over (4.21) is that it does not depend on the mutation rate.

In the case of the finite island model, $\bar{T} = ET_r$, and $\bar{T}_0 = 2NEt_s$, so using (4.15) and (4.12) we have

$$F_{ST} = \frac{(n-1)^2}{4Nmn^2} \bigg/ \left(1 + \frac{(n-1)^2}{4Nmn^2}\right)$$

and it follows that

$$F_{ST} = \left(1 + 4Nm \cdot \frac{n^2}{(n-1)^2}\right)^{-1} \tag{4.23}$$

a formula first derived by Takahata (1983); see also (5) in Takahata and Nei (1984). Changing our previous convention, we will now follow the usage in a different part of the literature by defining the rescaled migration rate $M_1 = Nm$ to be the number of migrating gametes per colony per generation. To help remember the definition note that M_1 has a 1 where $M = 4Nm$ has a 4. When the number of colonies is large we can replace $n^2/(n-1)^2$ by 1 and we have

$$M_1 = \frac{1}{4}\left(\frac{1}{F_{ST}} - 1\right) \qquad (4.24)$$

Example 4.12. Seielstad, Minch, and Cavalli-Sforza (1998) have made an interesting application of F_{ST} estimates to human history. They collected data on Y chromsome single nucleotide polymorphisms and compared them with published studies of autosomal and mitochondrial DNA variation. Equating F_{ST} with 1 minus the fraction of the variation within population they found the following:

	within populations	within continents	between continents	F_{ST}	M_1
mtDNA	0.814	0.061	0.125	0.186	4.38
autosomes	0.856	0.057	0.088	0.144	1.49
Y chromosome	0.355	0.118	0.527	0.645	0.55

To infer migration rates from this, we can use (4.24) for autosomes, but mtDNA and the Y chromosome are haploid and exist in only one of the two sexes so their effective population size is $1/4$ as large and we must use $F_{ST} = (1 + M_1)^{-1}$ instead of (4.24).

The results are given in the table. Comparing the inferred migration rate from mtDNA, which resides only in females, with that from the Y chromosome in males, we conclude that females migrate about eight times as often as males. At first, this may seem surprising since men travel more than women in many societies. However, the most important parameter from a genetic perspective is the distance between the birthplace of the parents and of the children. Marriages in agricultural economies where land is inherited by sons, and probably in early foraging economies as well, tend to be patrilocal; that is, the wife moves to join the husband. Thus women tend to be more mobile than men as far as transgenerational movements are concerned.

We turn now to the problem of estimating F_{ST} from data. When there are only two populations, $\bar{T} = (Et_s + Et_d)/2$ and $\bar{T}_0 = Et_s$, (4.22) becomes

$$F_{ST} \approx \frac{Et_d - Et_s}{Et_s + Et_d}$$

Using (4.24) we have

$$M_1 = \frac{1}{4}\left(\frac{Et_s + Et_d}{Et_d - Et_s} - 1\right) = \frac{Et_s}{2(Et_d - Et_s)}$$

Let Δ_s and Δ_d be the average number of pairwise difference for two individuals sampled from the same colony or form different colonies. Since $E\Delta_i = \theta Et_i$, we can estimate M by

$$\hat{M}_1 = \frac{\Delta_s}{2(\Delta_d - \Delta_s)}. \tag{4.25}$$

This estimator is closely related to but slightly different than others that have been introduced. Weir and Cockerham's (1984) $\hat{\theta}$ and Lynch and Crease's (1990) N_{ST} estimate $(Et_d - Et_s)/Et_s$ so one has

$$\hat{M} = \frac{1}{2} \cdot \left(\frac{1}{\hat{\theta}} - 1 \right)$$

Hudson, Slatkin, and Maddison (1992) estimate F_{ST} by $\hat{F} = 1 - \Delta_s/\Delta_d$ so they again have

$$\hat{M} = \frac{1}{2} \cdot \left(\frac{1}{\hat{F}} - 1 \right) \tag{4.26}$$

Example 4.13. To illustrate the estimation of F_{ST} and inferences about spatial structure, we will use data from Hamblin and Veuille (1999), who sequenced 809 bp of the *vermillion* locus in four African populations of *D. simulans*. Values of F_{ST} as computed by the formula of Hudson, Slatkin, and Maddison (1992) are given above the diagonal in the following table. Note that $F_{ST} \geq 0$ by definition but the comparison of Kenya and Tanzania produced a negative estimate.

	Kenya	Tanzania	Zimbabwe	Cameroon
Kenya		−0.033 (0.854)	0.178 (0.001)	0.107 (0.006)
Tanzania	∞		0.122 (0.007)	0.077 (0.014)
Zimbabwe	4.6	7.2		0.159 (0.014)
Cameroon	8.3	12	5.3	

To determine the significance of these values, Hamblin and Veuille (1999) used the method of Hudson, Boos, and Kaplan (1992). To explain the method, we consider the comparison of the 13 sequences from Kenya and the 10 from Zimbabwe. If the two populations were part of a homogeneously mixing whole, then the distribution of the computed value of F_{ST} does not change if we pick 13 of the 23 sequences at random (without replacement) to be the Kenya sample and call the remaining 10 the Zimbabwe sample. They repeated this procedure 1000 times. The number in parentheses below the estimate $\hat{F}$, 0.001 in this case, indicates the fraction of times the randomized data set gave a larger value of F_{ST}. In all cases except the Kenya and Tanzania comparison we can reject with confidence the assumption of homogeneous mixing. Using (4.26) now we can estimate the number of migrants per generation $\hat{M}$.

The results are given below the diagonal. In the case of the Kenya-Tanzania comparison, the estimated $\hat{M}$ is negative, so we write ∞.

Here we have followed the authors in converting F_{ST} into an estimate of the number of migrants Nm. However, Felsenstein (1982) argued that estimates of F_{ST} based on genetic markers reflect the cumulative effects of genetic drift and gene flow for groups of populations. Because of this averaging, they are not appropriate for estimates of gene flow between specific pairs within a network of populations or for instantaneous rates of gene flow. Whitlock and McCauley (1999) also made this point – while F_{ST} is an excellent measure of the extent of population structure, it is rare that F_{ST} can be translated into an accurate estimate of Nm.

5

Stepping Stone Model

"Since the mathematicians have invaded the theory of relativity, I do not understand it myself anymore." Albert Einstein

In the island model, migration occurs to all subpopulations with equal probability. However, in natural populations dispersal preferentially occurs between geographically close subpopulations. In this chapter, we will study a more realistic model of a spatially distributed population, the stepping stone model, which has what Wright (1943) called "isolation by distance." Much has been written about this model, but we will only cite the sources we use, so we should complete the picture by mentioning three historically important papers: Kimura (1953), Kimura and Weiss (1964), and Weiss and Kimura (1965).

In the discrete-time stepping stone model, the population is divided into groups that we will call colonies. Generation $n + 1$ is obtained from generation n in the following way. Consider a given individual in colony x. With probability μ, this individual mutates to a new type and, with probability $(1 - \mu)p(x, y)$, assumes the type of an individual chosen at random from the colony y (in generation n). All such mutations and choices are assumed to be independent for all individuals at all colonies.

5.1 d = 1, Exact results

We will first consider the case of an infinite sequence of colonies indexed by the integers, and we will further simplify by supposing that migration is symmetric and only occurs to the nearest neighbors:

$$p(x, x) = 1 - \nu \qquad p(x, x + 1) = p(x, x - 1) = \nu/2$$

Let $\psi(x, y)$ be the probability that two individuals, one chosen from colony x and one chosen from colony y, are identical by descent in equilibrium. (When $x = y$ we suppose that two distinct individuals are chosen.) In the case we

R. Durrett, *Probability Models for DNA Sequence Evolution*,
DOI: 10.1007/978-0-387-78168-6_5, © Springer Science+Business Media, LLC 2008

are considering, the probability $\psi(x, y)$ depends only on the difference $y - x$, so we consider instead $\phi(x) = \psi(0, x)$.

Theorem 5.1. *In the symmetric nearest neighbor stepping stone model on $\mathbf{Z}$, the probability of identity by descent for two lineages sampled from colonies that differ by x is*

$$\phi(x) = \frac{\lambda_2^{|x|}}{1 + 4N\mu + 4N\nu(1 - \lambda_2)} \tag{5.1}$$

where $\lambda_2 = \left(2 + \frac{2\mu}{\nu} - \sqrt{\frac{8\mu}{\nu} + \frac{4\mu^2}{\nu^2}} \right) \Big/ 2$.

If $\mu \ll \nu$, i.e., the mutation probability is much smaller than the migration probability, then the $4\mu^2/\nu^2$ under the square root in the definition of λ_2 can be ignored, as can the $2\mu/\nu$ on the outside, so $\lambda_2 \approx 1 - \sqrt{2\mu/\nu}$. In addition, $\mu \ll \sqrt{2\mu\nu} \approx \nu(1 - \lambda_2)$ so

$$\phi(0) \approx \frac{1}{1 + 4N\sqrt{2\mu\nu}} \tag{5.2}$$

$$\phi(x) \approx \phi(0)(1 - \sqrt{2\mu/\nu})^{|x|} \tag{5.3}$$

which agrees with formulas on page 89 of Malécot (1969).

Proof. Considering what happens in one generation and supposing that μ and ν are small enough so that we can ignore the probability of two events affecting the two individuals we are considering, we have that

$$\phi(x) = \nu\phi(x - 1) + (1 - 2\mu - 2\nu)\phi(x) + \nu\phi(x + 1) \quad \text{for } x \neq 0 \tag{5.4}$$

Rearranging, we have

$$0 = \phi(x - 1) - \left(2 + \frac{2\mu}{\nu} \right) \phi(x) + \phi(x + 1) \quad \text{for } x \neq 0$$

Restricting our attention to $x \geq 1$ in the last equation, we have a second-order difference equation whose general solution is $A\lambda_1^x + B\lambda_2^x$, where $\lambda_1 > \lambda_2$ are the roots of

$$0 = 1 - \left(2 + \frac{2\mu}{\nu} \right) \lambda + \lambda^2$$

The quadratic formula tells us that the roots are

$$\lambda_i = \frac{2 + \frac{2\mu}{\nu} \pm \sqrt{\left(2 + \frac{2\mu}{\nu} \right)^2 - 4}}{2} = \frac{2 + \frac{2\mu}{\nu} \pm \sqrt{\frac{8\mu}{\nu} + \frac{4\mu^2}{\nu^2}}}{2}$$

Noting that $(\lambda - a)(\lambda - b) = \lambda^2 - (a + b)\lambda + ab$, we see that $\lambda_1\lambda_2 = 1$ and we have $\lambda_1 > 1 > \lambda_2$. From this we see that for $x \geq 0$, $\phi(x) = B\lambda_2^x$, for otherwise the probability $\phi(x)$ would become unbounded as $x \to \infty$. Using the symmetry $\phi(-x) = \phi(x)$, it follows that

$$\phi(x) = B\lambda_2^{|x|} \qquad (5.5)$$

To determine the constant B we use the equation analogous to (5.4) for $x = 0$. Now $\phi(1) = \phi(-1)$ and the probability of coalescence of two lineages in the same colony on one step is $1/2N$, so if we ignore the occurrence of two events in one generation

$$\phi(0) = 2\nu\phi(1) + (1 - 2\mu - 2\nu - 1/2N)\phi(0) + 1/2N$$

Rearranging gives

$$(2\mu + 2\nu + 1/2N)\phi(0) - 2\nu\phi(1) = 1/2N \qquad (5.6)$$

Substituting in $\phi(x) = B\lambda_2^{|x|}$ we have

$$B\left[2\mu + 2\nu + \frac{1}{2N} - 2\nu\lambda_2\right] = \frac{1}{2N}$$

Solving gives

$$B = \frac{1/2N}{2\mu + 2\nu(1 - \lambda_2) + 1/2N}$$

Combining this with (5.5) we have proved (5.1). □

Ring of colonies

We are now ready to tackle the more complicated case of a ring of L colonies, which we think of as the integers modulo L. As for an infinite sequence of colonies, we consider symmetric nearest neighbor migration:

$$p(x, x) = 1 - \nu \qquad p(x, x + 1) = p(x, x - 1) = \nu/2$$

but now the addition is done modulo L so $p(L - 1, 0) = \nu/2$ and $p(0, L - 1) = \nu/2$. Again we let $\phi(x) = \psi(0, x)$.

Theorem 5.2. *In the symmetric nearest neighbor stepping stone model with a ring of L colonies, the probability of identity by descent for two lineages sampled from colonies that differ by x modulo L is*

$$\phi(x) = C(\lambda_1^{x-L/2} + \lambda_2^{x-L/2}) \qquad (5.7)$$

where $\lambda_1 > \lambda_2$ are $\left(2 + \frac{2\mu}{\nu} \pm \sqrt{\frac{8\mu}{\nu} + \frac{4\mu^2}{\nu^2}}\right) \Big/ 2$ and

$$1/C = (1 + 4N\mu)(\lambda_1^{-L/2} + \lambda_2^{-L/2}) \qquad (5.8)$$
$$+ 4N\nu[(1 - \lambda_1)\lambda_1^{-L/2} + (1 - \lambda_2)\lambda_2^{-L/2}]$$

Proof. If we define $\phi(L) = \phi(0)$, then by the reasoning that led to (5.4)

$$\phi(x) = \nu\phi(x-1) + (1 - 2\mu - 2\nu)\phi(x) + \nu\phi(x+1) \quad \text{for } 0 < x < L \quad (5.9)$$

This is the same second order difference equation, so it has solutions $A\lambda_1^x + B\lambda_2^x$ for the same values of λ_1 and λ_2. The solution of interest has the symmetry property $\phi(x) = \phi(L - x)$, so recalling $\lambda_2 = \lambda_1^{-1}$ we write

$$\phi(x) = C(\lambda_1^{x-L/2} + \lambda_2^{x-L/2})$$

To check this guess, we note that $\lambda_2 = 1/\lambda_1$ so

$$\phi(L - x) = C(\lambda_1^{L/2-x} + \lambda_2^{L/2-x}) = C(\lambda_2^{x-L/2} + \lambda_1^{x-L/2}) = \phi(x)$$

To compute C we note that (5.6) implies

$$\left(2\mu + 2\nu + \frac{1}{2N}\right) \cdot C(\lambda_1^{-L/2} + \lambda_2^{-L/2}) - 2\nu \cdot C(\lambda_1^{1-L/2} + \lambda_2^{1-L/2}) = \frac{1}{2N}$$

Multiplying by $2N/C$ on both sides

$$(1 + 4N\mu + 4N\nu) \cdot (\lambda_1^{-L/2} + \lambda_2^{-L/2}) - 4N\nu \cdot (\lambda_1^{1-L/2} + \lambda_2^{1-L/2}) = \frac{1}{C}$$

which is equal to the formula for $1/C$ in (5.8). $\square$

Two extreme situations

Suppose $\mu \ll \nu$, i.e., the mutation rate is much smaller than the migration rate. As a consequence of Theorem 5.2 we get the following qualitative result.

Theorem 5.3. *In the symmetric nearest neighbor stepping stone model with a ring of L colonies,*

- *when $L^2/\nu \gg 1/\mu$ there will be very little difference between the ring and the line.*
- *when $L^2/\nu \ll 1/\mu$ the stepping stone model behaves like a homogeneously mixing population.*

To interpret the last result, we note that the displacement kernel p has variance ν, so by the central limit theorem, it takes about L^2/ν steps to go a distance L. On the right-hand side, $1/\mu$ is the average amount of time it takes for a lineage to encounter a mutation, so when $L^2/\nu \gg 1/\mu$ it is very likely that a mutation will occur before the lineage has moved a distance L. In the other direction, when $L^2/\nu \ll 1/\mu$ it is very unlikely that a mutation will occur before the lineage has moved a distance that is a large multiple of L and hence traveled around the circle many times, forgetting its starting point.

Proof. Recalling that $\lambda_1 \approx 1 + \sqrt{2\mu/\nu}$, we see that λ_1^L is large when $L\sqrt{\mu/\nu}$ is large or equivalently when $L^2/\nu \gg 1/\mu$. If λ_1^L is large (and hence λ_2^L is small) then for $x/L \leq 1/2 - \epsilon$, $\lambda_2^{x-L/2} \gg \lambda_1^{x-L/2}$, so we can drop the smaller term from (5.7). Using the formula for C from (5.8) and multiplying top and bottom by $\lambda_2^{L/2} = \lambda_1^{-L/2}$ we have

$$\phi(x) \approx \frac{\lambda_2^x}{(1+4N\mu)(\lambda_1^{-L}+1) + 4N\nu[(1-\lambda_1)\lambda_1^{-L} + (1-\lambda_2)]}$$

$$\approx \frac{\lambda_2^x}{1 + 4N\sqrt{2\mu\nu}}$$

since $\lambda_1^{-L} \approx 0$, $\lambda_2 \approx 1 - \sqrt{2\mu/\nu}$, and $\mu \ll \sqrt{\mu\nu}$. This agrees with our previous result for the stepping stone model on an infinite linear grid given in (5.2) and (5.3).

When $L^2/\nu \ll 1/\mu$, $L\sqrt{\mu/\nu} \approx 0$ and $\lambda_1^x \approx 1 \approx \lambda_2^x$ for all $-L \leq x \leq L$, so using (5.7) and (5.8) we have

$$\phi(x) \approx \frac{2}{(1+4N\mu) \cdot 2 + 4N\nu \cdot 0} = \frac{1}{1 + 4N\mu}$$

the result for a homogeneously mixing population.   □

5.2 d = 1 and 2, Fourier methods

The analysis in the previous section, which is based on difference equations, only works in the one-dimensional nearest neighbor case. In this section, we will use a more sophisticated approach based on Fourier series that will allow us to get results in one and two dimensions for fairly general dispersal distributions. Our starting point is a result Malécot (1969) . Let $\psi(x,y)$ be the probability that two individuals, one chosen from colony x and one from colony y, are identical by descent in equilibrium. (When $x = y$ we suppose that two distinct individuals are chosen.)

To state the result we let $p^k(x,y)$ be the probability of going from x to y in k steps. Intuitively, p^k is just the kth power of the matrix $p(x,y)$. Formally, it can be defined inductively by $p^1(x,y) = p(x,y)$ and for $k \geq 2$

$$p^k(x,y) = \sum_z p(x,z)p^{k-1}(z,y)$$

Theorem 5.4. *In a stepping stone model with a general symmetric transition probability $p(x,y) = p(y,x)$ the probability of identity by descent satsifies*

$$\psi(x,y) = \frac{1 - \psi(0,0)}{2N} \sum_{n=1}^{\infty} (1-\mu)^{2n} p^{2n}(x,y) \tag{5.10}$$

Proof. Let $\psi_n(x, y)$ be the probability that two individuals, one chosen from x and one from y are identical by descent in generation n. Again when $x = y$ we suppose that two distinct individuals are chosen. By definition, $p(x, y)$ is the probability that an individual in colony x in generation $n + 1$ is the offspring of one at y in generation n. Recalling that the number of individuals in a colony is $2N$ and considering what happens on one step, we see that

$$\psi_{n+1}(x, y) = (1 - \mu)^2 \sum_{x', y'} \psi_n(x', y') p(x, x') p(y, y')$$
$$+ (1 - \mu)^2 \sum_z \frac{1 - \psi_n(z, z)}{2N} p(x, z) p(y, z)$$

The second term compensates for the two individuals choosing the same parent. When this event of probability $1/2N$ occurs the probability of identity by descent is 1, rather than $\psi_n(z, z)$ so this needs to be added to the $\psi_n(z, z) p(x, z) p(y, z)$ in the first term.

Setting $\psi_{n+1} = \psi_n = \psi$ to compute the equilibrium and iterating once, we have

$$\psi(x, y) = (1 - \mu)^2 \sum_z \frac{1 - \psi(z, z)}{2N} p(x, z) \, p(y, z)$$
$$+ (1 - \mu)^2 \sum_{x', y'} p(x, x') p(y, y')$$
$$\left\{ (1 - \mu)^2 \sum_z \frac{1 - \psi(z, z)}{2N} p(x', z) p(y', z) \right.$$
$$\left. + (1 - \mu)^2 \sum_{x'', y''} \psi(x'', y'') p(x', x'') p(y', y'') \right\}$$

Noting $\sum_{x'} p(x, x') p(x', z) = p^2(x, z)$, the above becomes

$$\psi(x, y) = (1 - \mu)^2 \sum_z \frac{1 - \psi(z, z)}{2N} p(x, z) p(y, z)$$
$$+ (1 - \mu)^4 \sum_z \frac{1 - \psi(z, z)}{2N} p^2(x, z) p^2(y, z)$$
$$+ (1 - \mu)^4 \sum_{x'', y''} \psi(x'', y'') p^2(x, x'') p^2(y, y'')$$

Iterating n times, letting $n \to \infty$ and noting that the last term tends to 0 when $\mu > 0$, we have

$$\psi(x, y) = \sum_{n=1}^{\infty} (1 - \mu)^{2n} \sum_z \frac{1 - \psi(z, z)}{2N} p^n(x, z) p^n(y, z)$$
$$= \frac{1 - \psi(0, 0)}{2N} \sum_{n=1}^{\infty} (1 - \mu)^{2n} p^{2n}(x, y)$$

where in the last step we have used $\psi(z, z) = \psi(0, 0)$ and $p^n(y, z) = p^n(z, y)$, which follows from the assumed symmetry.   □

Fourier analysis

We restrict attention now to $\mathbf{Z}^d$. The identity by descent probability $\psi(x, y)$ depends only on $y - x$, so it is enough to consider $\phi(x) = \psi(0, x)$. To compute ϕ, we introduce the Fourier transforms

$$\hat{\phi}(\theta) = \sum_x e^{i\theta \cdot x} \phi(x) \quad \text{and} \quad \hat{p}(\theta) = \sum_x e^{i\theta \cdot x} p(0, x)$$

where the sums are over $x \in \mathbf{Z}^d$ and $\theta \cdot x = \sum_i \theta_i x_i$ is the dot product of the two vectors. We will assume

p is irreducible, i.e., it is possible to get from any colony to any other

$p(0, x) = 1 - \nu$ and $p(0, x) = \nu q(x)$ with $q(0) = 0$

q has the same symmetries as $\mathbf{Z}^d$ (5.11)

q has finite third moment

Finite variance should be sufficient for the results. However, having a third moment allows us to more easily estimate some error terms.

Transforming Malécot's formula (5.10) and summing the geometric series, we have

$$\hat{\phi}(\theta) = \frac{1 - \phi(0)}{2N} \sum_{n=1}^{\infty} (1 - \mu)^{2n} \hat{p}^{2n}(\theta) = \frac{1 - \phi(0)}{2N} \frac{(1 - \mu)^2 \hat{p}^2(\theta)}{1 - (1 - \mu)^2 \hat{p}^2(\theta)} \quad (5.12)$$

d=1, line

In this subsection, we will derive and generalize results from Section 3 of Maruyama (1970b) and Section 6 of Maruyama (1970c), which he obtained by diagonalizing matrices. Sawyer (1977) has obtained similar results.

Theorem 5.5. *Consider a stepping stone model on $\mathbf{Z}$ satisfying (5.11), and define the characteristic length scale by $\ell = (\nu \sigma^2 / 2\mu)^{1/2}$. The identity by descent satisfies*

$$\phi(0) \approx \frac{1}{1 + 4N(2\mu\nu\sigma^2)^{1/2}} \quad (5.13)$$

$$\phi(x) \approx \phi(0) \exp(-|x|/\ell) \quad (5.14)$$

The first conclusion generalizes the result in (5.2). Since $y = 2\mu/\nu\sigma^2$ is small, $\exp(-y|x|) \approx (1 - y)^{|x|}$ and in the nearest neighbor case, the second result reduces to (5.3). We call ℓ the characteristic length scale because it gives the scale on which exponential decay of ϕ occurs.

Proof. Using the inversion formula for Fourier transforms on **Z** gives

$$\phi(x) = \int_{-\pi}^{\pi} \hat{\phi}(\theta) e^{-i\theta x} \frac{d\theta}{2\pi}$$

$$= \frac{1 - \phi(0)}{2N} \int_{-\pi}^{\pi} \frac{(1-\mu)^2 \hat{p}^2(\theta) e^{-i\theta x}}{1 - (1-\mu)^2 \hat{p}^2(\theta)} \frac{d\theta}{2\pi} \qquad (5.15)$$

At first, it may look odd that the unknown $1 - \phi(0)$ is on the right-hand side. This problem is easily remedied. Setting $x = 0$ in (5.15), we have $\phi(0) = A(1 - \phi(0))$ where

$$A = \frac{1}{2N} \int_{-\pi}^{\pi} \frac{(1-\mu)^2 \hat{p}^2(\theta)}{1 - (1-\mu)^2 \hat{p}^2(\theta)} \frac{d\theta}{2\pi}$$

and solving we have $\phi(0) = (1 + 1/A)^{-1}$. In the nearest neighbor case with $p(0) = 1 - \nu$, $p(1) = p(-1) = \nu/2$ we have

$$\hat{p}(\theta) = 1 - \nu(1 - \cos\theta)$$

since $(e^{i\theta} + e^{-i\theta})/2 = \cos\theta$. However, there is no need to restrict our attention to the nearest neighbor case. Our assumptions imply

$$\hat{p}(\theta) = (1 - \nu) + \nu\hat{q}(\theta) = 1 - \nu[1 - \hat{q}(\theta)]$$

so we can write

$$4NA = \pi^{-1} \int_{-\pi}^{\pi} \frac{(1-\mu)^2(1 - \nu[1 - \hat{q}(\theta)])^2}{1 - (1-\mu)^2(1 - \nu[1 - \hat{q}(\theta)])^2} \, d\theta$$

Since q has mean 0, variance σ^2, and finite third moment

$$1 - \hat{q}(\theta) = \sigma^2 \theta^2 / 2 + O(\theta^3).$$

Thus if μ and θ are small, the denominator should be well approximated by

$$1 - (1 - 2\mu)(1 - \nu\theta^2\sigma^2) = 2\mu + \nu\theta^2\sigma^2$$

The numerator ≈ 1, so changing variables $\theta = (2\mu/\nu\sigma^2)^{1/2} t$ we have

$$4NA \sim \pi^{-1} \int_0^\infty \frac{1}{2\mu + 2\mu t^2} \left(\frac{2\mu}{\nu\sigma^2}\right)^{1/2} dx$$

$$= \frac{1}{(2\mu\nu\sigma^2)^{1/2}} \int_{-\infty}^\infty \frac{1}{\pi(1 + t^2)} \, dt = \frac{1}{(2\mu\nu\sigma^2)^{1/2}}$$

where to evaluate the integral we have used the fact that the Cauchy distribution integrates to 1 (see e.g., p. 43 of Durrett (2005)). Combining our calculations gives

$$\phi(0) \approx \frac{1}{1 + 4N(2\mu\nu\sigma^2)^{1/2}}$$

To compute $\phi(x)$ we let

$$I = \int_{-\pi}^{\pi} \frac{(1-\mu)^2 \hat{p}^2(\theta) e^{-i\theta x}}{1 - (1-\mu)^2 \hat{p}^2(\theta)} \frac{d\theta}{\pi}$$

Repeating the calculation of A and again changing variables $\theta = (2\mu/\nu\sigma^2)^{1/2} t$ we have

$$I = \frac{1}{(2\mu\nu\sigma^2)^{1/2}} \int_{-\infty}^{\infty} \frac{1}{\pi(1+t^2)} \exp(-i(2\mu/\nu\sigma^2)^{1/2} xt) \, dt$$

$$= \frac{1}{(2\mu\nu\sigma^2)^{1/2}} \exp(-(2\mu/\nu\sigma^2)^{1/2}|x|)$$

by the formula for the Fourier transform of the Cauchy distribution. Now (5.15) implies

$$\phi(x) = \frac{1 - \phi(0)}{4N} I = \frac{1}{1 + 4N(2\mu\nu\sigma^2)^{1/2}} \exp(-(2\mu/\nu\sigma^2)^{1/2}|x|)$$

so we have

$$\phi(x) \approx \phi(0) \exp(-(2\mu/\nu\sigma^2)^{1/2}|x|)$$

and using $\ell = (\nu\sigma^2/2\mu)^{1/2}$ we have the second conclusion.   □

The results for identity by descent give us information about hitting times. Now, in order for the two lineages to coalesce, they must first come to the same colony at time T_0 and then coalesce at time t_0.

Theorem 5.6. *In the one dimensional stepping stone model on* **Z**, *if the migration rate ν and variance σ^2 are fixed, then as $|x| \to \infty$, $2\nu\sigma^2 T_0/|x|^2$ converges to the hitting time of 1 for a standard Brownian motion.*

Proof. Since T_0 and $t_0 - T_0$ are independent.

$$\phi(x) = E_x(1-\mu)^{2T_0} \cdot E_0(1-\mu)^{2t_0}$$

Since the second term is $\phi(0)$, comparing with (5.14) and recalling $\ell = (\nu\sigma^2/2\mu)^{1/2}$ shows

$$E_x(1-\mu)^{2T_0} \approx \exp(-(2\mu/\nu\sigma^2)^{1/2}|x|)$$

Now if μ is small $(1-\mu)^{2T_0} \approx \exp(-\mu[2t_0])$, so letting $\mu = \lambda\nu\sigma^2/|x|^2$ and $|x| \to \infty$ we have

$$E_x(\exp(-\lambda[2\nu\sigma^2 t_0/|x|^2]) \to \exp(-\sqrt{2\lambda})$$

Consulting (4.4) on page 391 of Durrett (2005) we see that the right-hand side is the Laplace transform of the hitting time of 1 for Brownian motion, and the proof is complete.   □

d = 2, plane

We will use our Fourier methods to derive asymptotics that are a mixture of Nagylaki (1974), Sawyer (1977), and Slatkin and Barton (1989).

Theorem 5.7. *Consider a stepping stone model on* $\mathbf{Z}^2$ *satisfying (5.11), and define the characteristic length scale by* $\ell = (\nu\sigma^2/2\mu)^{1/2}$. *The identity by descent satisfies*

$$\phi(0) \approx \left[1 + \frac{2\pi\nu\sigma^2 \cdot 2N}{\ln(\ell)}\right]^{-1} \tag{5.16}$$

$$\phi(x) \approx \frac{\phi(0)}{\ln(\ell)}[K_0(|x|/\ell) - K_0(|x|)] \tag{5.17}$$

where K_0 *is the modified Bessel function of the second kind of order 0.*

The Bessel function is defined by

$$K_0(x) = \int_0^\infty \cos(tx)/\sqrt{t^2 + 1}\, dt$$

However, for our purposes, it is more useful to know that

$$K_0(x) \approx \begin{cases} \ln(1/x) & \text{when } x \text{ is small} \\ e^{-x}\sqrt{\pi/2x} & \text{when } x \text{ is large} \end{cases} \tag{5.18}$$

Formula (5.17) is often given without the second factor $-K_0(|x|)$ (see, e.g., Slatkin and Barton 1989, page 1353), but that term is very important when $|x|$ is small. To see this, let $|x| \to 0$ in (5.17) and use the first formula in (5.18) to see that if we ignore the second term then we get the nonsensical result $\phi(x) \to \infty$. (Recall that $\phi(x)$ is a probability and hence must always be ≤ 1.) With the second term present

$$K_0(|x|/\ell) - K_0(|x|) \to \ln(\ell)$$

as $|x| \to 0$, and we have $\phi(x) \to \phi(0)$.

Proof. Using the inversion formula for Fourier transforms on $\mathbf{Z}^2$ gives

$$\phi(x) = \int_{[-\pi,\pi]^2} e^{-i\theta \cdot x} \hat{\phi}(\theta)\, \frac{d\theta}{4\pi^2}$$

Using this on (5.12), and letting $R(\theta) = \hat{p}^2(\theta)$, we have

$$\phi(x) = \frac{1 - \phi(0)}{2N} \int_{[-\pi,\pi]^2} \frac{(1-\mu)^2 R(\theta) e^{-i\theta \cdot x}}{1 - (1-\mu)^2 R(\theta)} \frac{d^2\theta}{4\pi^2} \tag{5.19}$$

Noting that $R(\theta)$ is the Fourier transform of a distribution with mean 0 and covariance $2\nu\sigma^2 I$, leads to (see, e.g., (3.8) on page 103 of Durrett 2005)

$$R(\theta) = 1 - \nu\sigma^2 |\ell|^2 + O(|\ell|^3)$$

Here, we are using the finite third moment to control the error term. Arguing as in $d = 1$, we define

$$A \equiv \frac{1}{2N} \int_{[-\pi,\pi]^2} \frac{R(\theta)}{1 - (1 - \mu)^2 R(\theta)} \frac{d^2\theta}{4\pi^2}$$

$$\approx \frac{1}{2N} \int_{[-\pi,\pi]^2} \frac{1}{2\mu + \nu\sigma^2 |\theta|^2} \frac{d^2\theta}{4\pi^2}$$

Changing variables $\theta = t/\ell$ where $\ell = (\nu\sigma^2/2\mu)^{1/2}$, we have

$$= \frac{1}{2N \cdot 2\mu} \cdot \frac{2\mu}{\nu\sigma^2} \int_{[-\pi\ell,\pi\ell]^2} \frac{1}{1 + |t|^2} \frac{d^2 t}{4\pi^2}$$

Shifting to polar coordinates the above

$$\approx \frac{1}{2N \cdot 2\mu} \cdot \frac{2\mu}{\nu\sigma^2} \int_0^{\pi\ell} \frac{r}{1 + r^2} \frac{dr}{2\pi}$$

Since the integrand $\sim 1/r$ for large r, it follows that

$$A \approx \frac{1}{2\pi\nu\sigma^2 \cdot 2N} \ln(\ell)$$

(5.16) now follows from the fact that $\phi(0) = (1 + 1/A)^{-1}$.

To compute $\phi(x)$, we let

$$I = \int_{[-\pi,\pi]^2} \frac{(1 - \mu)^2 R(\theta) e^{-i\theta \cdot x}}{1 - (1 - \mu)^2 R(\theta)} \frac{d\theta}{4\pi^2}$$

Repeating the calculation of A, and again changing variables $\theta = t/\ell$ with $\ell = (\nu\sigma^2/2\mu)^{1/2}$ we have

$$I \approx \frac{1}{\nu\sigma^2} \int_{[-\pi\ell,\pi\ell]^2} \frac{1}{1 + |t|^2} \exp(-ix \cdot t/\ell) \frac{d^2 t}{4\pi^2}$$

If we integrate over the whole space we get ∞, so inspired by Nagylaki (1974), we write the above as

$$\approx \frac{1}{2\pi\nu\sigma^2} \int \int \left(\frac{1}{1 + |t|^2} - \frac{1}{\ell^2 + |t|^2} \right) \exp(-ix \cdot t/\ell) \frac{d^2 t}{2\pi}$$

The reason for introducing the second term is that when $|t| \geq \ell$ the difference is $\leq \ell^2/|t|^4$ so it gets rid of the divergence. The new term is $\leq 1/\ell^2$ so it reduces the integral over $[-\pi\ell, \pi\ell]^2$ by at most 2π when $x = 0$. The quantity above is

$$= \frac{1}{2\pi\nu\sigma^2}[K_0(|x|/\ell) - K_0(|x|)]$$

the last bit of calculus coming from page 323 of Malécot (1967). In words, "the bidimensional Fourier transform of $(m^2 + x^2)^{-1}$ is $K_0(m|\xi|)$." To finish up we note that

$$\phi(x) = \frac{1 - \phi(0)}{2N}I = \frac{\phi(0)}{2NA}I \approx \frac{\phi(0)}{\ln(\ell)}[K_0(|x|/\ell) - K_0(|x|)]$$

which proves (5.17). □

5.3 d = 2, Coalescence times

In this section we will prove results for the coalescence time of a sample of size 2 from the two dimensional stepping stone model on the torus $\Lambda(L) = (\mathbf{Z} \bmod L)^2$ due to Cox and Durrett (2002). In contrast to the anlytical methods in the previous section, the tools will be probabilistic and will give a more intuitive picture of the coalescence time.

We will consider the Moran model in which each individual is replaced at rate 1. Ignoring mutations for the moment, with probability $1 - \nu$ an individual is replaced by a copy of an individual in the colony in which it resides. With probability ν it is replaced by a copy of one chosen at random from a nearby colony $y \neq x$ with colony y being chosen with probability $q(y - x)$, where the difference is computed componentwise modulo L.

•   Throughout this section, we assume that q satisfies the symmetry and irreducibility assumptions stated in (5.11), but we strengthen the finite third moment assumption to finite range.

To study the behavior of the stepping stone model, we will work backwards in time using a special case of the structured coalescent introduced in Section 4.5. Lineages in colony x jump at rate $1 - \nu$ to a randomly chosen individual within the colony and at rate $\nu q(x, y)$ to a randomly chosen individual in colony y. Consider first the genealogy of a sample of size 2 chosen at random from the population. As we work backwards, let T_0 be the amount of time required until the two lineages first reside in the same colony, and let t_0 be the total amount of time needed for the two lineages to coalesce to one.

5.3.1 Random walk results

We begin by considering T_0. Let X_t be the difference in the colony numbers of the two lineages (computed modulo L) at time t. Since the two sampled individuals were chosen randomly from the population, the distribution of X_0

is the uniform distribution on the torus $\Lambda(L)$, which we denote by π. Let P_π denote the distribution of the process $\{X_t, t \geq 0\}$ when X_0 has distribution π. The first thing to observe is that while the migration rate ν is very important for properties of the stepping stone model, it is a harmless time change for the torus random walk, X_t. Jumps occur at rate 2ν, so scaling time by 2ν we reduce to the case in which jumps occur at rate 1.

If X_t were a random walk on $\mathbf{Z}^2$, its distance from X_0 at time t would be $O(\sqrt{2\nu t})$. Here O, read as "of order," means we are giving the order of magnitude ignoring constants. This result on the movement of the random walk implies that random walkers starting from two randomly chosen locations on the torus will take at least $O(L^2/2\nu)$ units of time to hit. The next result shows that the amount of time required for T_0 is actually $O(L^2 \log L/2\nu)$.

Theorem 5.8. *For any $t > 0$, as $L \to \infty$,*

$$P_\pi\left(T_0 > \frac{L^2 \log L}{2\pi\nu\sigma^2} t\right) \to e^{-t}$$

Proof. The proofs of this and the other results in this section are too technical to give the details, so we will only describe the intuition behind the results. To begin to explain the answer, note that $P_\pi(X_t = 0) = 1/L^2$, so the expected amount of time the two lineages chosen at random are in the same colony up to time L^2 is

$$\int_0^{L^2} P_\pi(X_t = 0)\, dt = 1$$

The next step is to note that since jumps have variance σ^2 and occur at rate 2ν the local central limit theorem implies that for large t

$$P_0(X_t = 0) \sim \frac{1}{2\pi\sigma^2(2\nu t)}$$

where $a(t) \sim b(t)$ means $a(t)/b(t) \to 1$. Thus the expected amount of time the two lineages chosen from the colony at 0 are in the same colony up to time L^2 is

$$\int_0^{L^2} dt\, P_0(X_t = 0) \sim \frac{\log(L^2)}{2\pi(2\nu\sigma^2)} \equiv I$$

where $\equiv$ indicates that the last equality defines I. At first the reader might worry that we get ∞ from the lower limit. This is not a problem since $P_0(X_t = 0) \leq 1$. A second issue is that the asymptotic formula for $P_0(X_t = 0)$ does not hold for small t. However, the right-hand side $\to \infty$, so these values do not matter in the limit.

At this point we have shown:

- the expected amount of time the two lineages chosen at random are in the same colony up to time L^2 is 1

- if the two lineages are in the same colony then the expected amount of time they are in the same colony in the next L^2 units of time is I

As we will now explain, it follows from this that

$$P_\pi(T_0 \leq L^2) \approx \frac{1}{I} = \frac{2\pi\nu\sigma^2}{\log L} \tag{5.20}$$

To argue this formally, we note that by breaking things down according to the value of T_0

$$1 = \int_0^{L^2} P_\pi(X_t = 0)\, dt$$

$$= \int_0^{L^2} ds\, P_\pi(T_0 = s) \int_0^{L^2 - s} dt\, P_0(X_t = 0)$$

Replacing $L^2 - s$ by L^2 in the upper limit of the integral (which can be justified mathematically) the above implies

$$1 \approx P_\pi(T_0 \leq L^2) \int_0^{L^2} dt\, P_0(X_t = 0) \approx I P_\pi(T_0 \leq L^2)$$

To prove that the limit is exponential, Cox and Durrett (2002) show that X_t comes to equilibrium in time $o(L^2 \log L/\nu)$, so the limit distribution of $\tau = T_0/(L^2 \log L/\nu)$ must have the lack of memory property

$$P(\tau > s + t | \tau > s) = P(\tau > t)$$

that characterizes the exponential. □

Sampling at random from the entire population is mathematically convenient. However, in many genetic studies, samples are not taken from the population as a whole. For example, one of the samples in Sabeti et al. (2002) consists of 73 Beni individuals who are civil servants in Benin City, Nigeria. To model this type of local sample in our context, we assume that the n chromosomes are sampled at random from a $L^\beta \times L^\beta$ region. To begin to analyze this situation, let P_x denote the law of the difference of two walks when one starts in colony 0 and one in colony x. (If $x = 0$, we pick two distinct individuals from colony 0.)

Theorem 5.9. *If* $\lim_{L \to \infty} (\log^+ |x_L|)/\log L = \beta \in [0, 1]$, *then*

$$P_{x_L}\left(T_0 > \frac{L^{2\gamma}}{4\pi\nu\sigma^2}\right) \to \frac{\beta}{\gamma} \quad \text{for } \beta \leq \gamma \leq 1$$

$$P_{x_L}\left(T_0 > \frac{L^2 \log L}{2\pi\nu\sigma^2} t\right) \to \beta e^{-t}$$

The positive part $z^+ = \max\{z, 0\}$ is here in the formula so that $\log^+ 0 = 0$ not $-\infty$. The second result shows that the probability that a coalescence occurs at a time $\ll L^2 \log L$ is $\approx 1 - \beta$ and that when this does not occurs the limiting time is exponential as in Theorem 5.8. To explain the first result, we note that (i) by our explanation of the second result, we must get β when $\gamma = 1$, and (ii) since the random walk takes time $O(L^{2\beta}/\sigma^2\nu)$ to move a distance L^β, it is natural that the probability of a coalescence after this time tends to 1, which is the value for $\gamma = \beta$. The scaling in the first result is a little different from the one in Zähle, Cox, and Durrett (2005), but as we will see at the end of the next section, it makes things work a little better.

Proof. To derive the first result, we note that since the standard deviation of $X_t - X_0$ is $\sqrt{2\nu\sigma^2 t}$, hitting 0 before time $L^{2\beta}/(4\pi\nu\sigma^2)$ is unlikely, so the local central limit theorem implies that

$$\int_0^{L^{2\gamma}/(4\pi\nu\sigma^2)} P_{x_L}(X_t = 0)\, dt \approx \int_{L^{2\beta}/(4\pi\nu\sigma^2)}^{L^{2\gamma}/(4\pi\nu\sigma^2)} \frac{1}{2\pi(2\nu\sigma^2 t)}\, dt$$
$$= \frac{2(\gamma - \beta) \log L}{4\pi\nu\sigma^2}$$

The local central limit theorem implies that if X_t hits 0 by time $L^{2\gamma}/(4\pi\nu\sigma^2)$ it will spend an average of

$$\approx \int_0^{L^{2\gamma}/(4\pi\nu\sigma^2)} \frac{1}{2\pi\sigma^2(2\nu t)}\, dt = \frac{2\gamma \log L}{4\pi\nu\sigma^2}$$

time units at 0, so by the computation in the proof of (5.20),

$$\frac{2\gamma \log L}{4\pi\nu\sigma^2} P_{x_L}\left(T_0 \leq \frac{L^{2\gamma}}{4\pi\nu\sigma^2}\right) \approx \frac{2(\gamma - \beta) \log L}{4\pi\nu\sigma^2}$$

and it follows that

$$P_{x_L}\left(T_0 \leq \frac{L^{2\gamma}}{4\pi\nu\sigma^2}\right) \approx 1 - \frac{\beta}{\gamma}$$

To derive the second result from the first with $\gamma = 1$, we use the local central limit theorem to show that when $T_0 > L^2/(4\pi\nu\sigma^2)$, it is also likely that $T_0 > L^2\sqrt{\log L}/\nu\sigma^2$, at which time the distribution of X_t has become uniform over the torus, and the longer time behavior is as in Theorem 5.8. $\square$

5.3.2 Samples of size 2

Getting the two lineages to the same colony at time T_0 is only the first part of the coalescence time t_0. Since the two lineages are in the same colony at time T_0, to study $t_0 - T_0$, we need only be concerned with the distribution of t_0 under P_0. Since the stepping stone model is a special case of the general matrix migration model in which the colony size is constant and the migration symmetric, the next result follows from Theorem 4.10, and the fact that the difference of the two lineages jumps twice as fast in the Moran model.

Theorem 5.10. $E_0 t_0 = NL^2$.

This result, like the others in this section, is for the stepping stone model on the two dimensional torus $\Lambda(L)$ with dispersal distribution satisfying the conditions announced at the beginning of the section.

Comparing Theorems 5.8 and 5.10, we see that there are two extreme possibilities:

$$ET_0 = O(L^2 \log L / \nu) \ll O(NL^2) = Et_0 \quad \text{or} \quad ET_0 \gg Et_0$$

In the first case, the two lineages will come to the same colony in $o(NL^2)$ so the actual starting positions of the particles don't matter, and the limit distribution will have the lack of memory property. This next result is the strong migration limit for the stepping stone model.

Theorem 5.11. *If* $\lim_{L \to \infty} N\nu / \log L = \infty$, *then for any* $t > 0$, *as* $L \to \infty$,

$$\sup_{x \in \Lambda(L)} |P_x(t_0 > NL^2 t) - e^{-t}| \to 0$$

Conventional wisdom, see pages 125–126 of Kimura and Maruyama (1971), says that "marked local differentiation of gene frequencies can occur if $N\nu < 1$ where N is the effective size of each colony and ν is the rate at which each colony exchanges individuals with the four surrounding colonies." In the other direction, "if $N\nu > 1$ local differentiation is less pronounced and especially if $N\nu \geq 4$, the whole population tends to behave as a panmictic unit." As Theorem 5.11 and the next result show, $N\nu$ must be much larger than $\log L$ in order for the system to behave as if it were homogeneously mixing.

Theorem 5.12. *Suppose* $\lim_{L \to \infty} 2N\nu\pi\sigma^2 / \log L = \alpha \in [0, \infty)$. *If* x_L *satisfies* $\lim_{L \to \infty} (\log^+ |x_L|) / \log L = \beta \in [0, 1]$ *then, as* $L \to \infty$,

$$P_{x_L}\left(t_0 > (1 + \alpha)\frac{L^2 \log L}{2\pi\nu\sigma^2} t\right) \to \left(\beta + (1 - \beta)\frac{\alpha}{1 + \alpha}\right) e^{-t}$$

If we pick X_0 *at random according to* π, *the assumption holds with* $\beta = 1$ *so*

$$P_\pi\left(t_0 > (1 + \alpha)\frac{L^2 \log L}{2\pi\nu\sigma^2} t\right) \to e^{-t}$$

Proof. By the reasoning for Theorem 5.9, the probability that two lineages will enter the same colony before time $L^2 / 4\pi\nu\sigma^2$ is $\approx 1 - \beta$. Using the computation in Theorem 5.8 one can show that when they do, they will be in the same colony a geometrically distributed number of times with mean $(\log L) / \pi\nu\sigma^2$ before time $L^2 / 4\pi\nu\sigma^2$ and hence the probability of coalescence before time $L^2 / 4\pi\nu\sigma^2$ is approximately

$$\frac{1/2N}{1/(2N) + \pi\nu\sigma^2 / \log L} \to \frac{1}{1 + \alpha}$$

which gives the desired result. □

Effective population size

If $\alpha > 0$ we can use the fact that $\log L/(2N\nu\pi\sigma^2) \to 1/\alpha$ to write the second conclusion in Theorem 5.12 as

$$P_\pi\left(t_0 > NL^2\left(\frac{1+\alpha}{\alpha}\right)t\right) \to e^{-t} \tag{5.21}$$

Recalling that we are considering the Moran model in which time is scaled by the population size (without a factor of 2),

Theorem 5.13. *For the two dimensional stepping stone models on $\Lambda(L)$ considered in this section, the nucleotide diversity effective population size,*

$$N_e^\pi = NL^2\left(1 + \frac{1}{\alpha}\right)$$

This has the same form as N_e^π in the island model when the number of colonies $n = L^2$ is large and the scaled migration rate $M = \alpha$, see (4.16). Note that the stepping stone scaled migration rate

$$\alpha = 2N\nu\sigma^2 \cdot \frac{\pi}{\log L}$$

combines N, ν, and σ^2 into one composite parameter, compared to $M = 4Nm$ for the Wright-Fisher version of the island model.

Example 5.1. To illustrate the use of the formula for the effective population size, consider $L = 10$ with migration to the four nearest neighbors with equal probability so $\sigma^2 = 0.5$. If the local population size $N = 25$ and we choose $\nu = 0.1$ so that $4N\nu = 1$ then

$$\alpha = \frac{\pi(0.25)}{\log 10} = 0.341 \quad \text{and} \quad N_e^\pi = \frac{2500(1.341)}{0.341} = 9829$$

versus the actual population size of 2500.

5.3.3 Fixation indices F_{ST}

Using Theorems 5.10 and 5.12 with Slatkin's approximation given in (4.22), $F_{ST} \approx (\bar{T} - \bar{T}_0)/\bar{T}$, it follows that if $2\pi\sigma^2 N\nu/\log L \to \alpha \in (0, \infty)$ then

$$F_{ST} \approx \frac{\frac{L^2 \log L}{2\pi\nu\sigma^2}}{\frac{L^2 \log L}{2\pi\nu\sigma^2} + NL^2} \approx \frac{1}{1+\alpha} \tag{5.22}$$

This says that F_{ST} is close to 0 if and only if $N\nu \gg \log L$, the condition in (5.11) for the system to be homogeneously mixing. Crow and Aoki (1984) did a numerical study of F_{ST} for the nearest neighbor stepping stone model and found (see their page 6075) that F_{ST} is roughly proportional to

$\log n$, where $n = L^2$ is the number of colonies. In the cases they considered, $\log L/(2\pi\nu\sigma^2) \ll N$, so the first term in the denominator of (5.22) can be ignored, and we have

$$F_{ST} \approx \frac{\log L}{2N\pi\nu\sigma^2} = \frac{1}{4N\pi\nu\sigma^2}\log(L^2)$$

confirming their prediction.

As noted in (4.24) one can estimate the scaled migration rate $M = N\nu$ in the island model by

$$\widehat{M} = \frac{1}{4}\left(\frac{1}{F_{ST}} - 1\right)$$

Suppose the population being sampled has a stepping stone structure, but one uses the island model formula. Using (5.22) and $2N\pi\sigma^2\nu/\log L \approx \alpha$, we find that

$$\widehat{M} \approx \frac{\alpha}{4} \approx N\nu\sigma^2 \cdot \frac{\pi}{2\log L} \tag{5.23}$$

so the resulting estimate $\widehat{M}$ has a bias that depends on the size of the system. However, this result has no impact on the work of Seielstad et al. (1998) discussed in Section 4.5 since they used the ratio of estimates from mtDNA and Y chromosome data and when this is done the extra factor $\pi/2\log L$ cancels out.

Decay of F_{ST} with distance

If we compare two colonies that are separated by a displacement of x in the stepping stone model then, as in Section 4.7, we define

$$F_{ST}(x) = \frac{\bar{T}_x - \bar{T}_0}{\bar{T}_x + \bar{T}_0}$$

where $\bar{T}_x$ is the average coalescence time of two lineages sampled from colonies separated by a displacement of x. As in Rousset (1997) it is convenient to consider

$$\frac{F_{ST}(x)}{1 - F_{ST}(x)} = \frac{\bar{T}_x - \bar{T}_0}{2\bar{T}_0} \tag{5.24}$$

Slatkin (1991) considers 1 over this quantity because of its relationship to the migration estimate $\hat{M}$ given in (4.24), but as the reader will see, (5.24) is more convenient.

If we have a ring of L colonies with nearest neighbor migration at rate ν then by Theorem 4.10 we have $\bar{T}_0 = NL$ in the Moran model. The numerator in (5.24) is the amount of time that it takes for two lineages to come to the same colony. In Section 7.6 we will compute that $\bar{T}_x - \bar{T}_0 = (L - x)x/2\nu$, so

$$\frac{F_{ST}(x)}{1 - F_{ST}(x)} = \frac{x(L - x)}{4NL\nu}$$

When $x \ll L$ this is linear in agreement with Figure 1 in Rousset (1997). Remembering the factor of 2 difference in the Moran and stepping stone models, this agrees with formula (17) in Slatkin (1991).

If we have an $L \times L$ torus with migration at rate ν that has variance σ^2 then in the Moran model $\bar{T}_0 = NL^2$. Using Theorem 5.9 we see that

$$\bar{T}_x - \bar{T}_0 \sim (\log^+ |x|)L^2 \log L / (2\pi\nu\sigma^2)$$

so we have

$$\frac{F_{ST}(x)}{1 - F_{ST}(x)} = \frac{\log^+ |x| \log L}{4\pi\nu\sigma^2 N}$$

This is linear in $\log^+ |x|$ in agreement with Figure 2 in Rousset (1997) and is considerably simpler than the formula in (19b) of Slatkin (1991). Slatkin (1993) has done extensive simulations. Unfortunately he chose to plot $\log(\hat{M})$ versus $\log |x|$.

5.4 d = 2, Genealogies

It is not hard to extend the argument for (5.21) to show that if we sample n individuals at random from the population, then when time is written in units of $2N_e^\pi$ generations, the genealogy is that of Kingman's coalescent. Hence, for a random sample $N_e^c = N_e^\pi$.

The main result of Zähle, Cox, and Durrett (2005) shows that for a random sample from an $L^\beta \times L^\beta$ subregion after a nonlinear change of time scale the coalescent in the stepping stone model reduces to Kingman's coalescent. We begin with a result for the small time behavior of the coalescence time in the stepping stone model on the torus. We continue using the assumptions made at the beginning of the previous section.

Theorem 5.14. *If* $\lim_{L \to \infty} (\log^+ |x_L|) / \log L = \beta \in [0,1]$ *then for* $\beta \leq \gamma < 1$ *we have*

$$P_{x_L}\left(t_0 > \frac{L^{2\gamma}}{4\pi\nu\sigma^2}\right) \to \frac{\beta + \alpha}{\gamma + \alpha}$$

To make the connection with Theorem 5.12 note that

$$\beta + (1 - \beta)\frac{\alpha}{1 + \alpha} = \frac{\beta + \beta\alpha + \alpha - \alpha\beta}{1 + \alpha} = \frac{\beta + \alpha}{1 + \alpha}$$

which is the answer for $\gamma = 1$.

Proof. The first result in Theorem 5.9 implies that

$$P_x\left(T_0 > \frac{L^{2\gamma}}{4\pi\nu\sigma^2}\right) \to \frac{\beta}{\gamma}$$

Generalizing the computation in the proof of Theorem 5.12 one can show that when $T_0 \leq L^{2\gamma}/4\pi\nu\sigma^2$, the probability of coalescence before time $L^{2\gamma}/4\pi\nu\sigma^2$ is approximately

$$\frac{1/2N}{1/(2N) + \pi\nu\sigma^2/\log(L^\gamma)} \to \frac{\gamma}{\gamma + \alpha}$$

A little algebra gives

$$\frac{\beta}{\gamma} + \left(1 - \frac{\beta}{\gamma}\right)\frac{\alpha}{\gamma + \alpha} = \frac{\beta(\gamma + \alpha) + (\gamma - \beta)\alpha}{\gamma(\gamma + \alpha)} = \frac{\beta + \alpha}{\gamma + \alpha} \qquad \square$$

Let $h_L = (1+\alpha)(L^2 \log L)/(2\pi\nu\sigma^2)$ be the effective population size, which gives the time scale at large times.

Theorem 5.15. *If we change time so that*

$$\frac{L^{2\gamma}}{4\pi\nu\sigma^2} \to \log\left(\frac{\gamma + \alpha}{\beta + \alpha}\right) \qquad \text{for } \beta \leq \gamma \leq 1$$

$$\frac{L^2}{4\pi\nu\sigma^2} + h_L t \to \log\left(\frac{1 + \alpha}{\beta + \alpha}\right) + t$$

then the genealogy of a sample of size n chosen at random from an $L^\beta \times L^\beta$ region in stepping stone model on the two dimensional torus reduces to Kingman's coalescent.

Proof. It is straightforward to generalize an argument of Cox and Griffeath (1986) to show that when there are k lineages, the $\binom{k}{2}$ possible coalescences are essentially independent events, see pages 692–693 in Zähle, Cox, and Durrett (2005). The result then follows from the results for samples of size 2 in Theorems 5.12 and 5.14. $\qquad \square$

To get a feel for the time change note that if $s = L^{2\gamma}/(4\pi\nu\sigma^2)$ then

$$\gamma = \frac{\log(4\pi\nu\sigma^2) + \log(s)}{2 \log L}$$

Since $d\gamma/ds = 1/(2s \log L)$, the rate of coalescence at time s is

$$\frac{d}{ds} \log\left(\frac{\gamma + \alpha}{\beta + \alpha}\right) = \frac{1}{\gamma + \alpha} \cdot \frac{1}{2s \log L} \tag{5.25}$$

To see why we divide by $4\pi\nu\sigma^2$ in the small time regime, note that when $s = L^2/(4\pi\nu\sigma^2)$, $\gamma = 1$ and the coalescence rate becomes

$$\frac{2\pi\nu\sigma^2}{(1+\alpha)(L^2 \log L)} = \frac{1}{h_L}$$

the coalescence rate at large times, and the overall coalescent rate is continuous.

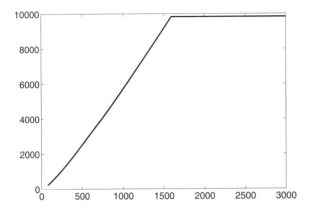

Fig. 5.1. 1 over the coalescence rate in Example 5.1, a 10×10 stepping stone model with colony size 25 and nearest neighbor migration, which has $N_e = 9829$.

Note that the coalescence rate given in (5.25) is decreasing in the first phase and occurs much more rapidly in the beginning. Inspired by Figure 4 of Wilkins (2004), we have graphed 1 over the coalescence rate versus time in our figure to show that 1 over the rate is roughly linear in time. This can be seen intuitively by noting that at time s the central limit theorem says that lineages will be spread over a region with radius $\sqrt{s}$ and hence area s, so the effective population size is of order s.

5.4.1 Simulation results

While Theorem 5.15 gives the limiting behavior of genealogies in the stepping stone model, it is not easy to use that result to compute quantities of interest for the limiting process. Thus to get numerical results we must turn to simulation. De and Durrett (2007) have done this. We begin with results for the site frequency spectrum. Because the coalescence rate is faster initially and then slows down to the usual rate, we should expect a reduction in the number of rare alleles. The graph on the left in figure 5.2 shows results from 350,000 simulations of a sample of size 40 taken from one subpopulation in (i) an island model with 100 colonies of size $N = 50$ ($\times$'s) and (ii) a 10×10 stepping stone model with colonies of size $N = 25$ (+'s) with scaled migration rate $4N\nu = 1$. The values are chosen to make the effective population size computed from (4.16) for the island model or from (5.13) for the stepping stone model close to 10,000. One can see that the island model and the stepping stone model both show the predicted skew compared to the prediction of $1/x$ for a homogeneously mixing population. In contrast, simulations in De and Durrett (2007) show that if we sample at random from the whole grid using the same parameters then the three site frequency spectra are almost identical. For the stepping stone model this confirms the result mentioned at the beginning of this section. For the island model this is predicted by Wakeley's many demes limit, Theorem 4.15.

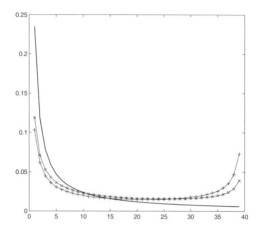

Fig. 5.2. Site frequency spectrum in the stepping stone ($\times$'s) and island ($+$'s) models, compared with the homogeneously mixing case for a sample from one colony.

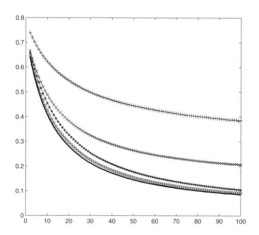

Fig. 5.3. Decay of r^2 in the stepping stone ($\times$'s) and island ($+$'s) models for samples from one subpopulation or at random compared with the homogeneously mixing case. The x-axis gives the distance in kilobases.

In the stepping stone model and the island model, the fact that the co-alescence rate is faster initially makes the decay of the linkage disequilibrium slower along a chromosome. The graph in Figure 5.3 shows results from 350,000 simulations of a sample of size 40 taken from (i) an island model with 100 colonies of size $N = 91$ ($\times$'s) and (ii) a 10×10 stepping stone model with colonies of size $N = 77$ (+'s) with scaled migration rate $4N\nu = 10$. Again the values are chosen to make the effective population sizes roughly 10,000. In the top two curves the sampled individuals are all from one subpopulation. Even though the migration rate is large, the decay of r^2 is much slower than in a homogeneously mixing population (solid line). The two other curves are for random samples and give results similar to the homogeneously mixing case.

5.5 d = 1, Continuous models

The title of this section is a little misleading since, as the reader will see, space is not continuous, but is instead a fine lattice. The word "continuous" refers to the fact that individuals are no longer organized into subpopulations. Wilkins and Wakeley (2002) considered a one dimensional stepping stone model in which there is one haploid individual at $0, 1/n, 2/n, \ldots 1$. Each individual produces a large number of offspring that are distributed according to a normal distribution with variance σ_m^2 (m is for migration) centered at the location of the individual and reflecting boundaries at the ends of the habitat. At each site one of these offspring is selected at random to become the adult at that location at the next generation.

The motivation for this research came from Bowen and Grant (1997) who collected data from the mitochondrial control region in five different sardine populations in five temperate upwelling zones off the coasts of Japan, California, Chile, Australia, and South Africa. Their one dimensional universe consists of temperate waters that connect the five regions. Rather than divide the ocean artificially into local populations they introduced the continuous model described above.

Wilkins and Wakeley (2002) obtained some analytic results that we will discuss later, and presented simulation results for $n = 100$ with $\sigma_m^2 = 0.0001$, 0.001 and 0.01. Seeing the three examples brings up the question: what relationship between n and σ_m^2 leads to interesting behavior? To address this question we will consider the one dimensional stepping stone model on the circle $\mathbf{Z}$ mod n. To be able to use Malecot's result given in (5.10), we will suppose that there is $N = 1$ diploid individual at each integer, for otherwise it would be impossible to define $\psi(0,0)$.

Theorem 5.16. *Consider the Wright-Fisher stepping stone model on $\mathbf{Z}$ mod n with dispersal distribution $p(0,x)$ a normal with mean 0 and variance σ^2 (rounded to the nearest integer). If $\sigma/\sqrt{n} \to \infty$ and $\sigma/n \to 0$ then the population behaves as if it is homogeneously mixing. That is, if t_0 is the coalescence* *time of two lineages sampled from colonies with a displacement of x, then $t_0/2n$ converges in distribution to a mean 1 exponential.*

We have assumed a normal distribution only to simplify the proof. As the next theorem will show, the exact form of the distribution is not important. The assumption $\sigma/n \to 0$ is a technicality needed for our proof. Matsen and Wakeley (2006) have shown that the conclusion holds if dispersal distance is uniform over $(-\sigma, \sigma)$ with $\sigma/n \to c > 0$.

Proof. Using the inversion formula on $\mathbf{Z}$ mod n with (5.12) gives

$$\phi(x) = \frac{1 - \phi(0)}{2n} \sum_{m=0}^{n-1} \frac{(1 - \mu)^2 \hat{p}^2 (2\pi m/n) e^{-i \cdot 2\pi m x/n}}{1 - (1 - \mu)^2 \hat{p}^2 (2\pi m/n)} \qquad (5.26)$$

which in the nearest neighbor case is (6.6) of Maruyama (1970c). To obtain the desired conclusion about the coalescence time we will take $\mu = \lambda/2n$. Since $\hat{p}(0) = 1$ the $m = 0$ in the term is

$$\frac{(1 - \mu)^2}{2\mu + \mu^2} \approx \frac{1}{2\mu}$$

If we can show that the other terms can be ignored, then we will have $\phi(x) \approx A(1 - \phi(0))$, with $A \approx 4N\mu$, so using the result first for $x = 0$ and then for a general x we have

$$\phi(x) \approx \frac{1}{1 + 4n\mu} = \frac{1}{1 + 2\lambda}$$

To obtain the desired conclusion we note that

$$\phi(x) = E_x \left(1 - \frac{\lambda}{2n} \right)^{2t_0}$$

so we have shown

$$E_x \exp(-2\lambda(t_0/2n)) \to \frac{1}{1 + 2\lambda}$$

which is the desired conclusion written in terms of Laplace transforms.

The normal distribution has $\hat{p}^2(\theta) = e^{-\theta^2 \sigma^2}$. Here we are ignoring the effect of rounding, which is valid if σ is large, and that dispersal occurs on the circle, which is valid by our assumption $\sigma/n \to 0$. If $\theta\sigma$ is small then $1 - e^{-\theta^2 \sigma^2} \approx \theta^2 \sigma^2$ so the denominator in the fraction in the sum can be approximated by

$$2\mu + \sigma^2 \left(\frac{2\pi m}{n} \right)^2 \qquad \text{when } m \le \epsilon n/\sigma$$

Since $\sigma^2/n \to \infty$ and $\mu = \lambda/2n$, the 2μ can be discarded for all $m \ge 1$, and we have

$$\sum_{m=1}^{\epsilon n/\sigma} \approx \frac{n^2}{(2\pi)^2\sigma^2} \sum_{m=1}^{\epsilon n/\sigma} \frac{1}{m^2} \leq C\frac{n^2}{\sigma^2} \ll \frac{1}{\mu}$$

By symmetry it suffices to now consider the sum from $\epsilon n/\sigma$ to $n/2$. Over this range $1 - (1 - \mu)^2(1 - e^{-(2\pi m)^2\sigma^2/n^2}) \geq \delta > 0$ so

$$\sum_{m=\epsilon n/\sigma}^{n/2} \leq \frac{1}{\delta} \sum_{m=\epsilon n/\sigma}^{n/2} e^{-(2\pi m)^2\sigma^2/n^2}$$

Changing variables $x = \sigma m/n$ gives a sequence of values with spacing σ/n, so comparing the sum with an integral gives that it is

$$\approx \frac{1}{\delta} \cdot \frac{n}{\sigma} \int_{\epsilon}^{\infty} e^{-(2\pi x)^2} \, dx$$

The integral is convergent, so we have

$$\sum_{m=\epsilon n/\sigma}^{n/2} \leq C\frac{n}{\sigma} \ll \frac{1}{\mu}$$

since $\sigma^2 \to \infty$, which completes the proof. $\square$

Our next goal is to show that the conclusion of Theorem 5.16 fails if $\sigma/\sqrt{n}$ has a finite limit. This implies that the interesting behavior happens when the range of the dispersal distribution is $O(\sqrt{n})$. In Wilkins and Wakeley's context, where the lattice spacing is $1/n$, this translates into a standard deviation $\sigma_m = O(1/\sqrt{n})$ or $\sigma_m^2 = O(1/n)$. In the three concrete examples mentioned above $n = 100$ so our $n\sigma_m^2$ is 0.01, 0.1 or 1.

For simplicity, Durrett and Restrepo (2007) considered the corresponding problem for haploid individuals on the integers $\mathbf{Z}$, with one individual sampled at 0 and one at $L_n = O(n)$. They used the Moran version of the stepping stone model in which each individual is replaced at rate 1, and when a replacement occurs at x, the new individual is a copy of the one at y with probability $q^n(y - x)$. They made the following assumptions about the dispersal distributions q^n:

1. symmetry: $q^n(z) = q^n(-z)$
2. the variance $\sum_{z\in\mathbf{Z}} z^2 q^n(z) = v_n^2 n$ with $v_n \to v \in (0,\infty)$
3. there is an $h > 0$, independent of n, so that $q^n(z) \geq h/\sqrt{n}$ for $|z| \leq n^{1/2}$
4. exponential tails: $q^n(z) \leq C\exp(-c|z|/\sqrt{n})$

These assumptions contain uniform, bilateral exponential, and normal distributions as special cases.

Theorem 5.17. *Let t_0 be the coalescence time of the individual at 0 and the one at L_n. If the positive numbers L_n have $L_n/vn \to x_0 \geq 0$ then $2t_0/n$ converges in distribution to $\ell_0^{-1}(v\xi/2)$, where ℓ_0 is the local time at 0 of a standard Brownian motion started from x_0 and ξ is independent with a mean 1 exponential distribution.*

It will take some time to explain what this result means. To begin, we note that if X_t^0 is the genealogy of the individual at 0 then $X^0(nt)/vn$ converges to a Brownian motion B_t. To explain the unusual scaling, note that $X^0(nt)$ has variance $nt \cdot v_n^2 n$, so the variance of $X^0(nt)/vn$ converges to t. Let X_t^1 be the genealogy of the individual at L_n. Until the two lineages coalesce the difference $(X^1(nt) - X^0(nt))/vn$ behaves like a Brownian motion starting at x_0 and run at rate 2. If $\delta > 0$ is fixed then in the limit $n \to \infty$ coalescence is not possible when $|X^1(nt) - X^0(nt)|/vn > \delta$, so it will only occur when the two lineages have spent enough time close to each other.

The local time of a Brownian motion at 0 is a measure of the amount of time B_t spends at 0. The mathematically simplest definition is that $\ell_0(t)$ is the increasing process that makes $|B_t| - \ell_0(t)$ a martingale. To see why this measures the time spent at 0, note that $|B_t|$, being a convex function of a martingale, is a submartingale. However $|x|$ is linear on $(-\infty, 0)$ and $(0, \infty)$ so $|B_t|$ is a martingale except on $\{t : B_t = 0\}$. This is a set of Lebesgue measure 0 but $\ell_0(t)$ provides a measure of the size of this set. Indeed, it is possible to define a family of local times $\ell_a(t)$ that measures the occupation time of a in such a way that for nice functions f (e.g., continuous and with compact support)

$$\int_0^t f(B_t)\,dt = \int f(a)\ell_a(t)\,da$$

A more concrete answer

To get a more explicit description of the distribution of the limits in Theorem 5.17 we would like to compute

$$P_x(\ell_0^{-1}(\xi/\lambda) > t) = P_x(\ell_0(t) < \xi/\lambda) = E_x \exp(-\lambda \ell_0(t))$$

Formula 1.3.7 in Borodin and Salaminen's (1996) tells us that

$$E_x(e^{-\lambda \ell_0(t)}; W_t \in dz) = \frac{1}{\sqrt{2\pi t}} e^{-(z-x)^2/2t}\,dz$$
$$- \frac{\lambda}{2} \exp((|z| + |x|)\lambda + \lambda^2 t/2)\mathrm{Erfc}\left(\frac{\lambda^2\sqrt{t}}{\sqrt{2}} + \frac{|z| + |x|}{\sqrt{2t}}\right)\,dz \quad (5.27)$$

where Erfc is the error function, i.e., the upper tail of the normal distribution.

Although this formula is explicit it is not easy to understand. Another approach to computing $u(t, x) = E_x \exp(-\lambda \ell_0(t))$, due to Maruyama (1971), see his (2.3), is to write a differential equation.

Lemma 5.1. $u(t, x)$ *satisfies the heat equation*

$$\frac{\partial u}{\partial t} = \frac{1}{2}\frac{\partial^2 u}{\partial x^2}$$

with the boundary condition $\frac{\partial u}{\partial x}(t, 0+) = \lambda u(t, 0)$.

Proof. When $B_t \neq 0$, $u(t, B_t)$ is a martingale so $u(t, x)$ satisfies the heat equation for $x \neq 0$. To determine the boundary condition at 0, we run Brownian motion until $\tau_h = \inf\{t : B_t \notin (-h, h)\}$ and use symmetry $u(t, x) = u(t, -x)$ to conclude that

$$u(t, 0) = E_0(e^{-\lambda \ell_0(\tau_h)} u(t - \tau_h, h); \tau_h \leq t) + O(P_0(\tau_h > t))$$

The strong Markov property implies that $\ell_0(\tau_h)$ is exponentially distributed. Let $D_\epsilon(\tau_h)$ be the number of downcrossings of $(0, \epsilon)$ by reflecting Brownian motion before it hits h. $D_\epsilon(\tau_h)$ is geometrically distributed with mean h/ϵ and $\lim_{\epsilon \to 0} \epsilon D_\epsilon(t) = \ell_0(t)$, see e.g., page 48 of Itô and McKean (1974), so $E_0 \ell_0(\tau_h) = h$ and

$$E_0(e^{-\lambda \ell_0(\tau_h)}) = \frac{1/h}{\lambda + 1/h} = \frac{1}{1 + \lambda h}$$

Using the explicit formula in (5.27) or the fact that $u(t, x)$ satisfies the heat equation with a bounded boundary condition on $[0, \infty) \times \{0\}$ shows $u(t, x)$ is Lipschitz continuous on $[0, T] \times [-K, K]$. Since τ_h has the same distribution as $h^2 \tau_1$, $|u(t - \tau_h, h) - u(t, h)| = O(h^2)$. Using this with $P_0(\tau_h > t) = o(h)$, we have

$$\frac{\partial u}{\partial x}(t, 0+) = \lim_{h \to 0} \frac{u(t, h) - u(t, 0)}{h}$$
$$= u(t, 0) \lim_{h \to 0} \frac{1 - E_0(e^{-\lambda \ell_0(\tau_h)})}{h} = \lambda u(t, 0)$$

which completes the proof.   □

Wilkins and Wakeley (2002) also take an approach using differential equations. They work on an interval $[0, 1]$ with reflecting boundary conditions at the ends. Suppose two lineages are sampled at $z_1(0)$ and $z_2(0)$. They are interested in computing the joint distribution $(z_1(t), z_2(t))$ on the event of no coalescence. They have the clever idea to change variables

$$x = 1 - |z_1 - z_2| \qquad y = z_1 + z_2 - 1$$

to map the problem into the triangle on the right side of the following square. $x = y$ corresponds to $z_1 = 1$ or $z_2 = 1$ while $x = -y$ corresponds to $z_1 = 0$ or $z_2 = 0$, so these boundaries have reflecting boundary conditions, and they can be removed by reflecting the function across the boundary lines to extend it to a function $U(x, y, t)$ defined on the square. To help explain this, we have indicated four points in the picture where the function will have the same value.

$x = 1$ corresponds to $z_1 = z_2$ the coalescing event, so Wilkins and Wakeley (2002) used the boundary condition:

$$-2\sigma^2 \frac{\partial U}{\partial x}(1, y, t) = \frac{1}{N} \frac{1}{\sqrt{2\pi(2\sigma^2)}} \int_0^1 U(1 - z, y, t) e^{-z^2/2(2\sigma^2)} \, dz$$

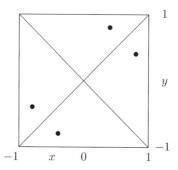

Fig. 5.4. Wilkins and Wakeley's change of variables.

To quote the authors: "In this equation, the flux across the boundary is set equal to the average probability over the next generation that the separation between the two lineages is $< 1/2N$." If we set $\sigma = c/\sqrt{N}$ then in the limit as $N \to \infty$ we have

$$-2c^2 \frac{\partial U}{\partial x} = U$$

another version of our boundary condition. Wilkins and Wakeley (2002) solved their equations numerically with an infinite sum of exponentially decaying sines and cosines and found good agreement with simulations.

5.6 d = 2, Continuous models

Wilkins (2004) has studied what he calls "isolation by distance in a continuous, two-dimensional habitat," i.e., using a two dimensional version of the model of Wilkins and Wakeley (2002) considered in the previous section. The results are similar to ones obtained earlier by Slatkin and Barton (1989), so we refer the treader to his paper for details.

Intuitively, if dispersal is uniform over an $L \times L$ square in the torus ($\mathbf{Z}$ mod n)^2 then the situation should be similar to a stepping stone model with colony size L^2 and migration probability $\nu = 1$. Matsen and Wakeley (2006) have considered the case in which $L/n \to c > 0$ and used results of Aldous (1989) and Diaconis and Stroock (1989) to show that the system behaves as if it is homogeneously mixing. Based on the analogy in the previous sentence, we should expect the result to be true under the much weaker assumption $l/\sqrt{\log n} \to \infty$. The next result proves this in a special case.

Theorem 5.18. *Consider the Wright-Fisher stepping stone model on $(\mathbf{Z} \bmod n)^2$ with dispersal distribution $p(0, x)$ a normal with mean 0 and covariance $\sigma^2 I$ (rounded to the nearest point on the lattice). If $\sigma/\sqrt{\log n} \to \infty$ and $\sigma/n \to 0$ then the population behaves as if it is homogeneously mixing. That is, if t_0 is the coalescence time of two lineages sampled from colonies with a displacement of x, then $t_0/2n$ converges in distribution to a mean 1 exponential.*

Proof. Using the inversion formula on $(\mathbf{Z} \bmod n)^2$ with (5.12) gives

$$\phi(x) = \frac{1 - \hat\phi(0)}{2n^2} \sum_{m_1=0}^{n-1} \sum_{m_2=0}^{n-1} \frac{(1 - \mu)^2 \hat{p}^2 (2\pi m/n) e^{-i \cdot 2\pi m \cdot x/n}}{1 - (1 - \mu)^2 \hat{p}^2 (2\pi m/n)} \tag{5.28}$$

where $m = (m_1, m_2)$ and $m \cdot x = m_1 x_1 + m_2 x_2$. In the nearest neighbor case with $K = \infty$ alleles this is (3.7) of Maruyama (1970d). To obtain the desired conclusion about the coalescence time we will take $\mu = \lambda/2n^2$. Since $\hat{p}(0) = 1$ the $m = 0$ in the term is

$$\frac{(1 - \mu)^2}{2\mu + \mu^2} \approx \frac{1}{2\mu}.$$

If we can show that the other terms can be ignored, then we will have $\phi(x) \approx A(1 - \hat\phi(0))$, with $A \approx 4n^2 \mu$, so using the result first for $x = 0$, and then for a general x we have

$$\phi(x) \approx \frac{1}{1 + 4n^2 \mu} = \frac{1}{1 + 2\lambda}.$$

To obtain the desired conclusion we note that

$$\phi(x) = E_x \left(1 - \frac{\lambda}{2n^2} \right)^{2t_0}$$

so we have shown

$$E_x \exp(-2\lambda(t_0/2n^2)) \to \frac{1}{1 + 2\lambda}$$

which is the desired conclusion written in terms of Laplace transforms.

The normal distribution has $\hat{p}^2(\theta) = e^{-|\theta|^2 \sigma^2}$, where $\theta = (\theta_1, \theta_2)$ and $|\theta|^2 = \theta_1^2 + \theta_2^2$. Here we are ignoring the effect of rounding, which is valid if σ is large, and that dispersal occurs on the torus, which is valid by our assumption $\sigma/n \to 0$. If $|\theta|\sigma$ is small then $1 - e^{-|\theta|^2 \sigma^2} \approx |\theta|^2 \sigma^2$ so the denominator in the fraction in the sum can be approximated by

$$2\mu + \sigma^2 \left(\frac{2\pi |m|}{n} \right)^2 \quad \text{when } |m| \le \epsilon n/\sigma$$

Since $\sigma^2 \to \infty$ and $\mu = \lambda/2n^2$, the 2μ can be discarded for all $m \ge 1$, and we have

$$\sum_{1 \le |m| \le \epsilon n/\sigma} \approx \frac{n^2}{(2\pi)^2 \sigma^2} \sum_{1 \le |m| \le \epsilon n/\sigma} \frac{1}{|m|^2}$$

Comparing with an integral and changing to polar coordinates

$$\sum_{1\le|m|\le\epsilon n/\sigma}\frac{1}{|m|^2}\le C\int_{1\le|x|\le\epsilon n/\sigma}\frac{1}{x^2}\,dx$$

$$=C\int_1^{\epsilon n/\sigma}\frac{1}{r^2}(2\pi r)\,dr\le C'\log n$$

so we have

$$\sum_{1\le|m|\le\epsilon n/\sigma}\le C''\frac{n^2\log n}{\sigma^2}\ll\frac{1}{\mu}$$

It suffices now to consider the sum over $\epsilon n/\sigma\le|m|\le n/2$. Over this range $1-(1-\mu)^2(1-e^{-(2\pi|m|)^2\sigma^2/n^2})\ge\delta>0$ so

$$\sum_{\epsilon n/\sigma}^{n/2}\le\frac{1}{\delta}\sum_{\epsilon n/\sigma\le|m|\le n/2}e^{-(2\pi m)^2\sigma^2/n^2}$$

Changing variables $x=\sigma m/n$ gives a grid of values with spacing σ/n, so comparing the sum with an integral gives that the right-hand side is

$$\approx\frac{1}{\delta}\cdot\frac{n^2}{\sigma^2}\int_{\epsilon\le|x|<\infty}e^{-(2\pi x)^2}\,dx$$

The integral is convergent, so we have

$$\sum_{\epsilon n\le|m|\le\sigma}^{n/2}\le C\frac{n^2}{\sigma^2}\ll\frac{1}{\mu}$$

since $\sigma^2\to\infty$, which completes the proof. □

6

Natural Selection

"It is not the strongest of the species that survives, nor the most intelligent, but rather the most responsive to change." Charles Darwin

In this chapter, we will consider various forms of natural selection and investigate their effects on the genealogy of a sample and observed patterns of genetic variability.

6.1 Directional selection

In this section, we will introduce selection into the Moran model considered in Section 1.5. As we said there, when we write the population size as $2N$ we are thinking of N diploid individuals, but the dynamics, which replace one copy of the locus at a time are more appropriate for $2N$ haploids. The haploid viewpoint is even more clear here, since we let 1 and $1 - s$ be the relative fitnesses of the two alleles, A and a. However, as the story unfolds in this chapter and the next, we will see that this situation is the same as a diploid model in which the relative fitnesses of A, Aa and aa are 1, $1 - s$, and $(1 - s)^2 \approx 1 - 2s$.

Let X_t be the number of A's at time t. Thinking of the fitnesses as the probability that an offspring of that type is viable, we can formulate the transition rates of the Moran model with selection as

$$i \to i + 1 \ \text{ at rate } b_i = 2N - i \cdot \frac{i}{2N}$$

$$i \to i - 1 \ \text{ at rate } d_i = i \cdot \frac{2N - i}{2N} \cdot (1 - s)$$

In words, a's are selected for possible replacement at total rate $2N - i$. The number of A's will increase if an A is chosen to be the parent of the new individual, an event of probability $i/2N$. The reasoning is similar for the second rate, but in this case the replacement only occurs with probability $1 - s$.

R. Durrett, *Probability Models for DNA Sequence Evolution*,
DOI: 10.1007/978-0-387-78168-6_6, © Springer Science+Business Media, LLC 2008

6.1.1 Fixation probability

Since there is no mutation in our model, the A allele will either be lost or become fixed in the population. Let $T_y = \min\{t : X_t = y\}$ be the first hitting time of y. Let $h(i) = P_i(T_{2N} < T_0)$ be the probability that the A allele becomes fixed when there are initially i copies. In the neutral case $h(i) = i/2N$ by (1.2).

Theorem 6.1. *In the Moran model with selection* $s > 0$

$$P_i(T_{2N} < T_0) = \frac{1 - (1-s)^i}{1 - (1-s)^{2N}} \tag{6.1}$$

When $i = 1$, the numerator is just s. If selection is strong, i.e., $2Ns$ is large, then $(1-s)^{2N} \approx 0$ and the probability of fixation of a new mutant is just s. When s is small, $(1-s) \approx e^{-s}$, so (6.1) can be written as

$$P_i(T_{2N} < T_0) \approx \frac{1 - e^{-is}}{1 - e^{-2Ns}} \tag{6.2}$$

The genealogical process for the Moran model is that of the Wright-Fisher model, but run twice as fast. If we replace s by $2s$ to compensate for this and let $p = i/2N$ the proportion in the population, we get the classic result of Kimura (1962) that the probability of fixation for the Wright-Fisher model is

$$P_{2Np}(T_{2N} < T_0) \approx \frac{1 - e^{-4Nsp}}{1 - e^{-4Ns}} \tag{6.3}$$

Proof. Births happen at rate b_i and deaths at rate d_i, so the probability a birth occurs before a death is $b_i/(b_i + d_i)$ and we have

$$h(i) = \frac{b_i}{b_i + d_i} h(i+1) + \frac{d_i}{b_i + d_i} h(i-1)$$

Multiplying on each side by $b_i + d_i$ and rearranging, we have

$$h(i+1) - h(i) = \frac{d_i}{b_i}(h(i) - h(i-1)) = (1-s)(h(i) - h(i-1))$$

Now $h(0) = 0$, so if we let $c = h(1)$ and iterate, it follows that

$$(\star) \qquad\qquad h(i+1) - h(i) = c(1-s)^i$$

Summing we have

$$h(j) = \sum_{i=0}^{j-1} c(1-s)^i = c\frac{1 - (1-s)^j}{s}$$

We must have $h(2N) = 1$ so $c = s/(1 - (1-s)^{2N})$ and the desired result follows. $\qquad\square$

For some calculations in this section, and later on, it is useful to give

Another derivation of (6.1). To motivate the computation, we begin by re-calling the proof of (1.2). Let $\tau = T_0 \wedge T_{2N}$. When $s = 0$, EX_t is constant in time, so we have

$$i = 2N \cdot P_i(X_\tau = 2N) + 0 \cdot P_i(X_\tau = 0)$$

Solving, we have $P_i(X_\tau = 2N) = i/2N$. When $s > 0$, $b_i/(b_i + d_i) = 1/(2 - s)$. A little calculation shows that

$$(1 - s)^{i+1}\frac{1}{2 - s} + (1 - s)^{i-1}\frac{1 - s}{2 - s}$$
$$= (1 - s)^i\frac{1 - s}{2 - s} + (1 - s)^i\frac{1}{2 - s} = (1 - s)^i$$

so, in this case, the value of $E(1 - s)^{X_t}$ stays constant in time. Reasoning as before,

$$(1 - s)^i = (1 - s)^{2N}P_i(X_\tau = 2N) + 1 \cdot [1 - P_i(X_\tau = 2N)]$$

Solving we have

$$P_i(X_\tau = 2N) = \frac{1 - (1 - s)^i}{1 - (1 - s)^{2N}}$$

in agreement with (6.1).   □

Subdivided populations

Maruyama (1970a, 1974) noticed that in some cases, the fixation proba-bility in the Moran model version of the matrix migration models described in Section 4.5, is the same as in a homogeneously mixing population. In the model under consideration here, there are n populations with haploid popula-tion sizes $2N_1, \ldots 2N_n$, and there are two alleles A and a with relative fitnesses 1 and $1 - s$.

- Each individual is subject to replacement at rate 1.
- An individual from population i is replaced by one chosen at random from population j with probability p_{ij}, where $\sum_j p_{ij} = 1$.
- The proposed new individual is accepted with probability equal to its fitness.

Theorem 6.2. *If the migration process satisfies the detailed balance condi-tion, i.e., $N_i p_{ij} = N_j p_{ji}$, then the fixation probability is the same as in a homogeneously mixing population with the same number of individuals.*

Proof. Let n_i be the number of A's in population i. If we consider only events that cause an a from population i to be replaced by an A from population j or vice versa, then changes in the number of A's occur at the following rates.

$$+1 \quad (2N_i - n_i)p_{ij}\frac{n_j}{2N_j}$$

$$-1 \quad n_j p_{ji}\frac{2N_i - n_i}{2N_i}(1 - s)$$

Our condition $N_i p_{ij} = N_j p_{ji}$ implies that the second rate is $(1 - s)$ times the first. Since this holds for all pairs i, j the overall rate of increase of A's is $(1 - s)$ times the rate of decrease and the desired result follows.   □

This result has recently been rediscovered by Lieberman, Hauert, and Nowak (2005).

6.1.2 Time to fixation

Our next goal is to compute the expected time to fixation, given that fixation occurs. Kimura and Ohta (1969a) did this using the diffusion approximation. However, as in Section 1.5, it is possible to do their computation without leaving the discrete setting, so we will take that approach here. Let $\tau = T_0 \wedge T_{2N}$ be the fixation time.

Theorem 6.3. *In the Moran model with selection $s > 0$, as $N \to \infty$,*

$$E_1(\tau|T_{2N} < T_0) \sim \frac{2}{s}\log N \tag{6.4}$$

where $a_N \sim b_N$ means $a_N/b_N \to 1$ as $N \to \infty$.

Proof. The proof begins in the same way as the proof for the neutral case given in Section 1.5. Let S_j be the amount of time spent at j before time τ and note that

$$E_i\tau = \sum_{j=1}^{2N-1} E_i S_j \tag{6.5}$$

Let N_j be the number of visits to j. Let $q(j) = (2 - s)j(2N - j)/2N$ be the rate at which the chain leaves j. Since each visit to j lasts for an exponential amount of time with mean $1/q(j)$ we have

$$E_i S_j = \frac{1}{q(j)}E_i N_j \tag{6.6}$$

If we let $T_j = \min\{t : X_t = j\}$ be the first hitting time of j and $R_j = \min\{t : X_t = j$ and $X_s \neq j$ for some $s < t\}$ be the time of the first return to j, then the reasoning for (1.36) gives

$$E_i N_j = \frac{P_i(T_j < \infty)}{P_j(R_j = \infty)} \tag{6.7}$$

To compute the quantities that enter into the last formula, we note that the average value of $(1-s)^{X_t}$ stays constant in time, so if $0 \le i \le j$

$$(1-s)^i = (1-s)^j P_i(T_j < T_0) + 1 \cdot [1 - P_i(T_j < T_0)]$$

and solving gives that for $0 \le i \le j$

$$P_i(T_j < T_0) = \frac{1-(1-s)^i}{1-(1-s)^j} \quad P_i(T_0 < T_j) = \frac{(1-s)^i - (1-s)^j}{1-(1-s)^j} \qquad (6.8)$$

Likewise, if $j \le i \le 2N$

$$(1-s)^i = (1-s)^j P_i(T_j < T_{2N}) + (1-s)^{2N} \cdot [1 - P_i(T_j < T_{2N})]$$

and solving gives that if $j \le i \le 2N$ then

$$P_i(T_j < T_{2N}) = \frac{(1-s)^i - (1-s)^{2N}}{(1-s)^j - (1-s)^{2N}}$$

$$P_i(T_{2N} < T_j) = \frac{(1-s)^j - (1-s)^i}{(1-s)^j - (1-s)^{2N}} \qquad (6.9)$$

From (6.8) and (6.9), it follows that

$$\begin{aligned} P_j(R_j = \infty) &= \frac{1}{2-s} P_{j+1}(T_{2N} < T_j) + \frac{1-s}{2-s} P_{j-1}(T_0 < T_j) \\ &= \frac{1}{2-s} \cdot \frac{(1-s)^j - (1-s)^{j+1}}{(1-s)^j - (1-s)^{2N}} \\ &\quad + \frac{1-s}{2-s} \cdot \frac{(1-s)^{j-1} - (1-s)^j}{1-(1-s)^j} \end{aligned} \qquad (6.10)$$

If $s > 0$ is fixed and N is large, $(1-s)^{2N} \approx 0$, so using (6.10)

$$P_j(R_j = \infty) \approx \frac{s}{2-s} + \frac{1-s}{2-s} \cdot \frac{s(1-s)^{j-1}}{1-(1-s)^j}$$

It will turn out that most of the answer will come from large values of j. In this case, the last formula simplifies to

$$P_j(R_j = \infty) \approx \frac{s}{2-s} \qquad (6.11)$$

If we take $i = 1$, then there is only one case to consider, so using (6.7) and (6.8) with the last formula, we have for large j

$$E_1 N_j = \frac{P_1(T_j < \infty)}{P_j(R_j = \infty)} \approx \frac{1-(1-s)^1}{1-(1-s)^j} \cdot \frac{2-s}{s} \approx 2-s$$

Using $q(j) = (2-s)(2N-j)j/2N$ with (6.6) now we have

$$E_1 S_j \approx \frac{2N}{j(2N - j)}$$

Turning now to properties of the conditioned process, we let

$$\bar{E}_i = E_i(\cdot | T_{2N} < T_0)$$

$h(x) = P_x(T_{2N} < T_0) = (1 - (1 - s)^x)/(1 - (1 - s)^{2N})$ so using (1.40)

$$\bar{E}_1 S_j = \frac{h(j)}{h(1)} \approx \frac{2N}{sj(2N - j)}$$

(6.11) and hence the last three formulas are only valid for large j. As the first part of the next computation will show, we can ignore the contribution from small j.

To evaluate the asymptotic behavior of $\sum_{j=1}^{2N-1} \bar{E}_1 S_j$ we will divide the sum into three parts. Letting $M = 2N/(\log N)$ and omitting the factor $1/s$.

$$\sum_{j=1}^{M} \frac{2N}{j(2N - j)} \approx \sum_{j=1}^{M} \frac{1}{j} \approx \log(2N/\log N) = \log(2N) - \log \log N \approx \log N$$

Note that since $\log N \to \infty$, we can ignore the contribution from small values of j. At the other end, changing variables $k = 2N - j$ shows

$$\sum_{j=2N-M}^{2N-1} \frac{2N}{j(2N - j)} = \sum_{k=1}^{M} \frac{2N}{k(2N - k)} \approx \log N$$

In the middle, we have

$$\sum_{j=M+1}^{2N-M-1} \frac{2N}{j(2N - j)} = \sum_{j=M+1}^{2N-M-1} \frac{1}{\frac{j}{2N}\left(1 - \frac{j}{2N}\right)} \cdot \frac{1}{2N}$$

$$\approx \int_{1/\log N}^{1-1/\log N} \frac{1}{u(1 - u)} \, du$$

To evaluate the integral, we note $1/u(1 - u) = 1/u + 1/(1 - u)$ so it is

$$= 2 \int_{1/\log N}^{1-1/\log N} \frac{1}{u} \, du = 2[\log(1 - 1/\log N) + \log \log N]$$

The first term tends to 0 as $N \to \infty$. The second is much smaller than $\log N$, so combining our computations, we have

$$\bar{E}_1 \tau = \sum_{j=1}^{2N-1} \bar{E}_1 S_j \approx \frac{2}{s} \log N$$

which proves (6.4).   □

6.1.3 Three phases of the fixation process

To obtain more insight into what is happening during the fixation of a favorable allele, we will now give a second derivation of (6.4). As in the proof just completed, we divide the values into three regimes.

1. While the advantageous A allele is rare, the number of A's can be approximated by a supercritical branching process.
2. While the frequency of A's is $\in [\epsilon, 1 - \epsilon]$ there is very little randomness and it follows the solution of the logistic differential equation.
3. While the disadvantageous a allele is rare, the number of a's can be approximated by a subcritical branching process.

Kaplan, Hudson, and Langley (1989) incorporated these three phases into their simulation study of the effects of hitchhiking.

Phase 1. Let i be the number of A's. If $i/2N$ is small, then

$$i \to i+1 \quad \text{at rate } b_i \approx i$$
$$i \to i-1 \quad \text{at rate } d_i \approx (1-s)i$$

This is a continuous time branching process in which each of the i individuals gives birth at rate 1 and dies at rate $1 - s$. Letting Z_t be the number of individuals at time t, it is easy to see from the description that

$$\frac{d}{dt} E Z_t = s E Z_t$$

so $E Z_t = Z_0 e^{st}$. A result from the theory of branching processes, see Athreya and Ney 1972), shows that as $t \to \infty$

$$e^{-st} Z_t \to W \tag{6.12}$$

The limit W may be 0, and will be if the branching process dies out, that is, $Z_t = 0$ for some t. However, on the event that the process does not die out $\{Z_t > 0 \text{ for all } t\}$, we have $W > 0$.

Let T_1 be the first time that there are $M = 2N/\log N$ A alleles. We want $M/2N \to 0$ slowly, but there is nothing special about this precise value. Using (6.12), we see that $Z_t \approx e^{st} W$ so when the mutation survives

$$\frac{2N}{\log N} \approx \exp(sT_1)W$$

and solving gives

$$T_1 \approx \frac{1}{s} \log\left(\frac{2N}{W \log N}\right) \approx \frac{1}{s} \log(2N)$$

Phase 2. Let T_2 be the first time that there are $2N - M$ A alleles, where $M = 2N/\log N$. As we will now show, during the second phase from T_1 to T_2

the process behaves like the solution of the logistic differential equation. Let X_t be the number of copies of the mutant allele at time t, and let $Y_t^N = X_t/2N$. Y_t^N makes transitions as follows:

$$i/2N \to (i+1)/2N \text{ at rate } b_i = 2N - i \cdot \frac{i}{2N}$$

$$i/2N \to (i-1)/2N \text{ at rate } d_i \approx (1-s)i \cdot \frac{2N-i}{2N}$$

When $Y_0^N = i/2N = y$, the infinitesimal mean

$$\frac{d}{dt} EY_t^N = b_i \cdot \frac{1}{2N} + d_i \cdot \left(-\frac{1}{2N}\right) = s\frac{2N-i}{2N} \cdot \frac{i}{2N} = sy(1-y)$$

while the infinitesimal variance

$$\frac{d}{dt} E(Y_t^N - y_0)^2 = (b_i + d_i) \cdot \frac{1}{(2N)^2} = (2-s)\frac{2N-i}{2N} \cdot \frac{i}{2N} \cdot \frac{1}{2N} \to 0$$

(If the terms infinitesimal mean and variance are unfamiliar, they are explained in Section 7.1).

In this situation, results in Section 8.7 of Durrett (1996) or in Section 7.4 of Ethier and Kurtz (1986), show that as $N \to \infty$, Y_t^N converges to Y_t, the solution of the logistic differential equation

$$dY_t = sY_t(1 - Y_t)$$

It is straightforward to check that the solution of this equation is

$$Y_t = \frac{1}{1 + Ce^{-t}}$$

where $C = (1-Y_0)/Y_0$. In the case of interest, $Y_0 = 1/\log(N)$, so $C \approx \log(N)$ and $Y_t = 1 - 1/(\log N)$ when

$$(\log N)e^{-t} = \frac{\log N}{\log N - 1} - 1 = \frac{1}{\log N - 1} \sim \frac{1}{\log N}$$

Solving, we find that $T_2 - T_1 \approx 2 \log \log N$.

Phase 3. To achieve fixation of the A allele mutation after time T_2, the $M = 2N/(\log N)$ a alleles must decrease to 0. The number of a alleles, Z_t, makes transitions

$$j \to j+1 \text{ at rate } d_{2N-j} \approx (1-s)j$$

$$j \to j-1 \text{ at rate } b_{2N-j} \approx j$$

That is, Z_t is a continuous time branching process in which each of the j individuals gives birth at rate $(1 - s)$ and dies at rate 1. By arguments in phase 1, $EZ_t = Z_0 e^{-st}$ so it takes about $(1/s)\log(2N)$ units of time to reach 0.

The times of the three phases were

Phase 1	$(1/s)\log(2N)$
Phase 2	$\log\log(2N)$
Phase 3	$(1/s)\log(2N)$

and we have a seond proof of (6.4). □

6.1.4 Ancestral selection graph

Krone and Neuhauser (1997), see also Neuhauser and Krone (1997), were the first to figure out how to study the genealogy of samples in models with selection. Departing from our usual approach in order to keep a closer contact with their work, we will consider a haploid population of N individuals that evolves according to the Moran model. We focus on one locus, and limit the discussion to the case of two alleles A and a with relative fitnesses 1 and $1-s$. Following Krone and Neuhauser, we consider only symmetric mutation: a's mutate to A's at rate u and A's mutate to a's at rate u.

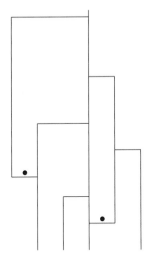

Fig. 6.1. A realization of the ancestral selection graph for a sample of size 4.

We first construct the genealogy ignoring mutation. The details may look a little strange, but the point of the construction is not to mimic reality, but simply to develop a scheme that will simulate the Moran model with selection. As in the ordinary case with $s = 0$, when we work backwards in time each individual in the population is subject to replacement at a total rate 1. However, this time there are two types of events. At rate $1 - s$, we have a replacement event that chooses a parent at random and always replaces the individual by the parent. At rate s, we choose a parent at random but replace the individual by the parent only if the parent is A. Since we will not know until the end of the genealogical computation whether or not replacement should occur, we must follow the lineages of both the individual and the parent and the result is branching in the genealogy. The picture in Figure 6.1 gives a possible outcome for a sample of four individuals. Here dots mark the edges that can only be used by A's.

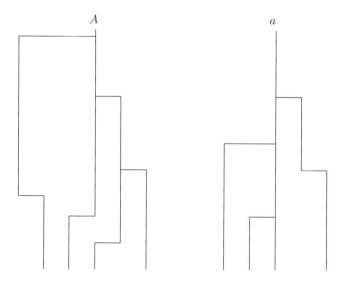

Fig. 6.2. The actual genealogy depends on the state of the ultimate ancestor.

If we speed up time by N and let $\sigma = Ns$ this results in a process in which coalescence occurs at rate $j(j - 1)/2$ and branching occurs at rate σj when there are j lineages. The time at which the genealogy reaches one individual is denoted T_{UA}, where UA stands for "ultimate ancestor." As the two pictures in Figure 6.2 show, depending on the state of the ultimate ancestor, it may or may not be the most recent common ancestor of the sample. Note also that the topology of the tree is different in the two cases.

The first thing to be proved is that the process ultimately will coalesce to one lineage. Krone and Neuhauser (1997) do this in their Theorem 3.2, but this can also be deduced from Theorem 3.10 for the ancestral recombination graph, since in each case there is a linear birth rate versus the coalescence death rate of $k(k-1)/2$.

To compute the time of the most recent common ancestor, one must work backwards to T_{UA} and then forwards from that time to see what lineages in the genealogy are real. For this reason, exact computations are difficult. To get around this difficulty, Krone and Neuhauser (1997) computed expected values to first order in σ, which corresponds to considering lineages with 0 or 1 mutation. Theorem 4.19 in Krone and Neuhauser (1997) shows that if σ is small and $\theta = 2Nu$ then the probability that two individuals chosen at random are identical by descent is

$$\frac{1}{1+\theta} - \frac{\theta(5+2\theta)}{4(1+\theta)^2(3+\theta)(3+2\theta)}\sigma + O(\sigma^2)$$

Again, when $\sigma = 0$ this reduces to the classic answer for the ordinary coalescent. When $\theta = 1$, this becomes

$$\frac{1}{2} - \frac{7}{320}\sigma + O(\sigma^2)$$

The coefficient of σ is small in this case. Figure 9 of Krone and Neuhauser (1997) shows it is < 0.025 for all $\theta \in [0, \infty)$.

To get results for σ that are not small, one has to turn to simulation. When σ gets to be 5–10, this becomes difficult because there are a large number of lineages created before the ultimate ancestor occurs. Slade (2000a,b) has recently developed some methods for trimming the tree in addition to proving some new theoretical results for the coalescent with selection.

6.2 Balancing selection

We begin this section by considering a Wright-Fisher model with diploid individuals and a locus with two alleles: A_1 and A_2. Let w_{ij} be the relative fitness of A_iA_j and assume that $w_{12} = w_{21}$. To be concrete, we will think of viability selection where $w_{ij} \leq 1$ is the probability an individual of genotype A_iA_j survives to maturity. Ignoring mutation, and considering the large population limit, if $x = k/2N$ is the current frequency of allele A_1, then the frequency of A_1 in the next generation is

$$x' = \frac{1}{\bar{w}}[w_{11}x^2 + w_{12}x(1-x)] \tag{6.13}$$

where $\bar{w} = w_{11}x^2 + w_{12} \cdot 2x(1-x) + w_{22}(1-x)^2$ is the average fitness of individuals in the population. To explain (6.13), note that if random mating occurs to produce M offspring then the number of individuals produced with A_i on their first chromosome and A_j on the second is Mx_ix_j. The number that survive to maturity is $Mx_ix_jw_{ij}$, so the fraction of A_iA_j individuals among the survivors is $x_ix_jw_{ij}/\bar{w}$. x' gives the fraction of survivors that have A_1 as their first chromosome.

Introducing $m_i = xw_{i1} + (1-x)w_{i2}$, we have $\bar{w} = xm_1 + (1-x)m_2$ and

$$x' = \frac{xm_1}{xm_1 + (1-x)m_2} \tag{6.14}$$

$m_i = xw_{i1} + (1-x)w_{i2}$ is called the *marginal fitness of A_i* since it is the fitness of A_i when paired with a randomly chosen allele. To have a fixed point of the iteration (6.14), we must have

$$\frac{m_1}{xm_1 + (1-x)m_2} = 1$$

or $m_1 = m_2$. Setting $xw_{11} + (1-x)w_{12} = xw_{12} + (1-x)w_{22}$ and solving gives

$$\bar{x} = \frac{w_{22} - w_{12}}{(w_{22} - w_{12}) + (w_{11} - w_{12})} \tag{6.15}$$

The fixed point $\bar{x} \in (0,1)$ if and only if $w_{22} - w_{12}$ and $w_{11} - w_{12}$ have the same sign. There are several cases to consider:

(i) A_1 *dominant over* A_2. $w_{11} > w_{12} > w_{22}$.

(ii) A_2 *dominant over* A_1. $w_{11} < w_{12} < w_{22}$.

(iii) *Underdominance.* $w_{11} > w_{12} < w_{22}$. Heterozygotes are less fit.

(iv) *Overdominance.* $w_{11} < w_{12} > w_{22}$. Heterozygotes are more fit.

Two inequalities that define our cases, so, ignoring the possbility of equality, this covers all of the cases. A fixed point $\bar{x} \in (0,1)$ exists in cases (iii) and (iv), but not in (i) and (ii).

Theorem 6.4. *Let x_n be the frequency of A_1 in generation n. In the four cases given above, we have the following behavior.*
(i) If $x_0 > 0$, $x_n \to 1$.
(ii) If $x_0 < 1$, $x_n \to 0$.
(iii) If $x_0 < \bar{x}$, $x_n \to 0$. If $x_0 > \bar{x}$, $x_n \to 1$.
(iv) If $x_0 \in (0,1)$, $x_n \to \bar{x}$.

Proof. In case (i), $m_1 > m_2$ so if $x \in (0,1)$

$$\frac{x'}{x} = \frac{m_1}{xm_1 + (1-x)m_2} > 1$$

Let x_n be the frequency in generation n. If $0 < x_0 < 1$, then x_n is increasing so x_n converges to a limit x_∞. To conclude that $x_\infty = 1$ we note that

$$x_{n+1} = \frac{m_1 x_n}{x_n m_1 + (1 - x_n) m_2}$$

so letting $n \to \infty$

$$x_\infty = \frac{m_1 x_\infty}{x_\infty m_1 + (1 - x_\infty) m_2}$$

Solving we have

$$x_\infty + (1 - x_\infty) \frac{m_2}{m_1} = 1$$

Since $m_2/m_1 < 1$, this implies $x_\infty = 1$. Case (ii) is similar to case (i).

To determine whether the fixed points are attracting or repelling in the last two cases, we note that

$$m_1 = w_{12} + x(w_{11} - w_{12})$$
$$m_2 = w_{22} + x(w_{12} - w_{22})$$

In case (iii), m_1 is an increasing function of x and m_2 is decreasing. The two functions are equal at $\bar{x}$, so

$$\text{if } x > \bar{x} \text{ then } m_1 > m_2 \text{ and } x' > x$$
$$\text{if } x < \bar{x} \text{ then } m_1 < m_2 \text{ and } x' < x$$

Since the limit of x_n will be a fixed point of the iteration, this implies that if $x_0 > \bar{x}$ then $x_n \to 1$, while if $x_0 < \bar{x}$ then $x_n \to 0$.

In case (iv), m_1 is a decreasing function of x and m_2 is increasing so

$$\text{if } x > \bar{x} \text{ then } m_1 < m_2 \text{ and } x' < x$$
$$\text{if } x < \bar{x} \text{ then } m_1 > m_2 \text{ and } x' > x$$

This is not quite enough to conclude that convergence to $\bar{x}$ for x' might overshoot the fixed point $\bar{x}$ and we could have a periodic orbit with period 2. To rule this out, we note that

$$x' - \bar{x} = \frac{m_1 x}{m_1 x + m_2 (1 - x)} - \frac{m_1 \bar{x}}{m_1 \bar{x} + m_2 (1 - \bar{x})}$$
$$= \frac{m_1 m_2 (x - \bar{x})}{(m_1 x + m_2 (1 - x))(m_1 \bar{x} + m_2 (1 - \bar{x}))}$$

so $x' - \bar{x}$ has the same sign as $x - \bar{x}$. From this we see that

$$\text{if } 0 < x_0 < \bar{x} \text{ then } x_n < \bar{x} \text{ is increasing}$$
$$\text{if } \bar{x} < x_0 < 1 \text{ then } \bar{x} < x_n \text{ is decreasing}$$

Since the limit of the x_n will be a fixed point of the iteration, in both cases we must have $x_n \to \bar{x}$ and the desired result follows.   □

In the case of overdominance, the selective advantage of the heterozygote maintains both alleles at positive frequencies, so this case is called *balancing selection*. If we write the fitnesses as

$$w_{11} = 1 - s_1 \quad w_{12} = 1 \quad w_{22} = 1 - s_2$$

then (6.15) shows that the fixed point is

$$\bar{x} = s_2/(s_1 + s_2) \tag{6.16}$$

Sickle-cell anemia

A classic case of overdominance is sickle-cell anemia in human beings. It is caused by an allele S that codes for a variant form of the β chain of hemoglobin. In persons of genotype SS, many blood cells assume a curved elongated shape and are removed from circulation, resulting in a severe anemia as well as pain and disability from the accumulation of defective cells in capillaries, joints, spleen, and other organs. The S allele is maintained at relatively high frequency because persons of genotype Ss, in which s is the nonmutant allele, have only a mild form of the anemia, but are resistant to malaria.

In regions of Africa in which malaria is common, the relative fitnesses of ss, Ss, and SS have been estimated as $w_{11} = 0.9$, $w_{12} = 1$, and $w_{22} = 0.2$. Substitution into (6.16) predicts s will have frequency

$$\frac{0.8}{0.1 + 0.8} = \frac{8}{9}$$

and hence that S has frequency $1/9 = 0.11$. This value is reasonably close to the average allele frequency of 0.09 across West Africa.

Mutation

Suppose mutation occurs after selection. Let μ_1 be the mutation probability $A_1 \rightarrow A_2$ and μ_2 be the mutation probability $A_2 \rightarrow A_1$. In this case

$$x' = (1 - \mu_1)\frac{xm_1}{xm_1 + (1 - x)m_2} + \mu_2\frac{(1 - x)m_2}{xm_1 + (1 - x)m_2} \tag{6.17}$$

Since $m_i = xw_{i1} + (1 - x)w_{i2}$, this is a cubic equation in x and is very messy to solve explicitly (see Theorem 8.10). To paraphrase, Ewens (1979), page 13: we have in mind the case where selective differences are 10^{-2} to 10^{-3}, while the mutation rates per locus are of order 10^{-5} to 10^{-6}, so the fixed point $\bar{x}$ for the recursion without mutation will be an excellent approximation for the equilibrium with mutation.

For a numerical example with selection at a low value and mutation high, suppose $w_{11} = 0.999$, $w_{12} = w_{21} = 1.0$, $w_{22} = 0.998$, and $\mu_1 = \mu_2 = 10^{-5}$. The fixed point without mutation is $\bar{x} = 2/3$. Iterating (6.17) starting from this value, we find that the equilibrium with mutation is 0.661848.

Genealogies

Kaplan, Darden, and Hudson (1988) were the first to study genealogies in models with balancing selection. The details are somewhat involved algebraically, but the idea is simple: this form of selection can be treated as if it were a two island model. Population i consists of the gametes with allele A_i, since in the absence of mutation two chromosomes can have the same parent only if they have the same allele. Mutation produces migration between populations.

Suppose that n genes are chosen at random from the 0th generation and let $Q(0) = (i, j)$ if the sample contains i A_1 alleles and j A_2 alleles.

Theorem 6.5. *For $t \leq 0$, let $Q(t) = (i(t), j(t))$ denote the number of lineages in generation t with alleles A_1 and A_2. If we assume the frequency of allele A_1 is at its equilibrium value $\bar{x}$, and if we ignore the posssibility of two events happening on one step, then $Q(t)$ is well approximated by a Markov chain with transtion probabilities*

$$P(Q(t-1) = (i+1, j-1)|Q(t) = (i,j)) = \frac{j\beta_1\bar{x}}{1-\bar{x}} \cdot \frac{1}{2N}$$

$$P(Q(t-1) = (i-1, j+1)|Q(t) = (i,j)) = \frac{i\beta_2(1-\bar{x})}{\bar{x}} \cdot \frac{1}{2N}$$

$$P(Q(t-1) = (i-1, j)|Q(t) = (i,j)) = \binom{i}{2} \frac{1}{2N\bar{x}}$$

$$P(Q(t-1) = (i, j-1)|Q(t) = (i,j)) = \binom{j}{2} \frac{1}{2N(1-\bar{x})}$$

The $t - 1$ may look odd on the left in the conditional probability, but it is not, since time is indexed by $t \leq 0$. To make the connection with the two-island model with populations of sizes $2N\bar{x}$ and $2N(1-\bar{x})$, we note that the forward migration probabilities $f_{1,2} = \beta_1/2N$ and $f_{2,1} = \beta_2/2N$ so the backward migration probabilities, using (4.9), are

$$m_{2,1} = \frac{2N\bar{x}\beta_1/2N}{2N\bar{x}\beta_1/2N + 2N(1-\bar{x})(1-\beta_2/2N)} \approx \frac{\beta_1}{2N} \frac{\bar{x}}{1-\bar{x}}$$

$$m_{1,2} = \frac{2N(1-\bar{x})\beta_2/2N}{2N(1-\bar{x})\beta_2/2N + 2N\bar{x}(1-\beta_1/2N)} \approx \frac{\beta_2}{2N} \frac{1-\bar{x}}{\bar{x}}$$

Proof. Let $X(t)$ be the frequency of allele A_1 in generation t. Let $m_i(t-1) = X_{t-1}w_{i1} + (1-X_{t-1})w_{i2}$ be the marginal fitness of A_i in generation $t-1$ and let

$$\bar{w}_{t-1} = X_{t-1}^2 w_{11} + 2X_{t-1}(1-X_{t-1})w_{12} + (1-X_{t-1})^2 w_{22}$$

be the mean fitness of the population in generation $t - 1$.

Let $f_{A_j}(A_k, t)$ denote the probability that a randomly chosen gene from generation t is of allelic type A_k and its parental gene from generation $t - 1$ is of allelic type A_j. By the calculation that led to (6.17),

$$f_{A_1}(A_1, t) = (1 - \mu_1) \frac{X_{t-1} m_1(t-1)}{\bar{w}_{t-1}}$$

$$f_{A_1}(A_2, t) = \mu_1 \frac{X_{t-1} m_1(t-1)}{\bar{w}_{t-1}}$$

$$f_{A_2}(A_2, t) = (1 - \mu_2) \frac{(1 - X_{t-1}) m_2(t-1)}{\bar{w}_{t-1}}$$

$$f_{A_2}(A_1, t) = \mu_2 \frac{(1 - X_{t-1}) m_2(t-1)}{\bar{w}_{t-1}}$$

To compute the transition probability of $Q(t)$, we let $f(A_i, t) = f_{A_1}(A_i, t) + f_{A_2}(A_i, t)$ be the probability of picking a gene of allelic type A_i in generation t regardless of the type of the parental gene. The probability that a sampled A_2 allele from generation t has an A_1 parental gene equals

$$\frac{f_{A_1}(A_2, t)}{f(A_2, t)} = \frac{\mu_1 X_{t-1} m_1(t-1)}{\mu_1 X_{t-1} m_1(t-1) + (1 - \mu_2)(1 - X_{t-1}) m_2(t-1)}$$

If we assume the allele frequencies are in equilibrium then $X_{t-1} \approx \bar{x}$ and hence $m_1(t-1) \approx m_2(t-1)$, so assuming $\mu_1 = \beta_1/2N$ and $\mu_2 = \beta_2/2N$ are small,

$$\frac{f_{A_1}(A_2, t)}{f(A_2, t)} \approx \frac{\bar{x} \frac{\beta_1}{2N}}{\bar{x} \frac{\beta_1}{2N} + (1 - \bar{x})\left(1 - \frac{\beta_2}{2N}\right)} \approx \frac{\beta_1 \bar{x}}{1 - \bar{x}} \cdot \frac{1}{2N}$$

Ignoring the possibility of two mutations on one step,

$$P(Q(t-1) = (i+1, j-1) | Q(t) = (i, j)) = \frac{j \beta_1 \bar{x}}{1 - \bar{x}} \cdot \frac{1}{2N}$$

A similar argument shows that

$$P(Q(t-1) = (i-1, j+1) | Q(t) = (i, j)) = \frac{i \beta_2 (1 - \bar{x})}{\bar{x}} \cdot \frac{1}{2N}$$

In the cases of interest, μ_1 and μ_2 are small, so most individuals pick their parents from their subpopulation. Ignoring the possibility of two coalescences or a coalescence and a mutation on one step we have

$$P(Q(t-1) = (i-1, j) | Q(t) = (i, j)) = \binom{i}{2} \frac{1}{2N\bar{x}}$$

$$P(Q(t-1) = (i, j-1) | Q(t) = (i, j)) = \binom{j}{2} \frac{1}{2N(1 - \bar{x})}$$

which completes the proof.   □

Adding recombination

As Hudson and Kaplan (1988) observed, it is not hard to generalize the setup above to include recombination. Suppose one is investigating the genealogy of a neutral locus B that is separated from A by recombination with probability $r = R/2N$ per generation, and that n genes are chosen at random from the 0th generation. Let $Q(0) = (i, j)$ if the sample contains i A_1 alleles and j A_2 alleles.

Theorem 6.6. *For $t \leq 0$, let $Q(t) = (i(t), j(t))$ denote the number of lineages in generation t with alleles A_1 and A_2. If we assume the frequency of allele A_1 is at its equilibrium value $\bar{x}$, and if we ignore the possibility of two events happening on one step, then $Q(t)$ is well approximated by a Markov chain with transition probabilities*

$$P(Q(t-1) = (i+1, j-1)|Q(t) = (i, j)) = \frac{j\bar{x}(\beta_1 + R(1-\bar{x}))}{1-\bar{x}} \cdot \frac{1}{2N}$$

$$P(Q(t-1) = (i-1, j+1)|Q(t) = (i, j)) = \frac{i(1-\bar{x})(\beta_2 + R\bar{x})}{\bar{x}} \cdot \frac{1}{2N}$$

$$P(Q(t-1) = (i-1, j)|Q(t) = (i, j)) = \binom{i}{2}\frac{1}{2N\bar{x}}$$

$$P(Q(t-1) = (i, j-1)|Q(t) = (i, j)) = \binom{j}{2}\frac{1}{2N(1-\bar{x})}$$

Proof. Recombination adds a term of the form

$$\frac{1}{\bar{w}_{t-1}}rX_{t-1}(1 - X_{t-1})w_{12}$$

to $f_{A_1}(A_2, t)$ and $f_{A_2}(A_1, t)$ and subtracts it from $f_{A_1}(A_1, t)$ and $f_{A_2}(A_2, t)$. Repeating the calculations in the previous proof and using $w_{ij} = 1 + O(1/N)$ we have

$$\frac{f_{A_1}(A_2, t)}{f(A_2, t)} \approx \frac{\mu_1 X_{t-1} m_1(t-1) + rX_{t-1}(1 - X_{t-1})w_{12}}{1 - X_{t-1}}$$

$$\approx \frac{\bar{x}\frac{\beta_1}{2N} + \bar{x}(1-\bar{x})\frac{R}{2N}}{1 - \bar{x}}$$

Combining this with a similar calculation for $f_{A_2}(A_1, t)/f(A_1, t)$ gives the indicated result. □

Coalescence times

Consider now a sample of size 2 of the B locus. Our first goal is to compute $M_{i,j}$, the mean coalescence time for two lineages when i start linked to the A_1 allele and j start linked to the A_2 allele. To simplify notation, we will let

$$\alpha_1 = \frac{(1 - \bar{x})(\beta_2 + R\bar{x})}{\bar{x}} \qquad \alpha_2 = \frac{\bar{x}(\beta_1 + R(1 - \bar{x}))}{1 - \bar{x}}$$

$$\gamma_1 = 1/\bar{x} \qquad \gamma_2 = 1/(1 - \bar{x})$$

Here α_i is the migration rate out of subpopulation i and γ_i is the coalescence rate within it. By considering what happens on the first event, it is easy to see that

$$M_{2,0} = \frac{1}{2\alpha_1 + \gamma_1} + \frac{2\alpha_1}{2\alpha_1 + \gamma_1} M_{1,1}$$

$$M_{0,2} = \frac{1}{2\alpha_2 + \gamma_2} + \frac{2\alpha_2}{2\alpha_2 + \gamma_2} M_{1,1}$$

$$M_{1,1} = \frac{1}{\alpha_1 + \alpha_2} + \frac{\alpha_1}{\alpha_1 + \alpha_2} M_{0,2} + \frac{\alpha_2}{\alpha_1 + \alpha_2} M_{2,0}$$

Plugging the first two equations into the third one, we have

$$(\alpha_1 + \alpha_2)M_{1,1} = 1 + \frac{\alpha_1}{2\alpha_2 + \gamma_2} + \frac{\alpha_1(2\alpha_2)}{2\alpha_2 + \gamma_2} M_{1,1}$$

$$+ \frac{\alpha_2}{2\alpha_1 + \gamma_1} + \frac{\alpha_2(2\alpha_1)}{2\alpha_1 + \gamma_1} M_{1,1}$$

Solving gives

$$M_{1,1} = \frac{1 + \frac{\alpha_1}{2\alpha_2 + \gamma_2} + \frac{\alpha_2}{2\alpha_1 + \gamma_1}}{\frac{\alpha_1 \gamma_2}{2\alpha_2 + \gamma_2} + \frac{\alpha_2 \gamma_1}{2\alpha_1 + \gamma_1}} \tag{6.18}$$

and then $M_{0,2}$ and $M_{2,0}$ can be computed from the preceding equations. To check this formula, we note that if $\gamma_1, \gamma_2 \gg \alpha_1, \alpha_2$, i.e., the coalescence rates are much larger than the migration rates, then

$$M_{1,1} \approx \frac{1}{\alpha_1 + \alpha_2}$$

In words, the two lineages wait an amount of time with mean $1/(\alpha_1 + \alpha_2)$ to come to the same population and then coalescence comes soon after that event.

In the other direction, if $\gamma_1, \gamma_2 \ll \alpha_1, \alpha_2$, then

$$M_{1,1} \approx \frac{1 + \frac{\alpha_1}{2\alpha_2} + \frac{\alpha_2}{2\alpha_1}}{\frac{\alpha_1 \gamma_2}{2\alpha_2} + \frac{\alpha_2 \gamma_1}{2\alpha_1}}$$

As we argued earlier for the fast migration limit of the island model, since migration is much faster than coalescence, we can pretend that before they hit the two lineages are independent and in equilibrium. A lineage is in the first population with probability $\alpha_2/(\alpha_1 + \alpha_2)$ and in the second with probability $\alpha_1/(\alpha_1 + \alpha_2)$, so the probability of coalescence on one step is approximately

$$\left(\frac{\alpha_2}{\alpha_1 + \alpha_2}\right)^2 \gamma_1 + \left(\frac{\alpha_1}{\alpha_1 + \alpha_2}\right)^2 \gamma_2 = \frac{\alpha_2^2 \gamma_1 + \alpha_1^2 \gamma_2}{(\alpha_1 + \alpha_2)^2}$$

Inverting this to get the mean coalescence time and then multiplying top and bottom by $1/2\alpha_1\alpha_2$, the mean is then

$$\frac{(\alpha_1 + \alpha_2)^2}{\alpha_2^2\gamma_1 + \alpha_1^2\gamma_2} = \frac{1 + \frac{\alpha_1}{2\alpha_2} + \frac{\alpha_2}{2\alpha_1}}{\frac{\alpha_1\gamma_2}{2\alpha_2} + \frac{\alpha_2\gamma_1}{2\alpha_1}}$$

Identity by descent

The same logic that led to (6.18) can be used to compute $h(i, j)$, the probability that two lineages are identical by descent when i starts in population 1 and j in population 2. Let u be the mutation rate and let $\theta = 4Nu$. By considering what happens at the first event,

$$(2\alpha_1 + \gamma_1 + \theta)h(2, 0) = 2\alpha_1 h(1, 1) + \gamma_1$$
$$(2\alpha_2 + \gamma_2 + \theta)h(0, 2) = 2\alpha_2 h(1, 1) + \gamma_2$$
$$(\alpha_1 + \alpha_2 + \theta)h(1, 1) = \alpha_1 h(0, 2) + \alpha_2 h(2, 0)$$

Plugging the first two equations into the third one

$$(\alpha_1 + \alpha_2 + \theta)\, h(1, 1) = \frac{\alpha_1(2\alpha_2)}{2\alpha_2 + \gamma_2 + \theta}h(1, 1) + \frac{\alpha_1\gamma_2}{2\alpha_2 + \gamma_2 + \theta}$$
$$+ \frac{\alpha_2(2\alpha_1)}{2\alpha_1 + \gamma_1 + \theta}h(1, 1) + \frac{\alpha_2\gamma_1}{2\alpha_1 + \gamma_1 + \theta}$$

which can be solved for $h(1, 1)$.

Example 6.1. Hudson and Kaplan (1988) used the last result to study Kreitman's (1983) data on the *Adh* locus. To examine the distribution of polymorphic sites along the chromosome, a "sliding window" method was used. Three different quantities were computed to characterize the variability in the window centered at each nucleotide site k: $\pi_{FS}(k)$, $\pi_{FF}(k)$, and $\pi_{SS}(k)$, the average number of pairwise differences between Fast and Slow sequences, between Fast sequences, and between Slow sequences. The region sequenced contained protein coding sequences as well as introns and other noncoding sequences. To take at least partial account of the different levels of constraints in these regions, the size of the window was varied to keep the number of possible silent changes in the window constant. The window size chosen corresponds to 50 base pairs in noncoding regions.

If each nucleotide is treated as an individual locus and if it is assumed that the allelic frequencies at position 2 of codon 192 are maintained by strong balancing selection, then the theory above can be used to calculate the expectation of $\pi_{FS}(k)$, $\pi_{SS}(k)$, and $\pi_{FF}(k)$. These calculations require that values be assigned to β_1, β_2, and $\bar{x}$, and that for each site i we must compute θ_i and R_i, the recombination rate between i and the location i_0 of the balanced polymorphism.

At sites where m of the three possible mutations is a silent change (does not change the amino acid), they assumed that the rate $\theta_i = m\theta_0/3$. The heterozygosity per nucleotide at silent sites, π, has been estimated to be 0.006 for a region that is 13 kb long that includes the Adh locus, so they set $\theta_0 = 0.006$. The mutations that change lysine to threonine and threonine back to lysine are the second-position transversions $A \to C$ and $C \to A$. Since $\theta_0 = 4Nu$, a plausible value for $\beta_1 = \beta_2 = \theta_0/6 = 0.001$, since these are $2N$ times the mutation rates. Since $\bar{x}$, the frequency of the Slow variant varies with geographic location (Oakeshott et al. 1982) it is not clear what value to assign to it. A more realistic model must take spatial structure into account, but Hudson and Kaplan simply set $\bar{x} = 0.7$, a value for a sample of $D.$ $melanogaster$ from Raleigh, North Carolina discussed in Kreitman and Aguadé (1986a).

Finally, they set $R_i = R_0|i - i_0|$. Recombination per base pair has been estimated for several regions of the $D.$ $melanogaster$ genome to be approximately 10^{-8} per generation in females (Chovnick, Gelbart, and McCarron 1977). There is no recombination in males. The neutral mutation rate has been estimated to be approximately 5×10^{-9} per year in many organisms. If we assume that $D.$ $melanogaster$ has four generations per year, then the ratio of the recombination rate to the neutral mutation rate per generation is approximately

$$\frac{10^{-8}/2}{(5 \times 10^{-9})/4} = 4$$

This implies that R_0 is approximately 4 times $\theta_0/2$, or 0.012.

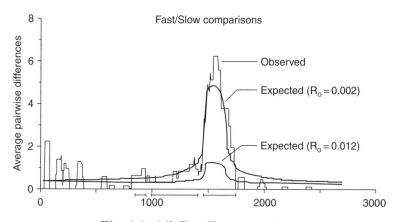

Fig. 6.3. Adh Fast-Slow comparison.

With all of the parameters estimated, they could compute the expected values of $\pi_{FS}(k)$, $\pi_{SS}(k)$, and $\pi_{FF}(k)$ by using the formulas for $h(1,1)$, $h(2,0)$, and $h(0,2)$ above. The next graph shows that in order to fit the FS data, the recombination rate needed to be lowered by a factor of 6 from what was

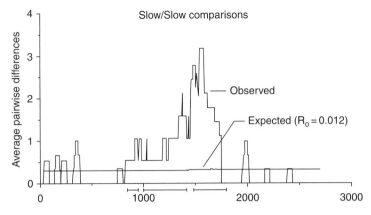

Fig. 6.4. Adh Slow-Slow comparison.

expected. Even after that was done, the fit was not good for the SS data, which showed a surprising amount of polymorphism in the SS near the locus under selection.

In his survey paper, Hudson (1991) redid the analysis using the divergence between *D. melanogaster* and *D. simulans* to estimate the mutation rate on a locus-by-locus basis. Now the recombination value that produces this fit is only a factor of 2 smaller than the a priori estimate.

6.3 Background selection

Mutation constantly introduces deleterious alleles. The action of selection to eliminate these deleterious alleles is called background selection. Following D. Charlesworth, B. Charlesworth, and Morgan (1995) and Hudson and Kaplan (1994, 1995, 1996), we will use a coalescent-based approach to asses its impact. To begin, we assume that the locus A is linked to one other locus at which deleterious mutations occur at rate u per individual per generation and that there is no recombination between the two loci. All deleterious mutations are assumed to have the same effect: $1 - sh$ in the heterozygous state and $1 - h$ in the homozygous state. However, individual deleterious mutations will have low frequencies, so we will ignore the possibility of homozygous mutants. Interaction between mutations is assumed to be multiplicative so that an individual heterozygous for i such mutations has fitness $(1 - sh)^i$.

Our first result was derived by B. Charlesworth, Morgan, and D. Charlesworth (1983), see their equation (3).

Theorem 6.7. *In the presence of background selection, the shape of the genealogy will be just like the neutral case except that the effective population size is changed to* $N_e = N e^{-u/2sh}$.

As a consequence, the site frequency spectrum will be the same as under the neutral model. To prepare for the statements of the next two results, note that if we let $\theta = 4N_e\mu$ and $\theta_0 = 4N\mu$, then we have $\theta/\theta_0 = e^{-u/2sh}$. If u is small then the nucleotide diversity $\pi \approx \theta$ so the reduction in the presence of background selection,

$$\pi/\pi_0 \approx e^{-u/2sh} \tag{6.19}$$

Proof. Following Kimura and Maruyama (1966) and Hey (1991), we formulate the dynamics of the process and the corresponding coalescent as follows. Let f_i be the frequency of gametes with i mutations, $m_k = \exp(-u/2)(u/2)^k/k!$ be the probability that a gamete experiences k new mutations, and $w_j = (1-sh)^j$ be the relative fitness of a gamete with j mutations. Assuming that selection acts before mutation, the proportion of the population that are gametes with j mutations produced by parental gametes with $i \leq j$ mutations is

$$p_{ij} = \frac{w_{ij}}{W} \quad \text{where} \quad w_{ij} = f_i w_i m_{j-i} \quad \text{and} \quad W = \sum_{ij} w_{ij}$$

Our first step is to check that

Theorem 6.8. *The Poisson $f_i = e^{-u/2sh}(u/2sh)^i/i!$ is a stationary distribution for the process.*

Proof. To do this, the algebra is simpler if we let $v = u/2$ to get rid of a number of 2's, and note that

$$\sum_{i=0}^{j} w_{ij} = \sum_{i=0}^{j} e^{-v/sh} \frac{(v/sh)^i}{i!}(1-sh)^i e^{-v} \frac{v^{j-i}}{(j-i)!}$$

$$= e^{-v(1+1/sh)} \frac{1}{j!} \sum_{i=0}^{j} \frac{j!}{i!(j-i)!} \left((v/sh)-v\right)^i v^{j-i}$$

$$= e^{-v(1+1/sh)} \frac{(v/sh)^j}{j!}$$

since

$$\sum_{i=0}^{j} \frac{j!}{i!(j-i)!} a^i b^{j-i} = (a+b)^j$$

Summing over j now, we have

$$W = \sum_{j=0}^{\infty} e^{-v(1+1/sh)} \frac{(v/sh)^j}{j!} = e^{-v}$$

so $\sum_{i=0}^{j} p_{ij} = \sum_{i=0}^{j} w_{ij}/W = e^{-v/sh}(v/sh)^j/j!$, and we have checked that the Poisson distribution is stationary. $\qquad\square$

By Bayes' theorem, the probability that a gamete with j mutations derives from one with $i \leq j$ mutations is

$$q_{ji} = \frac{p_{ij}}{\sum_{i=0}^{j} p_{ij}} = \frac{e^{-v/sh} \frac{(v/sh)^i}{i!} (1 - sh)^i \frac{v^{j-i}}{(j-i)!}}{e^{-v/sh} \frac{(v/sh)^j}{j!}}$$

$$= \frac{j!}{i!(j-i)!}(1 - sh)^i (sh)^{j-i} \qquad (6.20)$$

In words, q_{ji} is a binomial distribution.

When $j = 1$ and $i = 0$, $q_{10} = sh$. If $sh = 0.02$, as has been suggested by Crow and Simmons (1983), then as we work backwards in time, a geometrically distributed number of generations with mean 50 is needed for a sampled class 1 gamete to get to class 0. To estimate the amount of time needed for a sampled class j gamete to get to class 0, visualize the initial state as j white balls and that loss of mutation, with probability sh on each trial, results in a ball being painted green. Ignoring the possibility that two balls are painted on one turn, the total time to go from j to 0 then has mean

$$\leq \frac{1}{jsh} + \frac{1}{(j-1)sh} + \cdots + \frac{1}{2sh} + \frac{1}{sh} \approx \frac{\log(j+1)}{sh}$$

For most biologically reasonable values of sh, the sample size j, and population size N, $(\log(j+1))/sh \ll N$, so the time to get to class 0 is short compared to the coalescence time. Thus ignoring the time to get to class 0 and the possibility that the two lineages coalesce before getting to class 0, the coalescent process occurs as for the neutral model, with a population size of $2Ne^{-u/2sh}$ instead of $2N$. □

Adding recombination

We suppose now that there is recombination with probability R per generation between the selected locus and the neutral locus A.

Theorem 6.9. *Consider a sample of size 2 at the A locus. If the mutation rate u is small then, when time is measured in units of $2N$ generations, the time to coalescence of two lineages is roughly exponential with mean*

$$\lambda^{-1} \approx 1 - ush/2(R + sh)^2 \qquad (6.21)$$

Since the expected time under the neutral model is 1, $\pi/\pi_0 \approx \lambda^{-1}$.

Proof. If there are recombinations, then as we work backwards in time the march toward the 0 class is interrupted by recombinations. At recombination events, the selected locus is replaced by an independent copy drawn from the population and hence has a Poisson mean $u/2sh$ number of mutations. As (6.20) shows, in the absence of recombination, the number of mutations decreases according to the binomial distribution with a fraction $(1 - sh)$ of the mutations retained. From this, we see that if as we work backwards in time, the last recombination event happened k generations ago, an event of probability $R(1 - R)^k$, then the number of deleterious mutations will have a Poisson distribution with mean $(1 - sh)^k u/2sh$. Thus, in equilibrium the number of deleterious mutations is a geometric mixture of Poisson distributions.

Suppose now that u is small. In this case, there will always be 0 or 1 mutation, so the Poisson is replaced by the Bernoulli. Supposing R and sh are small, the probability of one mutation by the argument above is

$$\sum_{k=0}^{\infty} R(1 - R)^k \frac{u}{2sh}(1 - sh)^k = \frac{uR}{2sh(R + sh + Rsh)} \approx \frac{uR}{2sh(R + sh)}$$

Actually, in many cases R will not be small but if, say, $R = 0.1$ and $sh = 0.02$, then $R + sh + Rsh = 0.122$ while $R + sh = 0.120$, so the approximation is still reasonable.

Consider now a sample of size 2 at the A locus, and let $\mu(k)$ be the equilibrium probability that a lineage in the coalescent has k deleterious mutations. If we suppose that particle movement is more rapid than coalescence, then by the reasoning for the fast migration limit of the island model, Theorem 4.11, the probability of coalescence on one step is

$$\Lambda = \sum_k \frac{\mu(k)^2}{2N_k} \tag{6.22}$$

where $N_k = Ne^{-u/2sh}(u/2sh)^k/k!$ is the number of individuals in the population with k mutations. When u is small, we only have two classes to worry about: 0 and 1 mutation. In this case, $2N$ times the sum reduces to

$$\lambda = \frac{\left(\frac{uR}{2sh(R+sh)}\right)^2}{u/2sh} + \frac{\left(1 - \frac{uR}{2sh(R+sh)}\right)^2}{1 - u/2sh}$$

When y is small $(1 - y)^2 \approx 1 - 2y$ and $1/(1 - y) \approx 1 + y$. Thus, if u/sh is small, the above is

$$\approx \frac{uR^2}{2sh(R + sh)^2} + \left(1 - \frac{2uR}{2sh(R + sh)}\right) \cdot (1 + u/2sh)$$

If y and z are small $(1-y)(1+z) \approx 1-y+z$, so putting things over a common denominator, the above

$$\approx 1 + \frac{uR^2 - 2uR(R + sh) + u(R + sh)^2}{2sh(R + sh)^2} = 1 + \frac{ush}{2(R + sh)^2}$$

Since the per generation coalescence time is $\approx \lambda/2N$ the probability of no coalescence by time t is

$$\approx \left(1 - \frac{\lambda}{2N}\right)^{2Nt} \approx e^{-\lambda t}$$

and we have proved the desired result.   □

DNA segment

Consider now a continuous segment $[a, b]$ of DNA. Let $R(x)$ be the rate at which recombinations occur in $[0, x]$ and suppose that mutations in $[0, x]$ occur at rate $\int_0^x u(y)\, dy$.

Theorem 6.10. *The nucleotide diversity π compared to the predictions of the neutral theory, π_0, satisfies*

$$\frac{\pi}{\pi_0} \approx \exp\left(-\int_a^b \frac{u(x)sh}{2(sh + R(x))^2}\, dx\right) \tag{6.23}$$

Proof. Suppose first that the neutral locus A is followed by two selected loci B_1 and B_2 with the probability of a recombination between A and B_1 being R_1 and between B_1 and B_2 being $R_2 - R_1$. Let u_1 and u_2 be the mutation rates at the two loci. Combining the two loci into one locus with mutation rate $u = u_1 + u_2$ and then randomly allocating the mutations between the two loci with probabilities $p_i = u_i/(u_1 + u_2)$, it is easy to see that in the equilibrium of the process going forward, the numbers of deleterious mutations at the two loci are independent. To argue that this also holds for the process working backwards in time, we note that this is true when a recombination occurs between A and B_1 and all subsequent operations: thinning the number of mutations using the binomial or a recombination between B_1 and B_2 preserves this property.

The last argument is valid for any number of selected loci. We are now ready to divide up our segment of DNA into a lot of little pieces B_i and take a limit as the number of pieces goes to infinity. Let u_i be the mutation rate at B_i and let R_i be the recombination rate between A and B_i. Consider now a sample of size 2 at the A locus. When there are m loci the classes are denoted by vectors $\mathbf{k} = (k_1, k_2, \ldots, k_m)$ but the argument that leads to (6.22) stays the same. Let $\mu(\mathbf{k})$ be the equilibrium probability that a lineage in the coalescent has k_i deleterious mutations at locus i. If we suppose that particle movement is more rapid than coalescence, then the probability of coalescence on one step is approximately

$$\Lambda = \sum_{\mathbf{k}} \frac{\mu(\mathbf{k})^2}{2N_{\mathbf{k}}}$$

where $N_{\mathbf{k}}$ is the number of individuals in the population with k mutations and

$$N_{\mathbf{k}} = N \prod_{i=1}^{m} e^{-u_i/2sh}(u_i/2sh)^{k_i}/k_i!$$

$$\approx N \prod_{i=1}^{m} \left(\frac{u_i}{sh} 1_{\{k_i=1\}} + \left(1 - \frac{u_i}{sh}\right) 1_{\{k_i=0\}} \right)$$

Since μ is itself a product, it is easy to see that $2N$ times the sum is

$$\lambda = \prod_{i=1}^{m} \frac{\left(\frac{u_i R_i}{2sh(R_i+sh)}\right)^2}{u_i/2sh} + \frac{\left(1 - \frac{u_i R_i}{2sh(R_i+sh)}\right)^2}{1 - u_i/2sh}.$$

By a calculation in the previous proof, this is

$$\approx \prod_{i=1}^{m} 1 + \frac{u_i sh}{2(R_i + sh)^2} \approx \exp\left(\sum_{i=1}^{m} \frac{u_i sh}{2(R_i + sh)^2}\right).$$

To pass to the limit, let x be the distance from the neutral locus measured in kb and suppose that $u_i = \int_{B_i} u(x)\, dx$. In this case

$$\lambda \approx \exp\left(\int_a^b \frac{u(x)sh}{2(sh + R(x))^2}\, dx\right)$$

where a and b are the endpoints of the region under consideration. Since, when time is measured in units of $2N$ generations, the coalescence time is roughly exponential with mean λ^{-1}, we have proved the desired result. $\qquad\square$

Example 6.2. If we suppose $u(x) \equiv u$, $R(x) = r|x|$, $a = -L$, and $b = L$, then (6.23) reduces to

$$\frac{\pi}{\pi_0} \approx \exp\left(-2 \int_0^L \frac{ush}{2(sh + rx)^2}\, dx\right)$$

$$= \exp\left(\frac{2}{r} \left[\frac{ush}{2(sh + rL)} - \frac{u}{2}\right]\right) = \exp\left(-\frac{2uL}{2sh + 2rL}\right)$$

Writing $U = 2uL$ for the total deleterious mutation rate in the region and $R = 2rL$ for the total recombination rate, the above can be written as

$$\frac{\pi}{\pi_0} \approx \exp\left(-\frac{U}{2sh + R}\right) \qquad (6.24)$$

When $R = 0$, this reduces to the result in (6.19). Note that as we should expect, recombination lessens the impact of deleterious mutations.

(6.23) gives the effect of deleterious mutations on the variability at 0, but it is straightforward to extend this to calculate the effect at other places. Suppose we have a chromosome of length L and let $M(x)$ be the distance from 0 to x measured in Morgans, i.e., the expected number of recombinations in

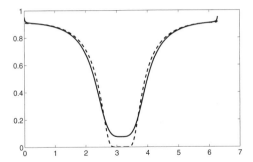

Fig. 6.5. Predicted variability under background selection for $sh = 0.0003$ (dashed line) and $sh = 0.03$ (solid line) when $M(x) = (x + \sin x)/12$

$[0, x]$ in one generation. In this case, the reduction in variability at y is given by

$$\frac{\pi}{\pi_0} \approx \exp\left(-\int_0^L \frac{u(x)sh}{2(sh + |M(x) - M(y)|)^2}\,dx\right) \qquad (6.25)$$

Example 6.3. Following Nordborg, B. Charlesworth, and D. Charlesworth (1996), we will now consider the case $M(x) = (x + \sin x)/12$ for $0 \le x \le 2\pi$. To explain their choice of this function we note that $M'(x) = (1 + \cos x)/12 = 0$ when $x = \pi$ so recombination is reduced near the centromere. The plot in Figure 6.5 shows what happens when $U = 0.1$ and $sh = 0.0003$ (dotted line) or $sh = 0.003$ (solid line). Note that near the centromere the weaker recombination causes a greater reduction in variability. The graphs turn up sharply toward the ends because at the telomeres contributions only come from one side.

Example 6.4. Hudson and Kaplan (1995) applied (6.25) to predict levels of variation on the third chromosome of *D. melanogaster*. The total diploid deleterious mutation rate has been estimated from mutation accumulation studies to be 1.0 or larger (Crow and Simmons 1983, Keightley 1994). Since there are approximately 5000 cytological bands in the *Drosophila* genome, they estimated u, the deleterious mutation rate per generation per band, to be $1.0/5000 = 2 \times 10^{-4}$. To estimate recombination as a function of distance, Hudson and Kaplan used information in Flybase to determine map distance as a function of physical position for 107 locations on the third chromosome.

The map gives the values of $M(x)$ at a discrete sequence of points x_i. From these they estimated the reduction in variability at x_k by $\pi/\pi_0 = e^{-G}$, where

$$G = \sum_i \frac{ush}{2} \cdot \frac{|x_{i+1} - x_i|}{(sh + |M(x_{i+1}) - M(x_k)|/2)(sh + |M(x_i) - M(x_k)|/2)}$$

To compare with (6.25), we first note that the $1/2$ in the denominator is to account for the fact that there is no recombination in male *Drosophila*.

Proof. Ignoring this 1/2, we can then derive their approximation as follows. Suppose without loss of generality that $M(x_k) \leq M(x_i) < M(x_{i+1})$. If we suppose that $M(x)$ is linear on $[x_i, x_{i+1}]$, then ignoring the constant factor $ush/2$ the contribution to the integral in (3.7) from this interval is

$$\int_0^1 \frac{(x_{i+1} - x_i)\, dv}{[sh + v(M(x_{i+1}) - M(x_k)) + (1-v)(M(x_i) - M(x_k))]^2}$$

$$= -\left.\frac{(x_{i+1} - x_i)/(M(x_{i+1}) - M(x_i))}{[sh + v(M(x_{i+1}) - M(x_k)) + (1-v)(M(x_i) - M(x_k))]}\right|_0^1$$

$$= -\frac{(x_{i+1} - x_i)/(M(x_{i+1}) - M(x_i))}{[sh + (M(x_{i+1}) - M(x_k))]}$$

$$+ \frac{(x_{i+1} - x_i)/(M(x_{i+1}) - M(x_i))}{[sh + (M(x_i) - M(x_k))]}$$

Adding the two fractions we have

$$= \frac{|x_{i+1} - x_i|}{(sh + |M(x_{i+1}) - M(x_k)|)(sh + |M(x_i) - M(x_k)|)} \qquad \square$$

Hudson and Kaplan compared the levels of variability at 17 loci on the third chromosome with predictions based on three different values of sh: 0.03, 0.02, and 0.005. The value of π_0 used, namely 0.014, was chosen to produce a good fit to the data as judged by eye. The fit given in the next figure is not very sensitive to the value of sh except near the centromere and is remarkably good except at the tips of the chromosomes.

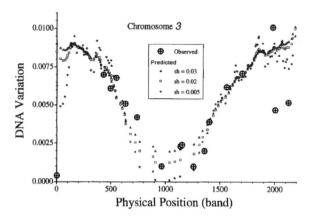

Fig. 6.6. Hudson and Kaplan's fit of background selection to *Drosophila* third chromosome data.

Example 6.5. Another example of this type of analysis occurs in Hamblin and Aquadro's (1996) study of *Gld*. To estimate $M(x)$ they used information about the following loci in *D. simulans*. Here x_i is the band number, $M(x_i)$ is the distance from the tip measured in Morgans, and nd means "not determined."

Locus	x_i	$M(x_i)$	Locus	x_i	$M(x_i)$
jv	175	0	*Rh3*	1089	nd
idh	285	0.064	*Aldox*	1299	0.754
Est-6	432	0.252	*rosy*	1403	nd
Pgm	567	0.381	*Men*	1413	0.877
ri	780	0.580	*boss*	1850	nd
Gld	996	0.590	*Ald*	1865	1.246
e	1059	0.600	*ca*	1990	1.300
H	1071	0.610	*Acph-1*	2000	1.340

Taking $sh = 0.02$, $u = 0.0002$, and choosing $\pi_0 = 0.11$ as Hamblin and Aquadro (1996) did gives the following prediction for π. We do not know the confidence intervals associated with the estimates of π, but with the exception of *boss*, the general pattern seems consistent with the background selection model.

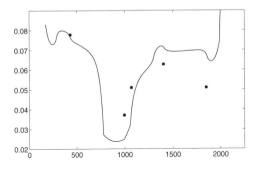

Fig. 6.7. Predicted versus under background selection versus observed levels of variablity at *Est-6, Gld, Rh3, rosy,* and *boss*.

6.4 Muller's ratchet

Asexual reproduction compels genomes to be inherited as indivisible blocks. This leads to an accumulation of deleterious mutations known as Muller's ratchet. Once the least mutated genome in the population carries at least one bad mutation, no genomes with fewer mutations can be found in future generations except as a result of a highly unlikely back mutation.

To see how fast the ratchet will click, we return to the calculations in the previous section which show that, assuming multiplicative fitness, the number of deleterious mutations in a randomly chosen individual is Poisson with mean u/sh where u is the genome wide deleterious mutation rate and sh is the selective disadvantage of a heterozygote with deleterious mutation. Suppose, as we did in the previous section, that $sh = 0.02$, a number that Gillespie (1998) supports with the observation that in *Drosophila*, the fitness disadvantage of flies with a recessive lethal mutation is 2–5%. If the genome wide deleterious mutation rate is $u = 1$ then the mutation free class has Ne^{-50} individuals, which will be zero for most sensible values of N. On the other hand if $u = 0.1$ and $N = 10^6$ then the answer is $10^6 e^{-5} = 6,738$.

The last calculation is not relevant to *Drosophila* because they reproduce sexually and so chromosomes undergo recombination. It does apply to the human Y chromosome, since it only exists in males, and with the exception of small "pseudo-autosomal" region near the chromosome ends does not undergo recombination. Sequencing of the nonrecombining or male-specific Y chromosome (MSY) by Skaletsky et al. (2003) revealed approximately 78 protein coding units, about 60 of which are members of nine MSY-specific gene families. In contrast, the finished sequence of the X chromosome described in Ross et al. (2005) contains more than 1000 genes. Combining this with Ohno's (1967) idea that the mammalian X and Y chromosomes are evolved from a pair of autosomes, we see that the Y chromosome has lost most of its genes over time.

At this point some readers may be wondering, what about mitochondrial DNA? The crucial difference is that mtDNA has about 16,000 nucleotides compared to 32 million on the Y chromosome. Thus, even if the mutation rate for mtDNA is 10 times larger, the ratio u/sh is 200 times smaller, and we will have $e^{-u/sh} \approx 1$.

6.4.1 Evolutionary advantages of recombination

One of the oldest hypotheses for the advantage of recombination, first articulated by Fisher (1930) and Muller (1932), is that it allows beneficial mutations that arise on different individuals to be placed together on the same chromosome. If no recombination occurs, one of the beneficial alleles is doomed to extinction, slowing the rate at which adaptive mutations are incorporated within a population.

The initial arguments were qualitative. Muller (1958) in his Gibbs Lecture to the American Math Society took the first steps in making the argument quantitative, see page 150. For mathematicians, this may seem like a fairly recent paper. However, in it Muller talks about "the brilliant recent theory of Watson and Crick" and he estimates the size of the human genome by noting

"the total mass of nucleotide material, or DNA, contained in one set of human chromosomes, such as would be found in a human sperm or egg nucleus just before they united in fertilization, is approximately 4×10^{-12} of a gram. Since the mass of one pair of nucleotides is about 10^{-21} of a gram there must be about 4×10^9 nucleotide pairs in the chromosome set."

In (1964) Muller gave a detailed argument. Using his notation and phrasing, we let some initial mutant be represented in the next generation by $(1+r)$ individuals, in the second successive generation by $(1+r)^2$, and so on. After g generations, the mutant will have given rise to a total of

$$1 + (1+r) + (1+r)^2 + \cdots + (1+r)^g = \frac{(1+r)^{g+1} - 1}{r} \quad \text{descendants.}$$

Suppose that the probability of a new advantageous mutation is $f = 1/F$. Then the number of generations needed, on the average, to generate an advantageous mutation in one of the descendants can be calculated by setting

$$\frac{(1+r)^{g+1} - 1}{r} = F$$

Solving, we get

$$g = \frac{\log(1 + rF)}{\log(1+r)} - 1 \tag{6.26}$$

For an example, if $r = 0.01$ and $F = 10^9$ then since $\log(1+x) \approx x$

$$g \approx \frac{\log(10^7)}{\log(1.01)} \approx \frac{7 \log(10)}{0.01} = 1612 \quad \text{generations.}$$

During this time, in the whole population, there will have been an average of gN/F advantageous mutations. That expression, then, represents the ratio of the amount of mutational accumulation in the sexual population to that in the asexual one, inasmuch as all of the gN/F advantageous mutations in the former have the opportunity to become concentrated, eventually, within the same descendants, by recombining as they multiply. If we continue with our example and set $N = 10^6$ then $gN/F = 1.612$. On the other hand if $r = 0.001$ then

$$g \approx \frac{\log(10^6)}{\log(1.001)} = 13,815 \quad \text{and} \quad gN/F = 13.815$$

and the evolution of the asexual population is 13.8 times as slow.

Crow and Kimura (1965) did a very similar analysis starting from the logistic growth of a mutation with selective advantage s:

$$p = \frac{p_0}{p_0 + (1 - p_0)e^{-st}}$$

Taking the initial frequency $p_0 = 1/N$ the solution is

$$p = \frac{1}{1 + (N - 1)e^{-st}}$$

Integrating

$$\int_0^g \frac{e^{st}}{e^{st} + N - 1} \, dt = \frac{1}{s} \log(e^{st} + N - 1) \Big|_0^g$$

$$= \frac{1}{s} \log\left(\frac{e^{sg} + N - 1}{N}\right)$$

Multiplying by N we see that the total number of individuals $= F$ when

$$F = \frac{N}{s} \log\left(\frac{e^{sg} + N - 1}{N}\right)$$

Rearranging gives $e^{sF/N} = (e^{sg} + N - 1)/N$ and hence

$$g = \frac{1}{s} \log[N(e^{sF/N} - 1) + 1]$$

Note that as $N \to \infty$, $N(e^{sF/N} - 1) \to sF$, so this answer reduces to the one in (6.26). The number of mutations in the whole population that occur in this time is

$$\frac{Ng}{F} = \frac{N}{sF} \log[N(e^{sF/N} - 1) + 1] \tag{6.27}$$

This formula is similar to Muller's, but has the advantage that the answer depends only on N and sF. When $N = 10^6$ and $sF = 10^7$, 10^6 the new answers are 2.38 and 14.4, compared with the previous 1.61 and 13.8.

Maynard Smith (1968) questioned these conclusions and proposed a counterexample. In his deterministic model for the infinite population limit, each mutant is originally at a low equilibrium frequency in the population maintained by a balance between recurrent mutation and natural selection. In this setting if the relative fitnesses of ab, Ab, Ab, AB are 1, $1 - H$, $1 - h$, $(1 - H)(1 - h)$ then the gene frequencies will satisfy $p_{ab}p_{AB} = p_{Ab}p_{aB}$. Thus there is no linkage disequilibrium initially and none will be created by the deterministic evolution.

Crow and Kimura (1969) replied by pointing out that if favorable mutations were initially rare, there would almost never be any individuals carrying two or more of them so the population would not be in a state of linkage equilibrium. Indeed, as Maynard Smith's own calculations showed, the probabilities are to a first approximation 1, μ_A, μ_B, $\mu_A\mu_B$. The mutation rates μ_A and μ_B are very small, so for most populations there will be no AB's. A lively debate ensued. See Felsenstein (1974) for an account.

6.4.2 Sex, epistasis, and Kondrashov

While it is fairly clear, based on the calculations above, that recombination is a good thing, the same cannot be said for the sexual reproduction which makes it possible. In 10 second sound bite form "asexual reproduction has an intrinsic twofold advantage over anisogamous sexual reproduction." where "anisogamous" refers to the situation in which the female produces relatively large gametes, such as eggs or seeds, and the male produces a large number of small and mobile gametes, such as sperm or seeds.

To explain the two-fold cost, note that in a population with a 1:1 sex ratio, each female produces an average of two offspring, one male and one female. Imagine now that a dominant mutation arises that allows *parthenogeneis*, i.e., females reproduce asexually. The genome carrying the mutation will persist in two copies in the next generation, and then double in subsequent generations until it takes over the population.

There must be some problem with this scenario, because sexual reproduction is predominant among eukaryotes. One possibility is that mutations that cause parthenogenesis either do not occur, or are so deleterious that they swamp the two-fold advantage. However, parthenogenesis does occur in the dandelion *Taraxacum officianale*, which probably lost sex only recently, so we must look somewhere else for an explanation.

Our calculations in the previous section assume that fitness is multiplicative, i.e., an individual with k mutations has fitness $(1-s)^k$. However this assumption is contradicted by a number of experimental studies. Mukai (1969) accumulated mutations on second chromosomes of *Drosophila melanogaster* for 60 generations. He found that the best fit for the fitness of an individual in generation k mutations was

$$w_k = 1 - 0.009813x - 0.00555x^2$$

where $x = 0.141k$ was the expected number of mutations in generation k. Under this fitness function, called positive epistasis, fitness drops off faster than in the additive or multiplicative cases.

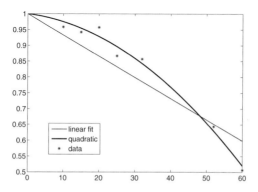

Fig. 6.8. Mukai's data showing the quadratic and linear fits to fitness.

Models with epistasis are very difficult to analyze so following Kondrashov (1988) we will consider truncation selection, i.e., the fitness of an individual with n mutations is

$$w_n = \begin{cases} 1 & \text{if } n \le k \\ 0 & \text{if } n > k \end{cases}$$

To compare asexual and sexual reproduction, we will compute the *mutation load*, which is the fraction of individuals that, in selection-mutation equilibrium, do not reproduce because of selection. Under truncation selection, an asexual species will, due to Muller's ratchet, eventually come to the point where all individuals have k mutations. In this case, if the deleterious mutation rate is U per genome per generation, then there will be a Poisson mean U number of mutations per reproduction, so the probability of at least one mutation is $1 - e^{-U}$. In words, the mutational load for an asexual species is

$$L_{asex} = 1 - e^{-U}$$

If $U \ge \log 2$ then $L_{asex} \le 1/2$ and there is potential for a two-fold advantage.

Theorem 6.11. *Under truncation selection, a sexually reproducing species has mutation load $L_{sex} \to 0$ as $k \to \infty$.*

Our account is inspired by Kondrashov (1988) and Gillespie (1998), but we use an argument by contradiction rather than numerical computation.

Proof. Suppose that the mean number of deleterious mutations per chromosome is μ before reproduction. If these are spread out along a reasonably large genome, then due to recombination, gametes will have roughly a Poisson mean $\lambda = \mu + U$ number of mutations. Those with $> k$ mutations will be eliminated by selection. Thus the distribution after selection will be

$$q_m = \frac{p_m}{\sum_{n=0}^{k} p_n} \quad \text{where} \quad p_m = e^{-\lambda} \frac{\lambda^m}{m!}$$

We will have equilibrium when the mean of q is μ.

This value of the mean $\mu(k)$ is not easy to find algebraically. To prove the theorem we will show

$$(k - \mu(k))/k^{1/2} \to \infty \quad \text{as } k \to \infty. \tag{6.28}$$

To see this is enough, note that when λ is large p_m is roughly a normal with mean λ and standard deviation $\sqrt{\lambda}$, so the fraction of the probability lost by truncation tends to 0.

To prove (6.28), let $C_k = (k - \lambda)/\sqrt{\lambda}$ so that $k = \lambda + C_k\sqrt{\lambda}$. Let X have the normal with mean λ and standard deviation $\sqrt{\lambda}$, truncated to removed values $> k$ and then renormalized to be a probability distribution. Let $A = [\lambda - C_k\sqrt{\lambda}, \lambda + C_k\sqrt{\lambda}]$. Using symmetry and a trivial inequality

$$E(X; X \in A) = \lambda P(X \in A)$$
$$E(X; X \in A^c) \leq (\lambda - C_k \sqrt{\lambda}) P(X \in A^c)$$

Adding the two inequalities we have

$$\mu \leq \lambda - C_k \sqrt{\lambda} P(X \in A^c)$$

If there is a sequence of values $k(n) \to \infty$ so that $C_{k(n)}$ stays bounded then $P(X \in A^c)$ stays bounded away from 0, a contradiction since $\lambda = \mu(k) + U$.
$\square$

The argument by contradiction does not provide any quantitative information about how large k has to be, so we will now consider a concrete example:

Example 6.6. Take $U = 1.4$, so $e^{-U} = 0.246$ and $L_{asex} = 0.754$. If we iterate the dynamics of truncation selection with $n = 6$, we find that the equilibrium has mean $\mu = 4.317$, $L_{sex} = 0.349 < L_{asex}/2$, and the distribution of the number of mutations per individual in the population is

0	1	2	3	4	5	6
0.0050	0.0288	0.0824	0.1572	0.2246	0.2569	0.2448

6.5 Hitchhiking

When a selectively favorable mutation occurs in a population and is subsequently fixed, the process will alter the frequency of alleles at closely linked loci. Alleles present on the chromosome on which the original mutation occurred will tend to increase in frequency, and other alleles will decrease in frequency. This is referred to as "hitchhiking" because an allele can get a lift in frequency from selection acting on a neighboring allele.

Following Maynard-Smith and Haigh (1974), we consider haploid individuals and begin by considering the behavior of the selected locus when alleles B and b have relative fitnesses $1 + s$ and 1. If we assume a very large population in which individuals of type B produce $1 + s$ times as many offspring as those of type b, then the fraction of individuals of type B in generation n, p_n, satisfies

$$p_{n+1} = \frac{(1+s)p_n}{(1+s)p_n + (1-p_n)} = \frac{1+s}{1+sp_n} p_n \qquad (6.29)$$

In n generations, B's will produce $(1+s)^n$ times as many offspring as b's will, so the general solution of the equation is

$$p_n = \frac{(1+s)^n p_0}{1 - p_0 + (1+s)^n p_0} \qquad (6.30)$$

Readers not convinced by the verbal argument can check that with this choice of p_n

$$\frac{(1+s)p_n}{1+sp_n} = \frac{(1+s)^{n+1}p_0}{1-p_0+(1+s)^np_0}\left(1+\frac{s(1+s)^np_0}{1-p_0+(1+s)^np_0}\right)^{-1}$$

$$= \frac{(1+s)^{n+1}p_0}{1-p_0+(1+s)^np_0+s(1+s)^np_0} = p_{n+1}$$

We will have $p_n \approx 1$ in (6.30) when $(1+s)^np_0 = C$ and C is large, that is, $n = \log(C/p_0)/\log(1+s)$. If $p_0 = 1/2N$ and s is small, so that $\log(1+s) \approx s$, this condition is $n = (\ln 2NC)/s$ in accord with our previous result in (6.4) on the duration of a selective sweep.

Consider now a second neutral locus with alleles A and a. Introducing Q_n and R_n to denote the conditional probabilities $P(A|B)$ and $P(A|b)$ in generation n we have

Genotype	AB	aB	Ab	ab
Fitness	$1+s$	$1+s$	1	1
Frequency	p_nQ_n	$p_n(1-Q_n)$	$(1-p_n)R_n$	$(1-p_n)(1-R_n)$

The main result of Maynard-Smith and Haigh (1974) is

Theorem 6.12. *Suppose that $Q_0 = 0$. In this case the frequency of the A allele after the sweep is*

$$Q_\infty = R_0(1-p_0)\sum_{n=0}^{\infty}\frac{(1-r)^nr}{1-p_0+p_0(1+s)^{n+1}} \tag{6.31}$$

To interpret this equation, we note that the initial frequency of the A allele is $R_0(1-p_0)$ while its frequency after the sweep is Q_∞. Thus, the sum on the right-hand side gives the factor by which it has been reduced.

Proof. Let r be the probability of a recombination between the A and B loci per generation. Our first goal is to show

$$Q_n - R_n = (1-r)^n(Q_0 - R_0) \tag{6.32}$$

Even though it will take a lot of algebra to derive this, the answer is intuitive. A recombination event makes the two loci independent so the difference between the two conditional probabilities is $(1-r)^n$ times the original difference.

Proof of (6.32). The first step is to compute Q_{n+1}. By considering the possible parent pairs we have

parents	expected number of AB offspring
AB,AB	$(1+s)p_nQ_n \cdot (1+s)p_nQ_n$
AB,aB	$(1+s)p_nQ_n \cdot (1+s)p_n(1-Q_n)$
AB,Ab	$(1+s)p_nQ_n \cdot (1-p_n)R_n$
AB,ab	$(1+s)p_nQ_n \cdot (1-p_n)(1-R_n) \cdot (1-r)$
Ab,aB	$(1-p_n)R_n \cdot (1+s)p_n(1-Q_n) \cdot r$

To explain the calculation, note that in the first case the offspring will always be AB. In the second and third, recombination has no effect. The offspring is AB $1/2$ of the time but there is a factor of 2 coming from the fact that the parents in the second case could be AB,aB or aB,AB. This $(1/2) \cdot 2$ occurs in the fourth and fifth cases as well. In the fourth, recombination must be avoided, while in the fifth it must occur to obtain the desired outcome. Adding up the first three rows with the part of the fourth that comes from the 1 in the $1 - r$ and then putting the rest in the second term, we have

$$(1+s)p_n Q_n \cdot ((1+s)p_n + 1 - p_n)$$
$$+ r(1+s)p_n(1-p_n)[R_n(1-Q_n) - Q_n(1-R_n)]$$

Dividing by the total number of offspring $(1 + p_n s)^2$, it follows that

$$p_{n+1}Q_{n+1} = \frac{(1+s)p_n Q_n(1+sp_n) + r(1+s)p_n(1-p_n)(R_n - Q_n)}{(1+sp_n)^2}$$

Rearranging gives

(a)
$$(1+sp_n)^2 p_{n+1}Q_{n+1} =$$
$$[(1+s)p_n]\{Q_n(1+sp_n) + r(1-p_n)(R_n - Q_n)\}$$

(6.29) implies that

$$(1+sp_n)^2 p_{n+1} = (1+s)p_n(1+sp_n)$$

Substituting this on the left-hand side of (a) and then dividing by $(1+s)p_n$, we have

(b)
$$(1+sp_n)Q_{n+1} = Q_n(1+sp_n) + r(1-p_n)(R_n - Q_n)$$

Considering Ab, we have

parents	expected number of Ab offspring
Ab,Ab	$(1-p_n)R_n \cdot (1-p_n)R_n$
Ab,ab	$(1-p_n)R_n \cdot (1-p_n)(1-R_n)$
Ab,AB	$(1-p_n)R_n \cdot (1+s)p_n Q_n$
Ab,aB	$(1-p_n)R_n \cdot (1+s)p_n(1-Q_n) \cdot (1-r)$
AB,ab	$(1+s)p_n Q_n \cdot (1-p_n)(1-R_n) \cdot r$

Adding things up as before, we get

(c) $$(1+sp_n)^2(1-p_{n+1})R_{n+1} = (1-p_n)R_n(1+sp_n)$$
$$+ r(1+s)p_n(1-p_n)(Q_n - R_n)$$

(6.29) implies that

$$(1+sp_n)^2(1-p_{n+1}) = (1+sp_n)^2 \cdot \frac{(1+sp_n) - (1+s)p_n}{1+sp_n} = (1+sp_n)(1-p_n)$$

Substituting this on the left-hand side of (c) and then dividing by $(1 - p_n)$, we have

$$(d) \qquad (1 + sp_n)R_{n+1} = R_n(1 + sp_n) + r(1 + s)p_n(Q_n - R_n)$$

Subtracting (d) from (b) and then dividing by $(1 + sp_n)$ we have

$$Q_{n+1} - R_{n+1} = Q_n - R_n - \frac{r(1 - p_n) + r(1 + s)}{(1 + sp_n)}(Q_n - R_n)$$
$$= (1 - r)(Q_n - R_n)$$

proving (6.32).   □

If we assume that initially $Q_0 = 0$ then

$$(e) \qquad\qquad Q_n - R_n = -R_0(1 - r)^n$$

Using (b) and (e), we have

$$Q_{n+1} - Q_n = \frac{r(1 - p_n)}{1 + sp_n}R_0(1 - r)^n$$

Using (6.30) twice, we have

$$1 - p_n = \frac{1 - p_0}{1 - p_0 + p_0(1 + s)^n} \qquad 1 + sp_n = \frac{1 - p_0 + p_0(1 + s)^{n+1}}{1 - p_0 + p_0(1 + s)^n}$$

Inserting this into the previous equation and summing, we have (6.31).   □

A second derivation

There is a different approach to this question, pioneered by Ohta and Kimura (1975) that leads to more insight. We assume that a favorable mutation B arises in the population at time $t = 0$ and is subsequently in the process of replacing allele b. As we have described in the first section of this chapter, the number of B's has a first phase in which it behaves like a supercritical branching process, a second in which the fraction of B's approximates the solution of the logistic differential equation, and a third in which the number of b's behaves like a subcritical branching process.

However, here we will simplify things by using the logistic differential equation to model the increase of p_t, the fraction of B alleles,

$$\frac{dp_t}{dt} = sp_t(1 - p_t) \tag{6.33}$$

We will call this the *logistic sweep model*. Using the notation introduced for Theorem 6.12, we have

Theorem 6.13. *Suppose that $Q_0 = 0$. In this case the frequency of the A allele after the sweep is*

$$Q_\infty = R_0(1 - p_0) \int_0^{2\tau} \frac{re^{-rt}}{(1 - p_0) + p_0 e^{st}} \, dt \qquad (6.34)$$

This is a close relative of the expression in (6.31):

$$Q_\infty = R_0(1 - p_0) \sum_{n=0}^{\infty} \frac{(1 - r)^n r}{1 - p_0 + p_0(1 + s)^{n+1}}$$

Proof. As one can easily check by differentiating, the solution to (6.33) is

$$p_t = \frac{p_0}{p_0 + (1 - p_0)e^{-st}} \qquad (6.35)$$

When $p_0 = \epsilon$ the solution reaches $1/2$ at the time τ when $\epsilon = (1 - \epsilon)e^{-s\tau}$, that is, when

$$\left(\frac{1 - \epsilon}{\epsilon}\right) e^{-s\tau} = 1$$

Solving gives $\tau = (1/s)\log((1/\epsilon) - 1) \approx -(1/s)\log(\epsilon)$. The solution in (6.35) has the symmetry property

$$p_{\tau+s} = 1 - p_{\tau-s} \qquad (6.36)$$

so it will reach $1 - \epsilon$ at time 2τ. The next graph gives a picture of p_t when $\epsilon = 1/200$ and $s = 0.02$. In this case $\tau = 264.67$.

To relate the selective sweep to population subdivision, we divide the population into two parts: chromosomes with the advantageous mutant B and those with the original allele b. Let $v(t)$ be the frequency of allele A in the B population and $w(t)$ be the frequency of allele A in the b population after t generations. Recombinations between chromosomes from the two populations occur at rate $rp_t(1 - p_t)$ so we have

$$\frac{dv}{dt} = r(1 - p_t)(w(t) - v(t))$$
$$\frac{dw}{dt} = rp_t(v(t) - w(t))$$

Subtracting the second equation from the first gives

$$\frac{d}{dt}(v(t) - w(t)) = -r(v(t) - w(t))$$

so we have

$$v(t) - w(t) = e^{-rt}(v(0) - w(0)) \qquad (6.37)$$

which is the analogue of (6.32). Using (6.37) and (6.35) in our pair of differential equations gives

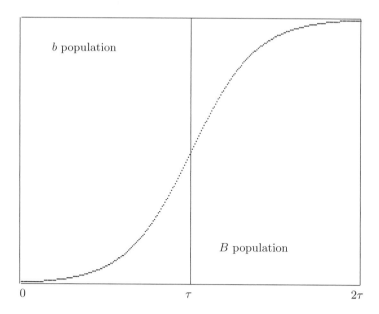

Fig. 6.9. Rise of the favored allele in a selective sweep. ϵ is the initial frequency, s is the selective advantage, and $\tau = (1/s) \log((1/\epsilon) - 1)$.

$$v(2\tau) = v(0) + r(w(0) - v(0)) \int_0^{2\tau} \frac{(1 - p_0)e^{-st-rt}}{p_0 + (1 - p_0)e^{-st}} \, dt \qquad (6.38)$$

$$w(2\tau) = w(0) + r(v(0) - w(0)) \int_0^{2\tau} \frac{p_0 e^{-rt}}{p_0 + (1 - p_0)e^{-st}} \, dt$$

When $v(0) = 0$ and $w(0) = R_0$, the first equation becomes (6.34) and the proof is complete. □

Let $v_1(t)$ be the solution given in (6.38) with $R_0 = 1$. A little thought reveals that $v_1(t)$ is the probability that the neutral locus of an individual in the B population at time t comes from the b population at time 0.

$$v_1(2\tau) = r \int_0^{2\tau} \frac{(1 - p_0)e^{-(r+s)t}}{p_0 + (1 - p_0)e^{-st}} \, dt$$

The integral is what Stephan, Wiehe, and Lenz (1992) call $I(\tau)$, except that they replace $1 - p_0$ by 1. To get an approximation to $v_1(2\tau)$, note that the expected number of times a lineage will cross from the B population to the b population during the sweep is

$$v_1(2\tau) \approx r \int_0^{2\tau} (1 - p_t) \, dt = r\tau = \frac{r}{s} \log(1/\epsilon) \qquad (6.39)$$

since the symmetry property (6.36) implies

$$\int_0^\tau (1 - p_{\tau-s} + 1 - p_{\tau+s})\, ds = \int_0^\tau 1\, ds$$

Since the number of crossings should have approximately a Poisson distribution, the probability of at least one is

$$1 - \exp(-(r/s)\log(1/\epsilon)) = 1 - \epsilon^{r/s} \tag{6.40}$$

The most natural value to use here is $\epsilon = 1/2N$, but we will see later in the discussion of (6.46) that it is better to set $\epsilon = 1/2Ns$.

It is not hard to show, see e.g., the appendix of Durrett and Schweinsberg (2004), that the result in (6.40) becomes exact when $N \to \infty$.

Theorem 6.14. *Under the logistic sweep model, if $N \to \infty$ and $r\log(2N)/s \to a$ then*

$$v_1(2\tau) \to 1 - e^{-a}.$$

The reader should note that since we assume $r\log(2N)/s \to a$ this is a little different from the rule of thumb that, see e.g., Nurminsky (2001), "hitchhiking of the neighboring neutral locus is efficient if $r < s$ and becomes negligible if $r \approx s$."

To judge the quality of the approximation, we consider the situation with a population size of $N = 10{,}000$ and selection coefficient $s = 0.01$ for various values of recombination $r = 0.0001$ ot 0.001. The curves compare the exact value from (6.31) with the approximation from (6.40). Some authors, see e.g., Fay and Wu (2000) page 1407, suggest the use of (6.39), but as the graph shows, this is not very accurate.

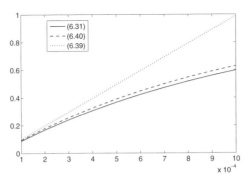

Fig. 6.10. Comparison of approximations for the probability a lineage does not escape from the selective sweep.

Site frequency spectrum

The results above show how the frequencies of linked neutral alleles are changed by a selective sweep. Using them we can determine the effect of hitchhiking on the site frequency spectrum.

Theorem 6.15. *Under the logistic sweep model, if $c = 1 - \epsilon^{r/s}$ then the site frequency spectrum after the sweep is*

$$\phi_c(x) = \left(\frac{\theta}{x} - \frac{\theta}{c}\right) 1_{[0,c]}(x) + \frac{\theta}{c} 1_{[1-c,1]}(x) \qquad (6.41)$$

where $1_{[a,b]}(x)$ is 1 on $[a, b]$ and 0 otherwise.

This is a result from Fay and Wu (2000), see their equations (5) and (6), which has been rewritten to allow for the possibility that $c > 1/2$ in which case the two regions overlap.

Proof. This is a fairly straightforward calculation but it is more easily done in terms of the distribution function, rather than in terms of the density function where one has to take into account how the two transformations stretch the space.

Returning to (6.38), and recalling that v_1 is the solution with $v(0) = 0$ and $w(0) = 1$ we have

$$v(2\tau) = v(0) + (w(0) - v(0))v_1(2\tau)$$
$$\approx v(0) + (w(0) - v(0))\left(1 - \epsilon^{r/s}\right)$$

From this, we see that if $w(0) = p_0$ is the frequency in the b population at time 0 then the frequency in the B population will be $v(0) = 1$ with probability p_0 and $v(0) = 0$ with probability $1 - p_0$, so the ending frequencies will be $(1 - c) + cp_0$ with probability p_0 and cp_0 with probability $1 - p_0$.

The second term in (6.41) comes from the fact that the new frequency $p' = (1 - c) + cp_0$ with probability p_0. In order for $p' \geq y > 1 - c$ we must have $p_0 \geq [y - (1 - c)]/c$. Taking into account the probability of this case is $p_0 = x$ we have

$$\int_{[y-(1-c)]/c}^{1} x \cdot \frac{\theta}{x} \, dx = \frac{\theta}{c}[c - y + (1 - c)] = \int_y^1 \frac{\theta}{c} \, dx$$

The first term in (6.41) comes from the fact that the new frequency $p' = cp_0$ with probability $1 - p_0$. In order for $p' \in [y, c]$ with $y < c$ we must have $p_0 \in [y/c, 1]$. Taking into account the probability of this case is $1 - p_0 = 1 - x$ we have

$$\int_{y/c}^{1} (1 - x) \cdot \frac{\theta}{x} \, dx = -\theta \log(y/c) - \theta \frac{c - y}{c} = \int_y^c \frac{\theta}{x} - \frac{\theta}{c} \, dx$$

Combining the two calculations gives the desired result. $\qquad \square$

Samples of size $n \geq 2$

Theorem 6.14 concerns the effect of a selective sweep on a single lineage. As Kaplan, Hudson, and Langley (1989) observed in their equation (16), the heterozygosity after the sweep H_∞ is related to that before the sweep H_0 by

$$\frac{H_\infty}{H_0} = p_{22} \tag{6.42}$$

where p_{22} is the probability that two lineages sampled from the B population at time 2τ are distinct at time 0. Since the time of the sweep is much smaller than the population size, two lineages will coalesce only if they both end up in the B population at time 2τ. That is,

$$p_{22} = 1 - (1 - v_1(2\tau))^2 = v_1(2\tau)(2 - v_1(2\tau)) \tag{6.43}$$

which is equation (14a) of Stephan, Wiehe, and Lenz (1992).

Using this we obtain formula (13) from Kim and Stephan (2000).

Theorem 6.16. *Under the logistic sweep model, the nucleotide diversity π in a population that experienced a selective sweep $2N\tau$ generations ago has*

$$E\pi = \theta[1 - (1 - p_{22})e^{-\tau}] \tag{6.44}$$

Proof. If T_s is the coalescence time in the presence of the sweep then $E\pi = \theta ET_s$. Let $T_0 \geq T_s$ be the coalescence time without the sweep. $T_s \neq T_0$ only if there is no coalescence by time τ, an event of probability $e^{-\tau}$, and the sweep causes the two lineages to coalesce, an independent event of probability $1 - p_{22}$. When these two events occur $T_s = \tau$. The lack of memory property of the exponential implies that $E(T_0 - \tau | T_0 > \tau) = 1$, so we have

$$ET_0 - ET_s = e^{-\tau}(1 - p_{22}) \cdot 1$$

which proves the desired result. $\square$

As Proposition 2 in Durrett and Schweinsberg (2004) shows, the reasoning that led to (6.43) implies

Theorem 6.17. *Under the logistic sweep model, if $N \to \infty$ with $r \ln(2N)/s \to a$ and $s(\ln N)^2 \to \infty$ then for $j \geq 2$*

$$p_{k,k-j+1} \to \binom{k}{j} p^j (1-p)^{k-j} \qquad \text{where } p = e^{-a}$$

and $p_{k,k} \to (1-p)^k + kp(1-p)^{k-1}$.

In the special case $k = 2$ we have

$$p_{22} = (1-p)^2 + 2p(1-p) = 1 - p^2$$

In words, the two lineages do not coalesce unless neither escapes from the sweep.

6.6 Better approximations

To evaluate the approximations proposed in the previous section, we turn to simulation, starting with a sample of size 2. Let $p2inb$ be the probability both lineages escapes the sweep and do not coalesce, let $p2cinb$ be the probability both lineage escapes the sweep but coalesce, and let $p1B1b$ be the probability one lineage escapes but the other does not. $p_{22} = p2inb + p1B1b$. Theorem 6.17 implies that if $r \log(2N)/s \to a$ then

$$p2inb \to p^2 \qquad p2cinb \to 0 \qquad p1B1b \to 2p(1-p) \qquad (6.45)$$

where $p = e^{-a}$ is the limiting probability that one lineage escapes from the sweep.

In the first set of simulations $N = 10{,}000$, $s = 0.1$, and the values of r have been chosen to make the approximate probabilities of escaping from the sweep ($pinb$) equal to 0.1 and 0.4 respectively. The first row in each group gives the probabilities obtained from the approximation in Theorem 6.17. The second row gives the result of integrating the differential equations for the probabilites. The third row gives the average of 10,000 simulation runs of the Moran model conditioned on fixation of the advantageous allele. The fourth row gives the values from the much more accurate Theorem 6.19 which will be stated in a moment.

		$pinb$	$p2inb$	$p2cinb$	$p1B1b$	p_{22}
	Th 6.17	0.1	0.01	0	0.18	0.19
$r = 0.00106$	logistic	0.09983	0.00845	0.03365	0.11544	0.12390
	Moran	0.08203	0.00620	0.01826	0.11513	0.12134
	Th 6.19	0.08235	0.00627	0.01765	0.11687	0.12314
	Th 6.17	0.4	0.16	0	0.48	0.64
$r = 0.00516$	logistic	0.39936	0.13814	0.09599	0.32646	0.46460
	Moran	0.33656	0.10567	0.05488	0.35201	0.45769
	Th 6.19	0.34065	0.10911	0.05100	0.36112	0.47203

Note that Theorem 6.17 does a good job of approximating the probabilities of escaping from the sweep, but the other approximations on the first row are poor. The logistic overestimates the true probability of escaping the sweep by about 20%, but does a good job of computing p_{22}, despite not being very accurate for the two probabilities make up p_{22}.

Results for $N = 100{,}000$, $s = 0.03$ are similar, but this time are based on only 1,000 simulations. Again the values of r have been chosen to make the approximate probabilities of escaping from the sweep ($pinb$) equal to 0.1 and 0.4 respectively.

		$pinb$	$p2inb$	$p2cinb$	$p1B1b$	$p22$
	Th 6.17	0.1	0.01	0	0.18	0.19
$r = .000259$	logistic	0.09989	0.00837	0.03034	0.11097	0.11935
	Moran	0.07675	0.00554	0.01545	0.11149	0.11704
	Th 6.19	0.07671	0.00553	0.01494	0.11249	0.11803
	Th 6.17	0.4	0.16	0	0.48	0.64
$r = .001256$	logistic	0.39826	0.13808	0.10204	0.31627	0.45437
	Moran	0.31846	0.09641	0.04581	0.35246	0.44888
	Th 6.19	0.32074	0.09790	0.04409	0.35750	0.45540

Stick breaking construction

We now begin the description of the more accurate approximation due to Durrett and Schweisnberg (2004). They describe the impact of a sweep on a sample of size k as a marked partition Π_s of $\{1, \ldots, k\}$. The integers i and j will be in the same block of the partition if and only if the ith and jth lineages in the sample coalesce during the sweep and the block will be marked if the individuals in the block are descended from the one who started the selective sweep.

In the approximation of Theorem 6.17, we made a partition Π_1 by flipping k coins with a probability $p = e^{-a}$ of heads. All of the integers with heads are combined into one marked block. They are the lineages that did not escpe from the sweep. The other integers are singletons in the partition. This partition has only one block with more than one integer, because in this approximation only the lineages that get trapped in the B population coalesce. The more accurate approximation allows multiple blocks to have more than one integer.

To define the new random partition Π_2, we will use a stick-breaking construction, which is essentially the same as the paintbox construction of Kingman (1978). The construction is simple to describe, but it will take us a while to explain the intuition behind it. The ingredients for the construction are as follows:

- Let $M = [2Ns]$, where $[m]$ is the greatest integer $\leq m$.
- Let ξ_l, $2 \leq l \leq M$, be independent Bernoulli random variables that are 1 with probability r/s and 0 otherwise.
- Let W_l, $2 \leq l \leq M$, be independent random variables with W_l having a beta$(1, l-1)$ distribution.
- For $2 \leq l \leq M$, let $V_l = \xi_l W_l$, and let $T_l = V_l \prod_{i=l+1}^{M}(1 - V_i)$.
- Let $T_1 = \prod_{l=2}^{M}(1 - V_l)$.

In words, $\prod_{i=l+1}^{M}(1 - V_i)$ is the length of the stick at time $\ell + 1$, and we break off a fraction V_ℓ of what remains. Now, divide the interval $[0, 1]$ into M subintervals (some of which may be empty) as follows. Let $a_{M+1} = 1$ and for $1 \leq l \leq M$, let $a_l = a_{l+1} - T_l$. Since $\sum_{l=1}^{M} T_l = 1$, we have $a_1 = 0$. Let $I_l = [a_l, a_{l+1}]$.

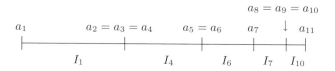

Fig. 6.11. Example of the stick breaking construction.

In the picture drawn $M = 10$, the Bernoulli variables ξ_2, ξ_3, ξ_5, ξ_8, $\xi_9 = 0$, so the corresponding $V_\ell = 0$.

To obtain a partition of $\{1, \ldots, k\}$, let $U_1, \ldots, U_k$ be i.i.d. random variables with a uniform distribution on $[0, 1]$. We declare i and j to be in the same block of the partition if and only if U_i and U_j are both in the interval I_l for some l. Since we wish also to keep track of which lineages are descended from the B population and which come from the b population, we will mark, with probability $s/(r(1 - s) + s)$, the block of the partition containing all of the i such that U_i is in I_1 to indicate that these lineages did not escape from the sweep. This defines the marked partition Π_2. The main result of Schweinsberg and Durrett (2005) is

Theorem 6.18. *There is a positive constant C so that for all N and marked partitions π,*

$$|P(\Pi_s = \pi) - P(\Pi_2 = \pi)| \leq C/(\log N)^2.$$

A similar result has been derived by Etheridge, Pfaffelhuber, and Wakolbinger (2006) by considering the diffusion limit with $2Ns \to \alpha$ and then letting $\alpha \to \infty$. Recent work of Pfaffelhuber, Haubold, and Wakolbinger (2006), Pfaffelhuber and Studeny (2007), and Eriksson, Fernström, Mehlig, and Sagitov (2007) has resulted in various improvements to the approximation or to associated simulation procedures.

In contrast the error in the approximation by Π_1 from Theorem 6.17 is of order $C/(\log N)$, since as we work backwards in time, with probability $\approx r/s$ there is a recombination at the last moment which takes all of the coalesced lineages to the b population. When $N = 10,000$, $\log N = 9.214$ so the $C/(\log N)^2$ error may not seem to be much of an improvement, but as the numbers in the tables indicate the new approximation is considerably more accurate.

Sketch of proof. Our approximation in Theorem 6.18 is based on the following ideas:

1. We ignore the possibility that a lineage experiences two recombinations during the sweep, taking it from the B population to the b population and back to the B population.

2. When the number of chromosomes with the B allele is much smaller than the population size, the number of individuals with the B allele can be approximated by a continuous-time branching process, in which each individual splits into two at rate 1 and dies at rate $1 - s$. We don't care about the lines of descent in the branching process that die out. Known results, see for example O'Connell (1993), imply that the lineages in a branching process that do not die out are themselves a branching process, in our case the Yule process, a continuous time branching process in which each particle splits into two at rate s. Since each lineage has an infinite line of descent with probability s, the number of such lineages at the end of the sweep is approximately $M = \lfloor 2Ns \rfloor$.

3. When there are $l \geq 2$ lineages in the Yule process, the time to the next birth is exponentially distributed with mean $1/sl$, and recombinations occur at rate lr, so the expected number of recombination events is r/s. We assume that the number of such events is always 0 or 1. The Bernoulli variables ξ_l, $2 \leq l \leq M$ tell us whether one occurs or not.

4. As time tends to infinity, the number of individuals in the Yule process divided by its mean converges to an exponential distribution with mean 1. See Theorem 1.15. This implies that when there are l lineages, the fraction of individuals at the end of the sweep that are descendants of a given lineage has roughly the same distribution as $\xi_1/(\xi_1 + \cdots + \xi_l)$ where ξ_i are independent exponentails. The ratio has a beta$(1, l-1)$ distribution. Thus, the W_l, $2 \leq l \leq M$ represent the fraction of descendants of an individual in the Yule process when there are l individuals.

5. If $\xi_l = 0$ there is no recombination and $V_l = 0$. If $\xi_l = 1$ there is a recombination that removes a fraction V_l of the remaining population, i.e., the fraction of individuals that recombine at time l is V_l. $T_1 = \prod_{l=2}^{M}(1 - V_l)$ is the fraction of the initial population that trace their ancestry back to the B population at the time when there was one lineage in the Yule process.

6. The probability a recombination happens before the first birth in the Yule process is $r(1 - s)/(r(1 - s) + s)$. In this case, no lineage for the neutral locus comes from the B population.   □

The procedure described above translates directly into a procedure for simulating genealogies for a linked neutral locus, which is much simpler than Kaplan, Hudson, and Langley (1989). However, in the case of one or two lineages one can compute the probabilities of interest analytically.

Theorem 6.19. *For the approximation Π_2, we have*

$$pinB = \frac{s}{r(1-s)+s}\prod_{l=2}^{M}\left(1-\frac{r}{sl}\right) \qquad pinb = 1 - pinB$$

$$p2inB = \frac{s}{r(1-s)+s}\prod_{l=2}^{M}\left(1-\frac{2r}{s(l+1)}\right)$$

$$p2cinb = \frac{r(1-s)}{r(1-s)+s}\prod_{l=2}^{M}\left(1-\frac{2r}{s(l+1)}\right)$$

$$+\sum_{i=2}^{M}\frac{2r}{si(i+1)}\prod_{l=i+1}^{M}\left(1-\frac{2r}{s(l+1)}\right)$$

$p1B1b = 2(pinB - p2inB)$ *and* $p_{22} = 1 - p2inB - p2cinb$.

Proof. Since W_l has a beta$(1, l-1)$ distribution, which has density function $(l-1)(1-x)^{l-2}$, integration shows that $E[W_l] = 1/l$ and $E[W_l^2] = 2/l(l+1)$. To calculate $pinB$, first note that $pinB = [s/(r(1-s)+s)]P(U_1 \in I_1)$ where U_1 is uniform on $(0,1)$. If U_1 is not in any of the intervals $I_{l+1}, \ldots, I_M$, then the probability, conditional on V_l, that $U_1 \in I_l$ is V_l. Therefore, for $2 \le l \le M$, we have

$$P(U_1 \in I_l | U_1 \notin I_{l+1} \cup \cdots \cup I_M) = E[V_l] = E[\xi_l]E[W_l] = \frac{r}{sl}$$

It follows that

$$P(U_1 \in I_1) = \prod_{l=2}^{M}P(U_1 \notin I_l | U_1 \notin I_{l+1} \cup \cdots \cup I_M) = \prod_{l=2}^{M}\left(1-\frac{r}{sl}\right)$$

which implies the first statement.

For the second, note that if U_2 is independent of U_1 and uniform on $(0,1)$, then

$$p2inB = \frac{s}{(r(1-s)+s)}P(U_1, U_2 \in I_1)$$

If U_1 and U_2 are not in any of the intervals $I_{l+1}, \ldots, I_M$, then the probability, conditional on V_l, that either U_1 or U_2 is in I_l is $1-(1-V_l)^2$. A little calculation shows

$$E[1 - (1-V_l)^2] = 2E[V_l] - E[V_l^2] = E[\xi_l](2E[W_l] - E[W_l^2])$$
$$= \frac{r}{s}\left(\frac{2}{l} - \frac{2}{l(l+1)}\right) = \frac{2r}{s(l+1)}$$

and the formula for $p2inB$ now follows by the same reasoning as the formula for $pinB$.

Finally, to obtain the formula for $p2cinb$, we note that

$$p2cinb = \frac{r(1-s)}{r(1-s)+s}P(U_1, U_2 \in I_1) + \sum_{i=2}^{M}P(U_1, U_2 \in I_i).$$

From the calculation for $p2inB$, we know that the probability that U_1 and U_2 are not in any of the intervals $I_{i+1}, \ldots, I_M$ is

$$\prod_{l=i+1}^{M} \left(1 - \frac{2r}{s(l+1)}\right)$$

This formula when $i = 1$ gives $P(U_1, U_2 \in I_1)$. Conditional on the event that U_1 and U_2 are not in any of the intervals $I_{i+1}, \ldots, I_M$, the probability that U_1 and U_2 are both in I_i is

$$E[V_i^2] = E[\xi_i]E[W_i^2] = 2r/[si(i+1)].$$

By combining these observations, we obtain the desired formula for $p2cinb$. The last two relationships follow easily from the definitions. □

The expressions in Theorem 6.19 are exact, but can be simplified without much loss of accuracy by using $1 - x \approx e^{-x}$. Consider for example $pinB$. Rewriting the first fraction and dropping the $r(1-s)$ from the denominator (recall $r = as/\log(2N)$), and using $M = [2ns]$,

$$pinB = \left(1 - \frac{r(1-s)}{r(1-s)+s}\right) \prod_{l=2}^{M} \left(1 - \frac{r}{sl}\right)$$

$$\approx \exp\left(-\frac{r(1-s)}{s} - \sum_{l=2}^{M} \frac{r}{sl}\right)$$

$$= \exp\left(-\frac{r}{s}\left[-s + \sum_{l=1}^{M} \frac{1}{l}\right]\right)$$

$$\approx \exp\left(-\frac{r}{s}\left[\ln(2Ns) + \gamma - s\right]\right) \tag{6.46}$$

where $\gamma = \lim_{k\to\infty} \sum_{j=1}^{k} \frac{1}{j} - \ln k \approx 0.57721$ is Euler's constant. If one ignores the $\gamma - s$, this corresponds to taking $\epsilon = 1/2Ns$ in (6.40). This provides an explanantion for why Kim and Stephan like this choice of ϵ in Theorem 6.41.

Stephan, Wiehe, and Lenz (1992) found a different approximation for p_{22} by writing differential equations for the second moments of the allele freqeuncy and then approximately solving the equations. Their result, which is given in (17) of their paper, is

$$p_{22} \approx \frac{2r}{s}\alpha^{-2r/s} \int_{1/\alpha}^{1} z^{-(2r/s)-1}e^{-z}\, dz \tag{6.47}$$

where $\alpha = 2Ns$ and we have taken $\epsilon = 1/2Ns$. This does not look much like the approximation from Theorem 6.19, in which $M = [2Ns]$:

$$p_{22} = 1 - \sum_{i=1}^{M} \frac{2r}{si(i+1)} \prod_{l=i+1}^{M} \left(1 - \frac{2r}{s(l+1)}\right)$$

but the numerical values are similar. To compare the two approximations, we use a case considered in Kim and Stephan (2002); we take $N = 200,000$ and $s = 0.001$, and use values of r that, assming a per nucleotide recombination probability of 10^{-8} correspond to 1, 4, 7, and 10 kilobases.

r	(6.47)	Logistic	Theorem 6.19	Theorem 6.17
0.00001	0.098790	0.098363	0.087375	0.227393
0.00004	0.338821	0.331240	0.305675	0.643684
0.00007	0.513202	0.494525	0.470897	0.835672
0.00010	0.640356	0.610135	0.595986	0.924214

The column labeled logistic gives the values obtained by solving the differential equations exactly. The approximation in (6.47) is always slightly larger than the value computed from the logistic, while the more accurate approximation from Theorem 6.19 is somewhat smaller. We have listed the rather awful approximations from Theorem 6.17 to show that the differences between the last three columns are not as large as they might first appear.

Using (6.44) as an approximation for p_{22} gives an approximation for the expected number of pairwise differences. To compare with Figure 2 in Kim and Stephan (2002) we again take $N = 200,000$, $s = 0.001$, and assume a recombination probability of 10^{-8} per nucleotide per generation. Setting $\theta = 0.01$ assuming that the time since the sweep occured at 20,000 is $\tau = 0.005$ in units of $2N$ generations, we have the results plotted in Figure 6.12. The upper line uses the Kim and Stephan approximation of p_{22}. The thicker line below comes from solving the differential equation. The third dotted curve is Durrett and Schweinsberg's result from Theorem 6.19.

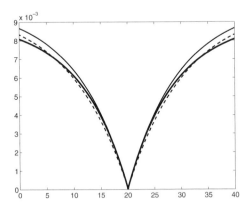

Fig. 6.12. Nucleotide diversity versus position in kilobases along the chromosome

6.7 Recurrent sweeps

Begun and Aquadro (1992) observed that in *Drosophila melanogaster* there is a positive correlation between nucleotide diversity and recombination rates. This observation may be explained by repeated episodes of hitchhiking caused by the fixation of newly arising advantageous mutations, because the average size of the region affected depends on the ratio s/r. Thus if selection intensities are similar across the genome, loci in regions of low recombination will be affected by more sweeps per unit time. Formula (6.23) shows that background selection has a greater impact in regions of low recombination, so as B. Charlesworth, Morgan and D. Charlesworth (1993) have suggested, it also provides an explanation.

To try to distinguish between the two explanations one must focus on other features of the data, such as the site frequency spectrum. Since background selection can be modeled as a reduction in the effective population size, it does not change the distribution of allele frequencies. However, the fixation of advantageous alleles is expected to result in an increase in the number of low frequency alleles because selective sweeps reduce variability, and then new mutations have low frequencies. In this section, we will try to quantify the effects of recurrent selective sweeps.

6.7.1 Nucleotide diversity

To consider in detail the impact of selective sweeps on genetic variability, we will now consider recurring selected substitutions that occur according to a Poisson process at rate ν per nucleotide per generation. If the physical distance of a selected substitution from the neutral region is m nucleotides, then its recombinational distance is ρm, where ρ is the recombination rate per nucleotide per generation. Recall p_{22} is the probability that two lineages do not coalesce during the selective sweep. Writing $p_{22}(r, s, N)$ to display the dependence of the probability on the population parameters, we can see that selective sweeps cause coalescence at an additional rate (per $2N$ generations) of

$$\lambda = 2N \cdot 2\nu \int_0^\infty 1 - p_{22}(\rho x, s, N)\, dx$$

$$= 4N\frac{\nu}{\rho} \int_0^\infty 1 - p_{22}(y, s, N)\, dy \tag{6.48}$$

so the time to coalescence has an exponential distribution with rate $1 + \lambda$. A coalescence caused by a sweep forces the state of the neutral allele to be equal in the two sampled individuals, so the heterozygosity under repeated sweeps is $1/(1 + \lambda)$ times that under the neutral theory. In symbols

$$H = \frac{H_{neu}}{1 + \lambda} \tag{6.49}$$

This is equation (21) of Stephan, Wiehe and Lenz (1992).

Stephan, Wiehe, and Lenz (1992) developed an approximation of p_{22}, see (6.47), which allowed Wiehe and Stephan (1993) to conclude that

$$2 \int_0^\infty 1 - p_{22}(y, s, N)\, dy \approx sI$$

where $I \approx 0.075$ is a constant. Using this in (6.48) we have $\lambda \approx I\alpha\nu/\rho$ where $\alpha = 2Ns$. Inserting this result into (6.49) we have

$$H \approx H_{neu} \cdot \frac{\rho}{\rho + \nu\alpha I} \tag{6.50}$$

From this, it follows that

$$\frac{1}{H} = \frac{1}{H_{neu}} + \frac{\nu\alpha I}{H_{neu}} \cdot \frac{1}{\rho} = \beta_1 + \beta_2 \cdot \frac{1}{\rho} \tag{6.51}$$

Wiehe and Stephan (1993) used this relationship to fit data on 17 *Drosophila melanogaster* loci from Begun and Aquadro (1992). Taking H to be the nucleotide diversity π, they found $\beta_1 = 125.54$ and $\beta_2 = 5.04 \times 10^{-7}$, which corresponds to $H_{neu} = 0.008$.

To interpret the parameters, we note that if we assume the per nucleotide mutation rate is 1×10^{-9}, then since $0.008 = H_{neu} = 4N_e\mu$, N_e is estimated to be 2×10^6. Assuming an average selective effect of 0.01, then $\alpha = 2N_es = 4 \times 10^4$ and

$$\nu = \frac{\beta_2 H_{neu}}{I\alpha} = \frac{(5.04 \times 10^{-7}) \cdot 0.008}{0.075 \cdot 4 \times 10^4} = 1.34 \times 10^{-12}$$

Comparing this to the mutation rate 10^{-9}, we see that $1/(1.34 \times 10^{-3})$, or 1 out of every 746 mutations, are driven to fixation by selection. Aquadro, Begun, and Kindahl (1994) repeated these calculations for 15 gene regions of the third chromosome of *Drosophila* with almost the same result.

The change of variables in (6.51) is called the Linweaver-Burk transformation. This procedure is generally used in biochemistry to fit models of the Michaelis-Menten type to data. This procedure has two problems: the estimates have large variance, and β_2 is a combination of H_{neu} and $\alpha\nu$ so it is not possible to provide confidence limits for the estimate of $\alpha\nu$. Stephan (1995) instead used the equation

$$\pi = H_{neu} - I\alpha\nu\frac{\pi}{\rho}$$

and reanalyzed the data of Aquadro, Begun, and Kindahl (1994) but the parameter estimates were almost identical to those in the first study. The next figure compares the fitted curve to the data.

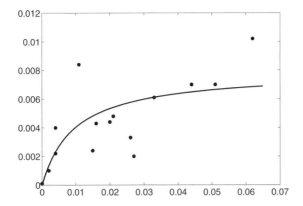

Fig. 6.13. Stephan's (1995) fit to data of Aquadro, Begun, and Kindahl (1994).

6.7.2 Genealogies

Using results from Section 6.4, we can obtain an approximation for the behavior of genealogies under recurrent selective sweeps. Let σ be a finite measure on $[-L, L] \times [0, 1]$. $\sigma([a, b] \times [c, d])$ gives the rate at which successful selective sweeps occur in the chromosome interval $[a, b]$ with selective advantage $c \leq s \leq d$. Let $r : [-L, L] \rightarrow [0, \infty)$ be a continuous function such that $r(0) = 0$ and r is nonincreasing on $[-L, 0]$ and nondecreasing on $[0, L]$. $r(x)$ gives the rate at which recombinations occur between 0 and x. Adding subscripts σ_N and r_N to indicate the dependence on the population size, Theorem 2.2 of Durrett and Schweinsberg (2005) shows

Theorem 6.20. *Suppose that, as* $N \rightarrow \infty$, *the measures* $2N\sigma_N$ *converge weakly to* σ *and the functions* $(\log 2N)r_N$ *converge uniformly to* r. *Let* η *be the measure on* $(0, 1]$ *such that*

$$\eta([y, 1]) = \int_{-L}^{L} \int_{0}^{1} 1_{\{e^{-r(x)/u} \geq y\}} \, \mu(dx \times du)$$

for all $y \in (0, 1]$. *Let* Λ *be the measure on* $[0, 1]$ *defined by* $\Lambda = \delta_0 + \Lambda_0$, *where* $\Lambda_0(dx) = x^2\eta(dx)$. *Then, as* $N \rightarrow \infty$, *the genealogy of the locus at 0 converges to the* Λ *coalescent defined by (4.1).*

The result in the paper cited has a u multiplying the indicator. That factor is missing because here our rates concern successful selective sweeps. The conclusion given above is a simple consequence of Theorem 6.17. A successful selective sweep that occurs at a recombination distance of r causes lineages to be trapped by the sweep with probability $p = e^{-r(\log 2N)/s}$.

We now derive the limiting Λ-coalescent in two natural examples.

Example 6.7. Selection at one locus. Consider the case in which we are concerned only with mutations at a single site, which we think of a rapidly evolving gene. To simplify, we assume that all mutations have the same selective advantage, but this assumption can easily be relaxed. Fix $\gamma > 0$, and let $\sigma_N = (\gamma/2N)\delta_{(z,s)}$ where $\delta_{(z,s)}$ is a point mass at (z,s) with $z \in [-L, L]$ and $s \in (0, 1]$. This means that beneficial mutations that provide selective advantage s appear on the chromosome at site z at times of a Poisson process with rate $\gamma/2N$. The measures $2N\sigma_N$ converge to $\sigma = \gamma\delta_{(z,s)}$. Assume that the recombination functions r_N are defined such that the sequence $(\log 2N)r_N$ converges uniformly to r, and let $\beta = r(z)$. Then, for all $y \in (0, 1]$, we have

$$\eta([y, 1]) = \int_{-L}^{L} \int_{0}^{1} 1_{\{e^{-r(x)/u} \geq y\}} \mu(dx \times du) = \gamma 1_{\{e^{-\beta/s} \geq y\}}.$$

Therefore, η consists of a mass γ at $p = e^{-\beta/s}$. It follows from Theorem 6.20 that the limiting coalescent process is the Λ-coalescent, where $\Lambda = \delta_0 + \gamma p^2 \delta_p$. Recalling the definition of a p-merger given in Example 4.1, we see that in addition to the mergers involving just two blocks, we have coalescence events at times of a Poisson process with rate γ in which we flip coins with probability p of heads for each lineage and merge the lineages whose coins come up heads.

Example 6.8. Mutations along a chromosome. It is also natural to consider the case in which favorable mutations occur uniformly along the chromosome. For simplicity, we will assume that the selective advantage s is fixed. Let dx denote Lebesgue measure on $(-\infty, \infty)$. (To treat this case, we have to first apply the result to $[-L, L]$ and then let $L \to \infty$.)

Suppose $\sigma_N = (2N)^{-1}(\gamma dx \times \delta_s)$, so the measures $2N\sigma_N$ converge to $\sigma = \gamma dx \times \delta_s$. To model recombinations occurring uniformly along the chromosome, we assume that the functions $(\log 2N)r_N$ converge uniformly to the function $r(x) = \beta|x|$, so the probability of recombination is proportional to the distance between the two sites on the chromosome. For all $y \in (0, 1]$, we have

$$\eta([y, 1]) = \gamma \int 1_{\{e^{-\beta|x|/s} \geq y\}} \, dx.$$

Since $e^{-\beta|x|/s} \geq y$ if and only if $|x| \leq -(s/\beta)(\log y)$, we have

$$\eta([y, 1]) = \frac{-2\gamma s \log y}{\beta}$$

$$-\frac{d}{dy}\eta([y, 1]) = \frac{2\gamma s}{\beta y}$$

By Theorem 6.20, the genealogies converge to a Λ-coalescent with $\Lambda = \delta_0 + \Lambda_0$ where Λ_0 has density $h(y) = cy$ with $c = 2\gamma s^2/\beta$.

To connect these parameters with the ones used by Braverman et al. (1995), suppose x is the distance along the chromosome in Morgans. Their Λ_r, which is the hitchhiking events per $2N$ generations per recombination unit, is $2N\gamma$, and $\alpha = 2Ns$ so $\gamma s = \Lambda_r \alpha$. In their simulations $2N = 10^8$, so $\beta = \log(2N) = 18.42$.

6.7.3 Segregating sites

Suppose the ancestral history of a sample of n individuals is given by a Λ-coalescent. Let λ_b be the total rate of all mergers when the coalescent has b blocks, see (4.2). Assume that, on the time scale of the coalescent process, mutations happen at rate $\theta/2$. Let Δ_{ij} be the number of sites at which the ith and jth sequences differ and let $\Delta_n = \binom{n}{2}^{-1} \sum_{i<j} \Delta_{ij}$ be the average number of pairwise differences. Since the average time for two lineages to coalesce is λ_2^{-1}

$$E\Delta_n = E\Delta_{ij} = \theta/\lambda_2 \qquad (6.52)$$

To compare with (6.49) we note that $\lambda_2 = 1 + \lambda$, where λ is the rate of coalescence due to selective sweeps.

To calculate the expected number of segregating sites, we note that any mutation in the ancestral tree before all n lineages have coalesced to 1 adds to the number of segregating sites. If, at some time, the coalescent has exactly b blocks, the expected time that the coalescent has b blocks is λ_b^{-1}. Let $G_n(b)$ be the probability that the coalescent, starting with n blocks, will have exactly b blocks at some time. Then

$$E[S_n] = \frac{\theta}{2} \sum_{b=2}^{n} b\lambda_b^{-1} G_n(b). \qquad (6.53)$$

Although we do not have a closed-form expression for $G_n(b)$, these quantities can be calculated recursively because the transition rates of the Λ-coalescent allow us to express $G_n(b)$ in terms of $G_k(b)$ for $k < n$.

We will now use (6.53) with Theorem 6.20 to compute the reduction in variability due to recurrent selective sweeps. Let ξ be a partition with b blocks and η a partition with $b - k + 1$ blocks obtained by collapsing k blocks of ξ into one in η. Recall from (4.1) that the jump rate

$$q_{\xi,\eta} = \int_0^1 x^{k-2}(1-x)^{|\xi|-k} \Lambda(dx)$$

Since there are $\binom{b}{k}$ possible such jumps, the rate at which the number of lineages decreases from b to $b - k + 1$ due to selective sweeps is

$$r_{b,b-k+1} = \binom{b}{k} \int_0^1 x^{k-2}(1-x)^{b-k} \Lambda_0(dx)$$

where $\Lambda_0 = \Lambda - \delta_0$.

In Example 6.8, Λ_0 has density cx, so

$$r_{b,b-k+1} = c\binom{b}{k} \int_0^1 x^{k-1}(1-x)^{b-k} \, dx$$

$$= c\frac{b!}{k!(b-k)!} \cdot \frac{(k-1)!(b-k)!}{b!} = \frac{1}{k}$$

The calculus can be done by remembering the definition of the beta distribution or integrating by parts to show that

$$\int_0^1 \frac{x^i}{i!} \cdot \frac{x^j}{j!}\, dx = \frac{1}{(i+j+1)!}$$

The total rate at which transitions occurs, when the partition has b blocks is $\lambda_b = \binom{b}{2} + \sum_{k=2}^b \frac{1}{k}$, so the probability that the next jump will take the chain from b to $b-j$ is

$$p_{b,b-j} = \begin{cases} \left(\binom{b}{2} + \frac{1}{2}\right)/\lambda_b & j = 1 \\ 1/(j+1)\lambda_b & 2 \le j \le b-1 \end{cases}$$

From this it follows that for $n > b$

$$G_n(b) = \sum_{j=1}^{n-b} p_{n,n-j} G_{n-j}(b)$$

Thus the values can be computed by starting with $G_b(b) = 1$ and then computing $G_n(b)$ for $n = b+1, b+2, \ldots$.

By the reasoning that led to (6.53) the expected total time in the tree for a sample of size n is

$$ET_n = \sum_{b=2}^n b\lambda_b^{-1} G_n(b)$$

Since $\lambda_b \ge \binom{b}{2}$ and $G_n(b) \le 1$, this is smaller than the neutral expectation.

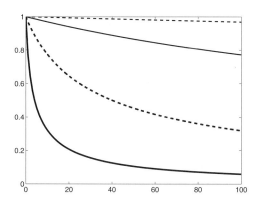

Fig. 6.14. Reduction in heterozygosity due to recurrent selective sweeps.

In Figure 6.14 we have plotted $ET_n/E_0 T_n$ for Example 6.8 as a function of distance along the chromosome measure in centiMorgans, when $c =$

$0.1, 1, 10, 100$. As noted earlier, in comparison to the parameters of Braverman et al. (1995)

$$c = \frac{2\gamma s}{\beta} = \frac{2\Lambda_r \alpha}{18.42} = 0.1085\Lambda_r \alpha$$

The graphs in their Figure 3 end at about $\Lambda_r = 0.001$ so our values of c correspond to their $\alpha = 10^3, 10^4, 10^5, 10^6$. Our analytical curves show good agreement with the values they found by simulation.

7

Diffusion Processes

"One cannot escape the feeling that these mathematical formulas have an independent existence and an intelligence of their own." H.R. Hertz

As we have mentioned several times, if we let the population size $N \to \infty$ then our processes become deterministic. In this section, we will see that if we let $N \to \infty$ and at the same time speed up time so that it runs at rate $O(N)$, then allele frequencies converge to limits called diffusion processes. This will allow us to obtain more detailed results about the models with selection introduced in the previous chapter. A rigorous treatment of diffusion processes requires a fair amount of mathematical sophistication and the details themselves could fill a book, see e.g., Durrett (1996). Here, we will content ourselves to state and explain the use of the main formulas useful for computation.

As an antidote to the mathematical skullduggery, we will give some anecdotes concerning the historical development of the use of diffusion process in genetics that, taking place in the 30s, 40s, and 50s, occurs in parallel to the development of a rigorous mathematical foundation for probability theory. Our first is a quote from Feller's (1951) Berkeley Symposium paper, which began the development of the mathematical machinery for treating the convergence of Markov chains to diffusion processes:

> "There exists a huge literature on the mathematical theory of evolution and statistical genetics, but existing methods and results are due almost entirely to R.A. Fisher and Sewall Wright. They have attacked individual problems with great ingenuity and an admirable resourcefulness, and had in some instances to discover for themselves isolated facts of the general theory of stochastic processes. However, as is natural with such pioneer work, it is not easy to penetrate to the mathematical core of the arguments to discover the explicit and implicit assumptions underlying the theory."

A footnote to the first sentence of the quote says: "See Fisher (1930), Wright (1939) and Wright (1942). It is difficult to give useful references to original

R. Durrett, *Probability Models for DNA Sequence Evolution*,
DOI: 10.1007/978-0-387-78168-6_7, © Springer Science+Business Media, LLC 2008

papers, since these are mostly highly technical and inaccessible to nonspecialists." I am sure that many biologists have similar feelings about the mathematics literature.

7.1 Infinitesimal mean and variance

To motivate the definition of a diffusion process, we begin by recalling that a continuous-time Markov chain X_t is defined by giving the rate $q(i,j)$ at which the chain jumps from i to j. That is, if we use P_i for the distribution of the process starting from i then

$$P_i(X_s = j) = q(i,j)s + o(s)$$

where $o(s)$, pronounced "little oh of s," is a quantity that when divided by s tends to 0 as $s \to 0$. If we define $q(i,i) = -\sum_{j \neq i} q(i,j)$ then the rows of the matrix sum to 0 and

$$P_i(X_s = i) = 1 + q(i,i)s + o(s)$$

Combining the last two formulas, it follows that if f is a bounded function then

$$E_i f(X_s) = (1 + q(i,i)s)f(i) + \sum_{j \neq i} q(i,j)sf(j) + o(s)$$

Rearranging we have

$$\frac{E_i f(X_s) - f(i)}{s} = \sum_j q(i,j)f(j) + o(1)$$

where $o(1)$ denotes a quantity that (when divided by 1) tends to 0 as $s \to 0$. Letting $s \to 0$

$$\frac{d}{ds} E_i f(X_s) \Big|_{s=0} = Qf(i) \tag{7.1}$$

where the right-hand side is the ith component of the product of the matrix $Q = q(i,j)$ and the vector $f(j)$. Q is called the *infinitesimal generator* of X_s.

To define diffusion processes, we will take an approach that is not intuitive, but is efficient.

Definition. A one dimensional diffusion process is a continuous Markov process with infinitesimal generator

$$Lf = \frac{1}{2}a(x)\frac{d^2}{dx^2}f + b(x)\frac{d}{dx}f$$

That is, we have $(d/dt)E_x f(X_t)|_{t=0} = Lf(x)$.

To see what this means, note that if we take $f(x) = x$ then $f'(x) = 1$ and $f''(x) = 0$ so

$$\frac{d}{dt} E_x X_t \Big|_{t=0} = b(x)$$

while if we fix x and define $f(y) = (y - x)^2$ then $f'(x) = 0$ and $f''(y) = 2$ so

$$\frac{d}{dt} E_x (X_t - x)^2 \Big|_{t=0} = a(x)$$

For this reason, $b(x)$ and $a(x)$ are called the *infinitesimal mean* and *infinitesimal variance*.

Taking $f(y) = (y - x)^4$ we have $f'(x) = 0$ and $f''(x) = 0$ so

$$\frac{d}{dt} E_x (X_t - x)^4 \Big|_{t=0} = 0$$

Since $(y - x)^4 \geq 0$, we have $E_x (X_t - x)^4 \geq \epsilon^4 P_x(|X_t - x| > \epsilon)$, and it follows that

$$\frac{1}{t} P_x(|X_t - x| > \epsilon) \to 0 \quad \text{as } t \to 0$$

It can be shown that this condition implies that the paths $t \to X_t$ are continuous. To see why we need this probability to be $o(t)$, recall that for continuous time Markov chains with jumps

$$P_i(X_t = j)/t \to q(i, j) \quad \text{as } t \to 0.$$

To explain the intuitive meaning of the coefficients $b(x)$ and $a(x)$ we will consider some examples.

Example 7.1. Deterministic motion. Suppose $X_0 = x$ and $dX_t/dt = b(X_t)$. A little calculus shows

$$f(X_t) - f(X_0) = \int_0^t \frac{d}{ds} f(X_s) \, ds$$
$$= \int_0^t f'(X_s) \frac{dX_s}{ds} \, ds = \int_0^t f'(X_s) b(X_s) \, ds$$

So if f' and b are continuous

$$\frac{f(X_t) - f(x)}{t} \to f'(x) b(x)$$

i.e., $Lf(x) = b(x) f'(x)$. Thus, when $a(x) = 0$, a diffusion process reduces to a differential equation.

Notation. In the next example and in what follows $B_t = B(t)$ and the second form will often be used when t has subscripts or a complicated formula.

Example 7.2. Brownian motion. Suppose $B(0) = x$ and for $0 = t_0 < t_1 < \ldots < t_n$, $B(t_1) - B(t_0)$, $B(t_2) - B(t_1)$, $\ldots B(t_n) - B(t_{n-1})$ are independent with $B(t_i) - B(t_{i-1})$ normally distributed with mean 0 and variance $\sigma^2(t_i - t_{i-1})$. Using Taylor's theorem, when t is small

$$f(B_t) - f(B_0) \approx f'(B_0)(B_t - B_0) + \frac{1}{2}f''(B_0)(B_t - B_0)^2$$

Taking expected values

$$E_x(f(B_t) - f(x)) \approx \frac{1}{2}f''(x)\sigma^2 t$$

so $Lf(x) = (\sigma^2/2)f''(x)$, i.e., $a(x) = \sigma^2$ and $b(x) = 0$. Thus, $a(x)$ measures the size of the stochastic fluctuations, or what biologists call random genetic drift.

Example 7.3. Stochastic differential equations. Let $\sigma(x) = \sqrt{a(x)}$. Intuitively, a diffusion process has for small t

$(\star)$ $$X_t - X_0 \approx b(X_0)t + \sigma(X_0)(B_t - B_0)$$

If b and σ are Lipschitz continuous, i.e., $|b(x) - b(y)| \le K|x - y|$ and $|\sigma(x) - \sigma(y)| \le K|x - y|$ then it can be shown that the integral equation

$$X_t - X_0 = \int_0^t b(X_s)\,ds + \int_0^t \sigma(X_s)dB_s$$

has a unique solution, where the second integral is defined to be the limit of approximating sums $\sum_i \sigma(X(s_{i-1}))(B(s_i) - B(s_{i-1}))$. The formalities involved in making the last sentence precise are considerable but the intuition in $(\star)$ is important: a diffusion process is a differential equation plus random fluctuations, which can be thought of as coming from a Brownian motion with a state dependent variance.

7.2 Examples of diffusions

In this section, we will introduce many of the examples from genetics that we will study. A formal proof of the convergence of Markov chains to limiting diffusions is somewhat complicated. Here, we will content ourselves to compute the limits of the infinitesimal mean and variance. Theoretical results which show that this is sufficient to conclude convergence can be found in Section 8.7 of Durrett (1996) or Section 7.4 of Ethier and Kurtz (1986).

Example 7.4. Wright-Fisher model with selection. There are two alleles, A and a. The fitness of A is 1 and fitness of a is $1 - s$ where $s \ge 0$. In the Wright-Fisher model, this can be implemented by declaring that, as we build up the state at time $t + 1$ by drawing with replacement from generation t, we always accept an A that is drawn, but we keep an a with probability $1 - s$. Here, selection acts on the individual chromosomes, so, in effect, we have a population of $2N$ haploid individuals. Later we will discuss the more complicated situation of selection acting on diploids.

If the frequency of allele A in generation 0 is $X_0 = x$ then a newly drawn ball will be kept with probability $x + (1 - x)(1 - s)$ so the expected frequency in the next generation will be

$$x' = \frac{x}{x + (1 - x)(1 - s)} = \frac{x}{1 - (1 - x)s} = x + x(1 - x)s + o(s)$$

since $1/(1 - y) = 1 + y + y^2 + \cdots$. The number of A's in the next generation N_1 will be binomial$(2N, x')$ so the frequency $X_1 = N_1/2N$ has

$$E(X_1 - X_0) = x(1 - x)s + o(s)$$

To take the diffusion limit, we want to write time in units of $2N$ generations, i.e., let $Y_t = X_{[2Nt]}$ where $[s]$ is the largest integer $\leq s$. Since time 1 for X corresponds to time $1/2N$ for Y, we want the change in the mean in one time step to be of order $1/2N$, so we let $\gamma = 2Ns$ and write

$$E(Y_{1/N} - Y_0) = x(1 - x)\gamma \cdot \frac{1}{2N} + o\left(N^{-1}\right)$$

where $o(N^{-1})$ is a term that when divided by N^{-1} tends to 0 as $N \to \infty$.

The variance of N_1 is $2Nx'(1 - x')$, and $\mathrm{var}\,(cZ) = c^2\,\mathrm{var}\,(Z)$ so the variance of X_1 is $x'(1 - x')/2N$. When $s = \gamma/2N$, $x' = x + o(1)$ and we have

$$\mathrm{var}\,(Y_{1/N} - Y_0) = x(1 - x) \cdot \frac{1}{2N} + o\left(N^{-1}\right)$$

Combining our calculations we see that the infinitesimal generator is

$$Lf = \frac{1}{2}x(1 - x)\frac{d^2}{dx^2}f + \gamma x(1 - x)\frac{d}{dx}f \tag{7.2}$$

In some papers in the biology literature time is not sped up and one sees

$$Lf = \frac{1}{4N}x(1 - x)\frac{d^2}{dx^2}f + sx(1 - x)\frac{d}{dx}f$$

Example 7.5. Wright-Fisher model with selection and mutation. As before, we have two alleles, A and a, with the fitness of A is 1 and fitness of a is $1 - s$. This time $a \to A$ with probability μ_1 and $A \to a$ with probability μ_2. In defining our process we will suppose that selection occurs first followed by mutation. For a concrete story, suppose that the fitnesses give the relative probabilities of the two types surviving long enough to reproduce, at which point a genetic mutation may occur. If the frequency of allele A in generation 0 is $X_0 = x$ then the frequency in the next generation will be

$$x'' = x' + \mu_1(1 - x') - \mu_2 x' \quad \text{where} \quad x' = x + x(1 - x)s + o(s)$$

Letting $Y_t = X_{[2Nt]}$, $\gamma = 2Ns$, $\beta_i = 2N\mu_i$, and noting that $x' = x + o(1)$ we see that

$$E(Y_{1/N} - Y_0) = \{x(1 - x)\gamma + \beta_1(1 - x) - \beta_2 x\} \cdot \frac{1}{2N} + o\left(N^{-1}\right)$$

Again $x'' = x + o(1)$ so we have

$$\text{var}\,(Y_{1/N} - Y_0) = x(1 - x) \cdot \frac{1}{2N} + o\left(N^{-1}\right)$$

Combining our calculations we see that the infinitesimal generator is

$$Lf = \frac{1}{2}x(1 - x)\frac{d^2}{dx^2}f + \{\gamma x(1 - x) + \beta_1(1 - x) - \beta_2 x\}\frac{d}{dx}f \qquad (7.3)$$

The source of each term is

binomial sampling	$\frac{1}{2}x(1 - x)\frac{d^2}{dx^2}f$
selection, $2Ns = \gamma$	$\gamma x(1 - x)\frac{d}{dx}f$
mutation, $2N\mu_i = \beta_i$	$\{\beta_1(1 - x) - \beta_2 x\}\frac{d}{dx}f$

Example 7.6. Moran model. There are two alleles, A and a. The fitness of A is 1 and fitness of a is $1 - s$. Mutations $a \to A$ occur at rate μ_1 and $A \to a$ occur at rate μ_2. For simplicity, we assume that mutations occur during the individual's life, not at birth, so adding the mutation rates to the transition rates from Section 6.1 we have

$$k \to k + 1 \quad \text{at rate} \quad (2N - k)\left(\frac{k}{2N} + \mu\right)$$

$$k \to k - 1 \quad \text{at rate} \quad k\left(\frac{2N - k}{2N}(1 - s) + \nu\right)$$

Let X_t be the fraction of individuals with the A allele. To derive the diffusion approximation we note that if $k/2N = x$

$$\frac{d}{dt}EX_t = \frac{1}{2N}\left[(2N - k)\left(\frac{k}{2N} + \mu_1\right) - k\left(\frac{2N - k}{2N}(1 - s) + \mu_2\right)\right]$$

$$= (1 - x)\mu_1 - x\mu_2 + x(1 - x)s$$

Letting $\beta_i = N\mu_i$, and $\gamma = Ns$ we see that the drift coefficient for the process run at rate N is

$$b(x) = (1 - x)\beta_1 - x\beta_2 + x(1 - x)\gamma$$

To compute the second order term, we note that after either an up jump or a down jump $(X_t - x)^2 = (1/2N)^2$, so

$$\frac{d}{dt}E(X_t - x)^2 = \frac{1}{(2N)^2}\left[(2N - k)\left(\frac{k}{2N} + \mu_1\right) + k\left(\frac{2N - k}{2N}(1 - s) + \mu_2\right)\right]$$

Since $\mu_1, \mu_2, s, \to 0$ and $k/2N = x$ we have

$$\frac{d}{dt}E(X_t - x)^2 = \frac{1}{2N}[2x(1 - x) + o(1)]$$

Thus for the process run at rate N the diffusion coefficient is

$$a(x) = x(1 - x)$$

Combining our calculations we see that the infinitesimal generator is again

$$Lf = \frac{1}{2}x(1 - x)\frac{d^2}{dx^2}f + [\gamma x(1 - x) + \beta_1(1 - x) - \beta_2 x]\frac{d}{dx}f \qquad (7.4)$$

In Section 1.5, we saw that the Moran model coalesces twice as fast as the Wright-Fisher model. To compensate for this, we sped up time by N rather than $2N$, in order to arrive at the same diffusion limit.

Example 7.7. General diploid selection model. We again have two alleles A and a but the fitnesses of diploid individuals are

$$\begin{array}{ccc} AA & Aa & aa \\ 1 - s_0 & 1 - s_1 & 1 - s_2 \end{array}$$

If the frequency of allele A in generation 0 is $X_0 = x$ then assuming random union of gametes and reasoning as in Section 6.2, the frequency in the next generation will be

$$x' = \frac{x^2(1 - s_0) + x(1 - x)(1 - s_1)}{x^2(1 - s_0) + 2x(1 - x)(1 - s_1) + (1 - x)^2(1 - s_2)}$$

$$= \frac{x - s_0 x^2 - s_1 x(1 - x)}{1 - x^2 s_0 - 2x(1 - x)s_1 - (1 - x)^2 s_2}$$

Ignoring terms with s_i^2 and $s_i s_j$ the above

$$\approx x - s_0 x^2 - s_1 x(1 - x) + s_0 x^3 + 2x^2(1 - x)s_1 + x(1 - x)^2 s_2$$

A little algebra now shows

$$x' - x \approx x(1 - x)[-s_0 x - s_1(1 - 2x) + s_2(1 - x)]$$
$$= x(1 - x)[s_2 - s_1 + x(2s_1 - s_0 - s_2)]$$

Letting $Y_t = X_{[2Nt]}$, $\gamma_i = 2Ns_i$, $\delta = \gamma_2 - \gamma_1$ and $\eta = 2\gamma_1 - \gamma_0 - \gamma_2$ then

$$E(Y_{1/N} - Y_0) = x(1-x)[\delta + \eta x] \cdot \frac{1}{2N} + o\left(N^{-1}\right)$$

Again $x' = x + o(1)$ so we have

$$\text{var}\,(Y_{1/N} - Y_0) = x(1-x) \cdot \frac{1}{2N} + o\left(N^{-1}\right)$$

Combining our calculations we see that the infinitesimal generator is

$$Lf = \frac{1}{2}x(1-x)\frac{d^2}{dx^2}f + x(1-x)[\delta + \eta x]\frac{d}{dx}f \qquad (7.5)$$

If there is mutation $a \to A$ at rate $\mu - 1$ and $A \to a$ at rate μ_2, and we let $\beta_i = 2N\mu_i$ then this adds a term of the form

$$\{\beta_1(1-x) - \beta_2 x\}\frac{d}{dx}f$$

There are several important special cases

Additive selection. $s_0 = 0$, $s_1 = s$, $s_2 = 2s$, and let $\gamma = 2Ns$. $\delta = \gamma_2 - \gamma_1 = \gamma$ and $\eta = 2\gamma_1 - \gamma_0 - \gamma_2 = 0$ so

$$\delta + \eta x = \gamma$$
$$b(x) = \gamma x(1-x) \qquad (7.6)$$

just as in our previous Wright-Fisher model with selection.

Balancing selection. $s_1 = 0$, so $\delta = \gamma_2$ and $\eta = -(\gamma_0 + \gamma_2)$. If we let $x_0 = \gamma_2/(\gamma_0 + \gamma_2)$ then

$$\delta + \eta x = (\gamma_0 + \gamma_2)(x_0 - x)$$
$$b(x) = (\gamma_0 + \gamma_2)x(1-x)(x_0 - x) \qquad (7.7)$$

From this we see that the drift is < 0 for $x > x_0$ and > 0 for $x < x_0$. In the symmetric case $\gamma_0 = \gamma_2 = \gamma$ so $x_0 = 1/2$ and

$$\delta + \eta x = \gamma(1 - 2x)$$
$$b(x) = \gamma x(1-x)(1 - 2x) \qquad (7.8)$$

A is dominant. Aa has the same fitness as AA. $s_0 = s_1 = 0$, $s_2 = s$. $\delta = \gamma_2 - \gamma_1 = \gamma$ and $\eta = 2\gamma_1 - \gamma_0 - \gamma_2 = -\gamma$

$$\delta + \eta x = \gamma(1 - x)$$
$$b(x) = \gamma x(1-x)^2 \qquad (7.9)$$

A is recessive. Aa has the same fitness as aa. $s_0 = 0$, $s_1 = s_2 = s$. $\delta = \gamma_2 - \gamma_1 = 0$ and $\eta = 2\gamma_1 - \gamma_0 - \gamma_2 = \gamma$.

$$\delta + \eta x = \gamma x$$
$$b(x) = \gamma x^2(1-x) \qquad (7.10)$$

To compare the drifts for additive selection versus dominant and recessive alleles we have graphed the solution of the differential equation $dX_t/dt = b(X_t)$ for the three drifts starting with $X_0 = 0.01$. To make the selective advantage of AA over aa the same in the three cases, we have taken $\gamma = 1, 2, 2$. Note that because of the extra factor of $(1 - x)$ the dominant case has more trouble getting to 1, while due to the extra factor of x, the recessive case has a hard time escaping from 0.

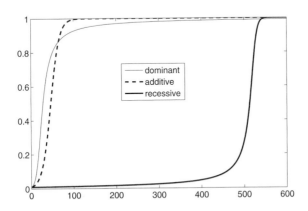

Fig. 7.1. Solution of $dX_t/dt = b(X_t)$ for the three drifts for additive selection, dominant alleles, and recessive alleles.

We can generate new examples of diffusions from old ones by

Theorem 7.1. Change of variables. *If h is increasing and has two continuous derivatives, then $Y_t = h(X_t)$ is a diffusion process with infinitesimal mean and variance*

$$\bar{a}(y) = a(x)h'(x)^2 \qquad \bar{b}(y) = Lh(x)$$

where $x = f^{-1}(y)$.

Proof. By calculus

$$\frac{d}{dx}f(h) = f'(h)h' \qquad \frac{d^2}{dx^2}f(h) = f''(h)(h')^2 + f'(h)h''$$

Using this in the definition of the generator

$$Lf(h) = \frac{1}{2}a(x)[f''(h)(h')^2 + f'(h)h''] + b(x)f'(h)h'$$

$$= \frac{1}{2}a(x)(h')^2 f''(h) + Lh(x)f'(h)$$

which gives the result. $\qquad\qquad\qquad\qquad\qquad\qquad\qquad\qquad\qquad\qquad\square$

Example 7.8. Fisher's transformation. Let X_t be the Wright-Fisher model with no mutation or selection. Fisher (1922) discovered a remarkable transformation, although he did not get the answer right the first time. See pages 88-89 in Fisher (1930) and pages 119-120 in Wright (1931). Let $h(x) = \cos^{-1}(1-2x)$. This maps $[0,1] \to [0,\pi]$. Recalling from calculus that

$$(f^{-1})'(x) = 1/f'(f^{-1}(x))$$

$$\frac{dh}{dx} = \frac{-2}{-\sin(\cos^{-1}(1-2x))}$$

$$\frac{d^2h}{dx^2} = \frac{-2\cos(\cos^{-1}(1-2x))}{\sin^2(\cos^{-1}(1-2x))} \cdot \frac{-2}{-\sin(\cos^{-1}(1-2x))}$$

To simplify the last expression we draw a picture

$$y = \sqrt{1-(1-2x)^2} = 2\sqrt{x(1-x)}$$

The last caclulation shows $\sin(\cos^{-1}(1-2x)) = 2\sqrt{x(1-x)}$ so $\theta_t = h(X_t)$ has generator

$$\frac{1}{2}\frac{d^2}{d\theta^2} - \frac{1}{2}\cot(\theta)\frac{d}{d\theta}$$

The infinitesimal variance is now constant, but a drift (which Fisher missed in his first attempt) has been introduced. We leave it as an exercise for the reader to check that $\sin^{-1}(\sqrt{x})$ also results in constant variance.

7.3 Transition probabilities

In discrete time, a Markov chain is defined by giving its transition probability $p(i,j)$. For a continuous-time Markov chain or a diffusion process, the transition probability $p_t(x,y) = P(X_t = y|X_t = x)$ must be computed by solving one of two differential equations. In the case of a continuous time Markov chain, it follows from the Markov property and the definition of the generator in (7.1) that

$$\frac{d}{ds}E_i f(X_s)\Big|_{s=t} = E_i Q f(X_t) \tag{7.11}$$

From this we get

$$\frac{d}{dt}\sum_j p_t(i,j)f(j) = \sum_k p_t(i,k)\sum_j q(k,j)f(j)$$

Since this holds for all f we must have

$$\frac{d}{dt}p_t(i,j) = \sum_k p_t(i,k)q(k,j) \tag{7.12}$$

or in matrix notation $(d/dt)p_t = p_t Q$. This is *Kolmogorov's forward equation*. It can also be derived by letting I be the identity matrix and using the Markov property to write

$$\frac{1}{h}(p_{t+h}(i,j) - p_t(i,j)) = \sum_k p_t(i,k)\frac{1}{h}(p_h(k,j) - I(k,j))$$
$$\rightarrow \sum_k p_t(i,k)q(k,j)$$

Here, we have broken the time interval $[0,t+h]$ into $[0,t]$ and $[t,t+h]$, with the small piece on the forward end. If we instead break it into $[0,h]$ and $[h,t+h]$ we get

$$\frac{1}{h}(p_{t+h}(i,j) - p_t(i,j)) = \frac{1}{h}\sum_k (p_h(i,k) - I(i,k))p_t(k,j)$$
$$\rightarrow \sum_k q(i,k)p_t(k,j) \tag{7.13}$$

or in matrix notation $(d/dt)p_t = Qp_t$. This is *Kolmogorov's backward equation*.

Consider now a diffusion process. Imitating (7.13) we can write

$$\frac{1}{h}(p_{t+h}(x,y) - p_t(x,y)) = \frac{1}{h}\left(\int p_h(x,z)p_t(z,y)\,dz - p_t(x,y)\right)$$
$$= \frac{1}{h}(E_x p_t(X_h,y) - p_t(x,y))$$

If we let $f(x) = p_t(x,y)$ for fixed y then the last quantity is $(1/h)(E_x f(X_h) - f(x))$, so recalling the definition of the generator and letting $h \rightarrow 0$ we have

$$\frac{d}{dt}p_t(x,y) = \frac{1}{2}a(x)\frac{d^2}{dx^2}p_t(x,y) + b(x)\frac{d}{dx}p_t(x,y) \tag{7.14}$$

Because we broke us the interval into $[0,h]$ and $[h,t+h]$, this is *Kolmogorov's backward equation*. Another reason is that the derivatives occur in the backward variable x. To get the forward equation note that as in (7.11)

$$\frac{d}{ds}E_x f(X_s)\Big|_{s=t} = E_x Lf(X_t) \tag{7.15}$$

So we have

$$\frac{d}{dt}\int p_t(x,y)f(y)\,dy = \int p_t(x,y)Lf(y)\,dy$$
$$= \int p_t(x,y)\left[\frac{1}{2}a(y)\frac{d^2}{dy^2}f(y) + b(y)\frac{d}{dy}f(y)\right]dy$$

To turn this into an equation for $p_t(x, y)$ we suppose that f is 0 outside $[\delta, 1-\delta]$ for some $\delta > 0$ and integrate by parts twice to get

$$\int \left[\frac{1}{2}\frac{d^2}{dy^2}(a(y)p_t(x, y)) - \frac{d}{dy}(b(y)p_t(x, y)) \right] f(y)\, dy$$

Since this holds for all f we have *Kolmogorov's forward equation*

$$\frac{d}{dt}p_t(x, y) = \frac{1}{2}\frac{d^2}{dy^2}(a(y)p_t(x, y)) - \frac{d}{dy}(b(y)p_t(x, y)) \qquad (7.16)$$

where the derivatives occur in the forward variable y. This is not as nice as the backward equations since it does not make sense unless $b(y)$ is differentiable and $a(y)$ is twice differentiable.

(7.16) is called the Fokker-Planck equation by physicists, due to work of Fokker in 1914 and Planck in 1917. The first rigorous mathematical derivation was given by Kolmogorov in 1931. The formula makes its first appearance in the biology literature in Wright (1945).

Defining the *adjoint operator* L^*

$$L^*f = \frac{1}{2}\frac{d^2}{dy^2}(a(y)f(y)) - \frac{d}{dy}(b(y)f(y))$$

we can write the two equations as

$$\frac{d}{dt}p_t(x, y) = L_x p_t(x, y) \qquad \frac{d}{dt}p_t(x, y) = L_y^* p_t(x, y)$$

where the subscript indicates the variable of $p_t(x, y)$ where the operator acts. In comparison, the two equations for continuous time Markov chains are

$$\frac{d}{dt}p_t(i, j) = \sum_k Q(i, k)p_t(k, j) \qquad \frac{d}{dt}p_t(i, j) = \sum_k p_t(i, k)Q(k, j)$$

Again Q acts on different variables in the two cases, but we don't need the formalities of defining the adjoint matrix. We just shift the matrix to the other side.

Only on rare occasions can one solve the differential equations given above to determine the transition probability.

Example 7.9. Brownian motion. Suppose $a(x) = \sigma^2$, $b(x) = 0$. In this case the backward equation (7.14) is

$$\frac{d}{dt}p_t(x, y) = \frac{\sigma^2}{2}\frac{d^2}{dx^2}p_t(x, y)$$

From the definition of the process in the previous section we know that

$$\sqrt{2\pi}p_t(x, y) = (t\sigma^2)^{-1/2}e^{-(y-x)^2/2\sigma^2 t}$$

Differentiating we find

$$\frac{d}{dt} = \left(-\frac{1}{2}t^{-3/2}\sigma^{-1} + t^{-1/2}\sigma^{-1}\frac{(y-x)^2}{2\sigma^2 t^2} \right) e^{-(y-x)^2/2\sigma^2 t}$$

$$\frac{d}{dx} = t^{-1/2}\sigma^{-1} \cdot \frac{y-x}{\sigma^2 t} e^{-(y-x)^2/2\sigma^2 t}$$

$$\frac{d^2}{dx^2} = \left(-t^{-3/2}\sigma^{-3} + t^{-1/2}\sigma^{-1}\frac{(y-x)^2}{\sigma^4 t^2} \right) e^{-(y-x)^2/2\sigma^2 t}$$

which shows that $p_t(x,y)$ satisfies the stated differential equation.

Example 7.10. Ornstein-Uhlenbeck process. Suppose $a(x) = \sigma^2$ and $b(x) = -\alpha x$. This is a model for the velocity of a particle with a random acceleration and experiences friction forces proportional to its velocity. The transition probability $p_t(x,y)$ is a normal with mean $u(x,t) = xe^{-\alpha t}$ and variance $v(t) = \sigma^2 \int_0^t e^{-2\alpha r}\, dr$, so

$$\sqrt{2\pi}p_t(x,y) = v(t)^{-1/2}e^{-(y-u(x,t))^2/2v(t)}$$

Let u_x and u_t be the partial derivatives of u, and note that $u_{xx} = 0$. Differentiating we find

$$\frac{d}{dt} = -\frac{1}{2}v(t)^{-3/2}v'(t)e^{-(y-u(x,t))^2/2v(t)}$$

$$+ v(t)^{-1/2} \cdot \frac{y-u(x,t)}{v(t)}u_t(x,t)e^{-(y-u(x,t))^2/2v(t)}$$

$$+ v(t)^{-1/2} \cdot \frac{(y-u(x,t))^2}{2v(t)^2}v'(t)e^{-(y-u(x,t))^2/2v(t)}$$

$$\frac{d}{dx} = v(t)^{-1/2} \cdot \frac{y-u(x,t)}{v(t)}u_x(x,t)e^{-(y-u(x,t))^2/2v(t)}$$

$$\frac{d^2}{dx^2} = v(t)^{-3/2}(-u_x(x,t)^2)e^{-(y-u(x,t))^2/2v(t)}$$

$$+ v(t)^{-1/2}\frac{(y-u(x,t))^2}{v(t)^2}u_x(x,t)^2 e^{-(y-u(x,t))^2/2v(t)}$$

Let $f_1, f_2, \ldots f_6$ denote the right hand sides. Since $v'(t) = \sigma^2 e^{-2\alpha t}$ and $u_x(x,t) = e^{-\alpha t}$, we have $f_1 = (\sigma^2/2)f_5$ and $f_3 = (\sigma^2/2)f_6$. Since $u_t(x,t) = -\alpha x e^{-\alpha t}$, we have $f_2 = -\alpha x f_4$. Combining these results we see that

$$\frac{d}{dt}p_t(x,y) = -\alpha x\frac{d}{dx}p_t(x,y) + \frac{\sigma^2}{2}\frac{d^2}{dx^2}p_t(x,y)$$

which verifies that $p_t(x,y)$ is the desired transition probability.

In most cases, one cannot find an explicit expression for the transition probability. Kimura (1955) was able to express the transition probability for the Wright-Fisher model as an infinite series of Gegenbauer polynomials. However, for our purposes, the following qualitative result is more useful. A function $f(x)$ on [0,1] is said to be *Hölder continuous* if there is a $\delta > 0$ and $C < \infty$ so that $|f(x) - f(y)| \leq C|x - y|^{\delta}$.

Theorem 7.2. *Suppose that the coefficients a and b are Hölder continuous on $[0,1]$, and $a(x) > 0$ on $(0,1)$. Then for any $\delta > 0$ there is an ϵ so that $p_t(x, y) \geq \epsilon$ when $x, y \in [\delta, 1 - \delta]$.*

7.4 Hitting probabilities

For our genetics models, we want to be able to compute the probability an allele becomes fixed in the population. Here, and throughout this chapter, we will first consider the analogous problems for discrete and continuous time Markov chains on $\{0, 1, \ldots 2N\}$. Let $T_k = \min\{n : X_n = k\}$ be the time of the first visit to k, and let $h(i) = P_i(T_{2N} < T_0)$. To compute $h(i)$ we note that if $0 < i < 2N$ then breaking things down according to what happens on the first step

$$h(i) = \sum_j p(i, j)h(j) \tag{7.17}$$

Introducing P for the transition matrix and I for the identity matrix, we can write (7.17) as $h = Ph$ or $(P - I)h = 0$. The second formula may look a little odd now, but soon it will seem natural. To compute $h(i)$, we first need a technical result.

Theorem 7.3. *Let $\tau = T_0 \wedge T_{2N}$. In discrete or continuous time, if it is possible to reach 0 and 2N from each $0 < i < 2N$ then $\sup_{0 < i < 2N} E_i \tau < \infty$.*

Proof. Our assumption implies that there are $\epsilon > 0$ and $M < \infty$ so that $P_i(\tau \leq M) \geq \epsilon$ for all $0 < i < 2N$. The Markov property implies $P_i(\tau > kM) \leq (1 - \epsilon)^k$, so

$$E_i \tau = \int_0^\infty P_i(\tau > t)\, dt \leq M \sum_{k=0}^\infty (1 - \epsilon)^k = M/\epsilon \qquad \square$$

Theorem 7.4. *Suppose that it is possible to reach 0 and 2N from each $0 < i < 2N$. $h(i) = P_i(T_{2N} < T_0)$ is the unique solution of $(P - I)h = 0$ with $h(0) = 0$ and $h(2N) = 1$.*

Proof. Theorem 7.3 implies $\sup_{0 < i < 2N} E_i \tau < \infty$. (7.17) implies that $Eh(X_{n \wedge \tau})$ is constant, since for any jump that starts at a point $0 < i < 2N$, the expected value after a jump is the same as before. The irreducibility condition implies that $P_i(\tau < \infty) = 1$. Letting $n \to \infty$, which can be justified since h is a bounded function,

$$h(i) = E_i h(X_\tau) = P_i(T_{2N} < T_0)$$

since $h(0) = 0$ and $h(2N) = 1$. $\qquad \square$

The next result is the continuous-time analogue of Theorem 7.4.

Theorem 7.5. *Suppose that it is possible to reach 0 and 2N from each* $0 < i < 2N$. $h(i) = P_i(T_{2N} < T_0)$ *is the unique solution of* $Qh = 0$ *for* $0 < i < 2N$ *with* $h(0) = 0$ *and* $h(2N) = 1$.

Proof. Theorem 7.3 implies $\sup_{0<i<2N} E_i \tau < \infty$. To check that the equation is satisfied, we note that

$$\frac{d}{ds} E_i h(X_s) \Big|_{s=t} = E_i Q h(X_t) = 0$$

so $E_i h(X_{t \wedge \tau})$ is constant in time, and we can repeat the argument from discrete time to conclude $h(i) = E_i h(X_\tau) = P_i(T_{2N} < T_0)$. □

Turning to the case of a diffusion process, let $T_a = \inf\{t : X_t = a\}$. Again, we begin with a technical result.

Theorem 7.6. *Let* $y < z$ *and* $\tau_{y,z} = T_y \wedge T_z$. *Suppose that it is possible to reach* y *and* z *from each* $y < x < z$. *Then* $\sup_{x \in (y,z)} E_x \tau_{y,z} < \infty$.

Proof. Pick $w \in (y, z)$. Pick M large enough so that $P_w(T_y \le M) \ge \epsilon > 0$ and $P_w(T_z \le M) \ge \epsilon > 0$. By the argument in discrete time, it is enough to show

$$\sup_{x \in (y,z)} P_x(\tau_{y,z} > M) \le 1 - \epsilon$$

By considering the first time the process starting from w hits z and using the Markov property, it follows that if $w < x < z$ then

$$P_w(T_z \le t) = E_w(P_x(T_z \le t - T_x); T_x \le t)$$
$$\le P_x(T_z \le t) P_w(T_x \le t) \le P_x(T_z \le t)$$

A similar argument shows that for $y < x < w$, $P_x(T_y \le t) \ge P_w(T_y \le t)$. and the desired result follows. □

Theorem 7.7. *Let* $y < z$. *Suppose that it is possible to reach* y *and* z *from each* $y < x < z$. $h(x) = P_x(T_z < T_y)$ *is the unique solution of* $Lh = 0$ *for* $y < x < z$ *with* $h(y) = 0$ *and* $h(z) = 1$.

Proof. Theorem 7.6 implies that $\sup_{x \in (y,z)} E_x \tau_{y,z} < \infty$, so h is well defined. To check that the equation is satisfied, we note that the Markov property implies that

$$\frac{d}{ds} E_x h(X_s) \Big|_{s=t} = E_x L h(X_t) \qquad (7.18)$$

$Lh = 0$ implies $E_x h(X_{t \wedge \tau_{x,y}})$ is constant in time, so we can argue as before that $h(x) = E_x h(X_\tau) = P_x(T_1 < T_0)$. □

Comparing the last three theorems shows that hitting probabilities satisfy

$$(P - I)h = 0 \quad \text{discrete time Markov chain}$$
$$Qh = 0 \quad \text{continuous time Markov chain}$$
$$Lh = 0 \quad \text{diffusion process}$$

so the three operators $P - I$, Q, and L are analogous.

Diffusion hitting probabilities

Based on Theorem 7.7, we want to solve

$$L\phi = \frac{1}{2}a(x)\frac{d^2}{dx^2}\phi + b(x)\frac{d}{dx}\phi = 0$$

ϕ is called the *natural scale* for the diffusion process because $\phi(X_t)$ is a martingale. To solve this equation, we let $\psi = \phi'$ and note that

$$\frac{1}{2}a(x)\psi' + b(x)\psi = 0 \quad \text{or} \quad \psi' = \frac{-2b(x)}{a(x)}\psi$$

As one can check by differentiating, this equation is solved by

$$\psi(y) = \exp\left(\int^y \frac{-2b(z)}{a(z)}\,dz\right)$$

where the lack of a lower limit indicates that we can choose any convenient value, or what is the same, use any antiderivative of $-2b(z)/a(z)$. ϕ can be obtained by ψ by integrating:

$$\phi(x) = \int^x \psi(y)\,dy$$

To have the boundary conditions $h(y) = 0$, $h(z) = 1$ satisfied

$$P_x(T_z < T_y) = \frac{\phi(x) - \phi(y)}{\phi(z) - \phi(y)}$$

$$P_x(T_z > T_y) = \frac{\phi(z) - \phi(x)}{\phi(z) - \phi(y)} \tag{7.19}$$

The second equation follows from $P_x(T_z > T_y) = 1 - P_x(T_z < T_y)$.

Turning to special cases:

Example 7.11. Martingale diffusions. Suppose $b(x) = 0$, $\psi'(x) = 0$ and hence $\phi(x) = x$. X_t is a martingale and we have

$$P_x(T_z < T_y) = \frac{x - y}{z - y} \qquad P_x(T_z > T_y) = \frac{z - x}{z - y}$$

In the next three examples, the mutation rates are zero and we are considering a special case of the general selection model so $a(x) = x(1 - x)$, $b(x) = x(1 - x)(\delta + \eta x)$ and hence

$$-\frac{2b(x)}{a(x)} = -2(\delta + \eta x)$$

Example 7.12. Additive selection. In this case, by (7.6) $-2b(x)/a(x) = -2\gamma$ so

$$\psi(y) = e^{-2\gamma y} \quad \text{and} \quad \phi(y) = [1 - e^{-2\gamma y}]/2\gamma \tag{7.20}$$

and the hitting probabilities are

$$P_x(T_1 < T_0) = \frac{1 - e^{-2\gamma x}}{1 - e^{-2\gamma}} \tag{7.21}$$

which agrees with (6.3). When $x = 1/2N$, using $\gamma/2N = s$ and $1 - e^{-2s} \approx 2s$ we have

$$P_{1/2N}(T_1 < T_0) \approx \frac{2s}{1 - e^{-2\gamma}} \approx 2s$$

when γ is large. The next figure shows the hitting probabilities when $\gamma = 0, 2, 5, 10$.

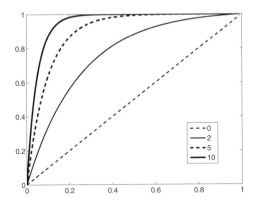

Fig. 7.2. Hitting probabilities for additive selection.

Example 7.13. Dominant advantageous allele. In this case (7.9) implies that $-2b(x)/a(x) = -2\gamma(1 - x)$, so $\psi(x) = e^{-\gamma(2x - x^2)}$, and

$$\phi(x) = \int_0^x e^{-\gamma(2y - y^2)} \, dy$$

Example 7.14. Recessive advantageous allele. In this case (7.10) implies that $-2b(x)/a(x) = -2\gamma x$, so $\psi(x) = e^{-\gamma x^2}$, and

$$\phi(x) = \int_0^x e^{-\gamma y^2}\, dy$$

Figure 7.3 shows $P_x(T_1 < T_0) = \phi(x)/\phi(1)$ when $\gamma = 0, 2, 5, 10$.

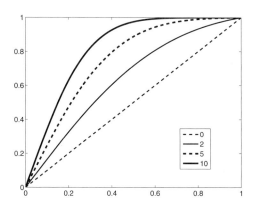

Fig. 7.3. Recessive advantageous allele hitting probabilities

The integrand is a constant multiple of the normal density with mean 0 and variance $1/2\gamma$, so if γ is large

$$\phi(1) \approx \frac{1}{2}\sqrt{\frac{\pi}{\gamma}}$$

If x is small $\phi(x) \approx x$ so if $x = 1/2N$

$$\frac{\phi(1/2N)}{\phi(1)} \approx \frac{1}{N}\sqrt{\frac{\gamma}{\pi}} = \sqrt{\frac{2s}{\pi N}}$$

which is (15) of Kimura (1962). This is larger than the neutral fixation probaiblity $1/2N$, but smaller than the $2s/(1-e^{-\gamma})$ for additive selection.

Example 7.15. Symmetric balancing selection. By (7.8) $-2b(x)/a(x) = -2\gamma$ $(1-2x)$ so

$$\psi(y) = e^{-2\gamma y(1-y)} \quad \text{and} \quad \phi(x) = \int_0^x \psi(y)\, dy$$

Figure 7.4 shows $P_x(T_1 < T_0) = \phi(x)/\phi(1)$ when $\gamma = 0, 2, 5, 10$.

If γ is large then most of the contribution to $\phi(1)$ comes from values within $O(1/2\gamma)$ of the boundary so

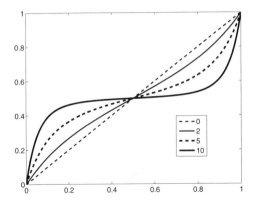

Fig. 7.4. Balancing selection hitting probabilities.

$$\phi(1) \approx 2 \int_0^\infty e^{-2\gamma y}\, dy = \frac{1}{\gamma}$$

If $x = c/(2\gamma)$ then

$$\phi(x) \approx \int_0^x e^{-2\gamma y}\, dy = \frac{1 - e^{-2\gamma x}}{2\gamma} = \frac{1 - e^{-c}}{2\gamma}$$

and we have

$$P_{c/(2\gamma)}(T_1 < T_0) \approx \frac{1}{2}(1 - e^{-c})$$

Using $P_{1-x}(T_0 < T_1) = P_x(T_1 < T_0)$, we see that if $x \gg 1/\gamma$ and $(1 - x) \gg 1/\gamma$ then $P_x(T_1 < T_0) \approx 1/2$.

Up to this point we have ignored the hypothesis "Suppose that it is possible to reach y and z from each $y < x < z$." It follows from Theorem 7.2 that this holds for $0 < y < z < 1$ whenever the coefficients a and b are Hölder continuous on $[0, 1]$, and $a(x) > 0$ on $(0, 1)$, which is true in all of our examples. In all of the results above we can make the derivation rigorous by first computing $P_x(T_y < T_z)$ and then letting $y \to 0$ and $z \to 1$. To show that problems can occur even in natural examples, we consider

Example 7.16. Wright-Fisher model with mutation. For simplicity we assume that there is no selection, so

$$Lf = \frac{1}{2}x(1 - x)\frac{d^2}{dx^2}f + (\beta_1(1 - x) - \beta_2 x)\frac{d}{dx}f$$

In this case, we have

$$\psi(x) = \exp\left(\int^x -2b(y)/a(y)\, dy\right)$$

$$= \exp\left(\int^x -\frac{2\beta_2}{1-y} - \frac{2\beta_1}{y}\, dy\right)$$

$$= x^{-2\beta_1}(1-x)^{-2\beta_2}$$

and $\phi(x) = \int_{1/2}^x y^{-2\beta_1}(1-y)^{-2\beta_2}\, dy$. $\phi(0) = -\infty$ if $\beta_1 \geq 1/2$ so taking $0 < y < x < z < 1$,

$$P_x(T_y < T_z) = \frac{\phi(z) - \phi(x)}{\phi(z) - \phi(y)} \to 0 \quad \text{as } y \to 0$$

and the process cannot get to 0. In words the mutation rate is so strong that the allele frequency cannot reach 0.

Likewise $\phi(1) = \infty$ if $\beta_2 \geq 1/2$ and $0 < y < x < z < 1$,

$$P_x(T_z < T_y) = \frac{\phi(x) - \phi(y)}{\phi(z) - \phi(y)} \to 0 \quad \text{as } z \to 1$$

so the process cannot get to 1. Of course if β_1 and β_2 are both $\geq 1/2$, $P_x(T_0 < T_1)$ is meaningless. We will return to this issue in Section 7.9, when we consider the boundary behavior of diffusion processes.

7.5 Stationary measures

Stationary distributions for Markov processes are important because they represent equilibrium states and are (under mild regularity conditions) the limiting distribution as time $t \to \infty$. In discrete time a nonnegative solution of

$$\sum_i \pi(i)p(i,j) = \pi(j) \tag{7.22}$$

is called a *stationary measure*. A solution with $\sum_i \pi(i) = 1$ is called a *stationary distribution*. If (7.22) holds then

$$\sum_i \pi(i)p^n(i,j) = \sum_{i,k} \pi(i)p(i,k)p^{n-1}(k,j) = \sum_k \pi(k)p^{n-1}(k,j)$$

and it follows by induction that $P_\pi(X_n = i) = \pi(i)$. Results from Markov chain theory imply that if there is a stationary distribution π, and $p(i,j)$ is irreducible and aperiodic (terms defined in Section 4.5) then $p^n(i,j) \to \pi(j)$ as $n \to \infty$.

To see what the condition for stationarity should be in continuous time, we note that if E_π is the expected value starting at π then

$$\frac{d}{dt} E_\pi(f(X_t))\bigg|_{t=0} = \sum_i \pi(i)\sum_j q(i,j)f(j) = \sum_j \left(\sum_i \pi(i)q(i,j)\right)f(j)$$

In order for this to be 0 for all f we must have

$$(\pi Q)(j) = \sum_i \pi(i) q(i, j) = 0 \quad \text{for all } j \tag{7.23}$$

To see that this is sufficient, note that the forward equation implies

$$\frac{d}{dt} \sum_i \pi(i) p_t(i, j) = \sum_{i,k} \pi(i) Q(i, k) p_t(k, j) = 0$$

Results from Markov chain theory imply that if there is a stationary distribution π and $p_t(i, j)$ is irreducible then $p_t(i, j) \to \pi(j)$ as $n \to \infty$.

For a diffusion process we want

$$0 = \frac{d}{dt} E_\pi f(X_t) \Big|_{t=0} = \int \pi(x) L f(x) \, dx$$

If f is 0 outside $[\delta, 1 - \delta]$ for some $\delta > 0$ then integrating by parts twice converts this into

$$\int L^* \pi(x) f(x) \, dx = 0$$

where L^* is the adjoint operator

$$L^* \pi = \frac{1}{2} \frac{d^2}{dx^2} (a(x) \pi(x)) - \frac{d}{dx} (b(x) \pi(x))$$

If this holds for all f then we must have

$$L^* \pi = 0 \tag{7.24}$$

To see that this is sufficient, note that the backward equation and integration by parts imply

$$\frac{d}{dt} \int \pi(x) p_t(x, y) \, dx = \int \pi(x) L_x p_t(x, y) \, dx = \int L^* \pi(x) p_t(x, y) \, dx = 0$$

From Theorem 7.2 and the theory of Harris chains, it follows that if there is a stationary distribution π, and $a(x) > 0$ for $x \in (0, 1)$ then $p_t(x, y) \to \pi(y)$ as $t \to \infty$.

Comparing the last three equations shows that the stationary measures satisfy

$$\begin{aligned} \pi(P - I) &= 0 & \text{discrete time Markov chain} \\ \pi Q &= 0 & \text{continuous time Markov chain} \\ L^* \pi &= 0 & \text{diffusion process} \end{aligned}$$

Again the three operators $P - I$, Q, and L are analogous, but as in Section 7.3, multiplying the matrices on the left by π corresponds to using the adjoint of the diffusion's generator.

Diffusion stationary measures

Theorem 7.8. *If $\psi(x)$ is the derivative of the natural scale then $m(x) = 1/a(x)\psi(x)$ is a stationary measure.*

m is sometimes called the *speed measure*, although as we will see in Section 7.7, this term is misleading. If $\int_0^1 m(x)\,dx < \infty$ then we can convert $m(x)$ into a stationary distribution by multiplying by a constant to make the integral equal to 1.

Proof. To solve $L^*\pi = 0$, it is convenient to note that since $\psi'(x)/\psi(x) = -2b(x)/a(x)$

$$\frac{1}{2}a(x)\psi(x)\frac{d}{dx}\left(\frac{1}{\psi(x)}\frac{d}{dx}f(x)\right)$$
$$= \frac{1}{2}a(x)\psi(x)\frac{1}{\psi(x)}\frac{d^2}{dx^2}f(x) + \frac{1}{2}a(x)\psi(x)\frac{-\psi'(x)}{\psi(x)^2}\frac{d}{dx}f(x) = Lf$$

Thus, if we let $m(x) = 1/a(x)\psi(x)$ then

$$Lf = \frac{1}{2m(x)}\frac{d}{dx}\left(\frac{1}{\psi(x)}\frac{d}{dx}f(x)\right) \tag{7.25}$$

Writing L in this form before we integrate by parts, it follows that

$$L^*m = \frac{d}{dx}\left(\frac{1}{\psi(x)}\left[\frac{d}{dx}\frac{1}{2m(x)}m(x)\right]\right) = 0 \qquad \square$$

Example 7.17. General diploid selection and mutation. In this case, the generator is

$$\frac{1}{2}x(1-x)\frac{d^2}{dx^2}f + \{x(1-x)[\delta + \eta x] + \beta_1(1-x) - \beta_2 x\}\frac{d}{dx}f$$

so we have

$$\psi(x) = \exp\left(\int^x -2b(y)/a(y)\,dy\right)$$
$$= \exp\left(\int^x -2[\delta + \eta y] + \frac{2\beta_2}{1-y} - \frac{2\beta_1}{y}\,dy\right)$$
$$= x^{-2\beta_1}(1-x)^{-2\beta_2}e^{-2\delta x - \eta x^2} \tag{7.26}$$

Since $a(x) = x(1-x)$ the stationary measure is

$$m(x) = \frac{1}{a(x)\psi(x)} = x^{2\beta_1 - 1}(1-x)^{2\beta_2 - 1}e^{2\delta x + \eta x^2} \tag{7.27}$$

In the case of additive selection, $\delta = \gamma$ and $\eta = 0$ so

$$m(x) = x^{2\beta_1 - 1}(1-x)^{2\beta_2 - 1}e^{2\gamma x} \tag{7.28}$$

If either $\beta_i = 0$ this is not integrable. If $\beta_1 = 0$ then there are no mutations from a to A, so 0 is an absorbing state. Likewise if $\beta_2 = 0$ then there are no mutations from A to a, so 1 is an absorbing state. If both $\beta_i > 0$ then there is a stationary distribution. In the case of no selection this is the beta distribution

$$\pi(x) = \frac{\Gamma(2\beta_1 + 2\beta_2)}{\Gamma(2\beta_1)\Gamma(2\beta_2)} x^{2\beta_1-1}(1-x)^{2\beta_2-1}$$

where $\Gamma(z) = \int_0^\infty t^{z-1} e^{-t}\, dt$ is the usual gamma function.

This formula can be found on page 123 of Wright's (1931) seminal paper on evolutionary theory. According to Will Provine's annotation of a collection of Wright's papers, Wright (1986), "this paper resulted in his admission to the National Academy of Science at a young age." It is interesting to note that the computation of the stationary distribution precedes the equation $L^*\pi = 0$, which first appears in the biology literature in Wright's (1945) work. In that paper, Wright says "Dr. A. Kolmogorov has recently been kind enough to send me a reprint of an important paper on this subject which was published in 1935, but which had not previously come to my attention." As one can see from the dates, the 10 year delay was likely due to World War II.

Suppose X has distribution π. Using this recursion and the fact that the constant makes $\int \pi(x)\, dx = 1$, we can compute

$$EX = \frac{\Gamma(2\beta_1 + 2\beta_2)}{\Gamma(2\beta_1)} \frac{\Gamma(2\beta_1 + 1)}{\Gamma(2\beta_1 + 2\beta_2 + 1)} = \frac{2\beta_1}{2\beta_1 + 2\beta_2}$$

$$EX^2 = \frac{\Gamma(2\beta_1 + 2\beta_2)}{\Gamma(2\beta_1)} \frac{\Gamma(2\beta_1 + 2)}{\Gamma(2\beta_1 + 2\beta_2 + 2)} = \frac{2\beta_1(2\beta_1 + 1)}{(2\beta_1 + 2\beta_2)(2\beta_1 + 2\beta_2 + 1)}$$

$$\mathrm{var}\,(X) = EX^2 - (EX)^2 = \frac{2\beta_1(2\beta_2)}{(2\beta_1 + 2\beta_2)^2(2\beta_1 + 2\beta_2 + 1)}$$

Using the first two formulas, we can compute the mean of the heterozygosity, i.e., the probability in equilibrium that two randomly chosen individuals are different:

$$E(2X(1-X)) = \frac{2\beta_1}{2\beta_1 + 2\beta_2}\left(1 - \frac{2\beta_1 + 1}{2\beta_1 + 2\beta_2 + 1}\right)$$

$$= 2 \cdot \frac{2\beta_1(2\beta_2)}{(2\beta_1 + 2\beta_2)(2\beta_1 + 2\beta_2 + 1)}$$

Reversibility

As we will now explain, the stationary measures of a one dimensional diffusion process have a very special property. Again, we begin by considering Markov chains. In discrete time the *detailed balance condition*:

$$\pi(i)p(i,j) = \pi(j)p(j,i) \tag{7.29}$$

implies that

$$\sum_i \pi(i)p(i,j) = \pi(j)\sum_i p(j,i) = \pi(j)$$

so π is a stationary distribution. In most cases, there is no π satisfying (7.29). However, if there is and we start the process from π, then the time reversal $Y_m = X_{n-m}$, $0 \le m \le n$ is a Markov chain with transition probability

$$p^*(i,j) = P_\pi(X_0 = j | X_1 = i) = \frac{P_\pi(X_0 = j, X_1 = i)}{P_\pi(X_1 = i)} = \frac{\pi(j)p(j,i)}{\pi(i)} = p(i,j)$$

Informally, a movie of a reversible process looks the same running forward or backwards in time.

For a continuous-time Markov chain, the detailed balance condition is

$$\pi(i)q(i,j) = \pi(j)q(j,i)$$

or, in equilibrium, the rate of jumps from i to j is the same as the rate of jumps from j to i. To extend the definition to diffusions, we define an inner product by

$$< f,g >_\pi = \sum_i f(i)\pi(i)g(i)$$

Given a linear operator R, we define the adjoint operator R^* with respect to π by

$$< f, Rg >_\pi = < R^* f, g >_\pi$$

Theorem 7.9. *If R is a matrix $r(i,j)$ then R^* is the matrix*

$$r^*(i,j) = \pi(j)r(j,i)/\pi(i).$$

In words, when R is a transition probability with stationary distribution π, R^* is the transition probability for the chain running backwards in time.

Proof. To check our proposed formula, we note that

$$< f, Rg >_\pi = \sum_j f(j)\pi(j)\sum_i r(j,i)g(i)$$

$$= \sum_i \left(\sum_j \frac{\pi(j)r(j,i)}{\pi(i)} f(j) \right) \pi(i)g(i)$$

$$= \sum_i \left(\sum_j r^*(i,j)f(j) \right) \pi(i)g(i) = < R^* f, g >_\pi \qquad \square$$

For a diffusion process, if we write

$$Lf = \frac{1}{2}a(x)\psi(x)\frac{d}{dx}\left(\frac{1}{\psi(x)}\frac{d}{dx}f(x) \right)$$

then using the speed measure $m(x) = 1/a(x)\psi(x)$ to define the inner product we have

$$< g, Lf >_m = \frac{1}{2} \int g(x) \frac{d}{dx} \left(\frac{1}{\psi(x)} \frac{d}{dx} f(x) \right) dx$$

If we assume that f and g vanish outside $[\delta, 1 - \delta]$, then integrating by parts twice shows that the above is

$$= \frac{1}{2} \int \frac{d}{dx} g(x) \cdot \frac{1}{\psi(x)} \frac{d}{dx} f(x) \, dx$$

$$= \frac{1}{2} \int \frac{d}{dx} \left(\frac{1}{\psi(x)} \frac{d}{dx} g(x) \right) f(x) \, dx = < Lg, f >_m$$

so L is self-adjoint with respect to m. As in the case of discrete state space, this implies that the transition probability has a symmetry property that resembles the detailed balance condition

$$p_t(x, y) = \frac{m(y)p_t(y, x)}{m(x)} \quad \text{or} \quad m(x)p_t(x, y) = m(y)p_t(y, x) \tag{7.30}$$

7.6 Occupation times

Let τ be the amount of time it takes for fixation or loss of an allele to occur. In addition to computing the probabilities of the two outcomes, we would like to determine the average time that this will take. For discrete models $\tau = T_0 \wedge T_{2N}$; for diffusions $\tau = T_0 \wedge T_1$. To compute $g(i) = E_i \tau$, it is convenient to consider a more general problem:

$$g(i) = E_i \sum_{0 \le m < \tau} f(X_m)$$

which reduces to the original question when $f \equiv 1$. If $0 < i < 2N$ then breaking things down according to what happens on the first step

$$g(i) = f(i) + \sum_j p(i, j)g(j) \quad \text{or} \quad E_i g(X_1) = g(i) - f(i) \tag{7.31}$$

The next result shows that this equation together with the boundary conditions $g(0) = 0$ and $g(2N) = 0$ are enough to identify $E_i \sum_{0 \le m < \tau} f(X_m)$.

Theorem 7.10. *Suppose that it is possible to reach 0 and 2N from each $0 < i < 2N$. $g(i) = E_i \sum_{0 \le m < \tau} f(X_m)$ is the unique solution of $(P - I)g = -f$ with $g(0) = 0$ and $g(2N) = 0$.*

Proof. By Theorem 7.3, our assumption implies $\sup_{0 < i < 2N} E_i \tau < \infty$, so g is well defined. To prove that the equation holds, we note that (7.31) implies that on $\{\tau > n\}$

$$E_i\left(g(X_{(n+1)\wedge\tau}) + \sum_{0\leq m<(n+1)\wedge\tau} f(X_m) \,\Big|\, \mathcal{F}_n\right)$$

$$= E_i\big(g(X_{(n+1)\wedge\tau})|\mathcal{F}_n\big) + \sum_{0\leq m<(n+1)\wedge\tau} f(X_m)$$

$$= g(X_{n\wedge\tau}) + \sum_{0\leq m<n\wedge\tau} f(X_m)$$

by (7.31). The last equality is trivial on $\{\tau \leq n\}$. Thus the expected value

$$E_i\left(g(X_{(n+1)\wedge\tau}) + \sum_{0\leq m<(n+1)\wedge\tau} f(X_m)\right)$$

is constant in time, so letting $n \to \infty$ and using $g(X_\tau) = 0$.

$$g(i) = E_i\left(g(X_\tau) + \sum_{0\leq m<\tau} f(X_m)\right) = E_i \sum_{0\leq m<\tau} f(X_m) \qquad \square$$

For a continuous-time Markov chain:

Theorem 7.11. *Suppose that it is possible to reach 0 and 2N from each $0 < i < 2N$. $g(i) = E_i \int_0^\tau f(X_s)\,ds$ is the unique solution of $Qg = -f$ for $0 < i < 2N$ with $g(0) = 0$ and $g(2N) = 0$.*

Proof. By Theorem 7.3, our assumption implies $\sup_{0<i<2N} E_i\tau < \infty$, so g is well defined. To prove that the equation holds, note that $Qg = -f$ for $0 < i < 2N$ implies

$$\frac{d}{dt}E_i\left(g(X_t) + \int_0^t f(X_s)\,ds\right) = E_i[Qg(X_t) + f(X_t)] = 0 \quad \text{when } t < \tau$$

so $E_i[g(X_{t\wedge\tau}) + \int_0^{t\wedge\tau} f(X_s)\,ds]$ is constant in time. If $g(0) = g(2N) = 0$ then letting $t \to \infty$ we have $g(i) = E_i \int_0^\tau f(X_s)\,ds$. $\qquad \square$

For a diffusion process:

Theorem 7.12. *Suppose that it is possible to reach y and z from each $y < x < z$. $g(x) = E_x \int_0^\tau f(X_s)\,ds$ is the unique solution of $Lg = -f$ for $0 < x < 1$ with $g(0) = 0$ and $g(1) = 0$.*

Proof. By Theorem 7.6, our assumption implies $\sup_{x\in(y,z)} E_x\tau < \infty$, so g is well defined. To prove that the equation holds, note that $Lg = -f$ for $y < x < z$ implies

$$\frac{d}{dt}E_x\left(g(X_t) + \int_0^t f(X_s)\,ds\right) = E_x[Lg(X_t) + f(X_t)] = 0 \quad \text{when } t < \tau$$

so $E_x[g(X_{t\wedge\tau}) + \int_0^{t\wedge\tau} f(X_s)\,ds]$ is constant in time. If $g(y) = g(z) = 0$ then letting $t \to \infty$ we have $g(x) = E_x \int_0^\tau f(X_s)\,ds$. $\qquad \square$

By now, the reader has probably learned that the three operators $P - I$, Q, and L are analogous, but, again, comparing the last three theorems shows that the occupation times satisfy

$$(P - I)g = -f \quad \text{discrete time Markov chain}$$
$$Qg = -f \quad \text{continuous time Markov chain}$$
$$Lg = -f \quad \text{diffusion process}$$

with $g = 0$ at the endpoints.

Exit times

We begin with two examples where $f \equiv 1$.

Example 7.18. Symmetric simple random walk. In this case we guess

$$g(i) = i(2N - i).$$

This obviously satisfies the boundary conditions $g(0) = 0$, $g(2N) = 1$. To check (7.31), we note that

$$\sum_j p(i,j)g(j) = (i+1)\frac{2N - i - 1}{2} + (i-1)\frac{2N - i + 1}{2}$$

$$= i(2N - i) + \frac{2N - i}{2} - \frac{i+1}{2} - \frac{2N - i}{2} + \frac{i - 1}{2}$$

$$= g(x) - 1$$

Example 7.19. Wright-Fisher model. In the case of no selection or mutation, inspired by a result of Kimura we guess

$$g(x) = -2[x \log x + (1-x) \log(1-x)]$$

$$g'(x) = -2\left[\log x + 1 - \log(1-x) + (1-x)\frac{1}{1-x} \cdot (-1) \right]$$

$$g''(x) = -2\left[\frac{1}{x} - \frac{1}{1-x} \cdot (-1) \right] = \frac{-2}{x(1-x)}$$

so $(1/2)x(1-x)g''(x) = -1$.

Example 7.20. Higher moments. As the proof will show, the next result is valid for continuous time Markov chains if we replace L by Q. This is Theorem 13.17 in Dynkin (1965).

Theorem 7.13. *If $f(x) = E_x \tau^k / k!$ then the solution of $Lg = -f$ is $g(x) = E\tau^{k+1}/(k+1)!$*

Proof. To begin, we recall that $E_x \tau^k / k! = \int_0^\infty \frac{u^{k-1}}{(k-1)!} P_x(\tau > u)\, du$ and write

$$g(x) = E_x \int_0^\tau E_{X_s}(\tau^k / k!)\, ds = E_x \int_0^\infty \int_0^\infty 1_{(\tau > s)} \frac{u^{k-1}}{(k-1)!} P_{X_s}(\tau > u)\, du\, ds$$

The Markov property implies that if $\mathcal{F}_s$ is the σ-field generated by the process up to time s then

$$P_x(\tau > t > s | \mathcal{F}_s) = 1_{(\tau > s)} P_{X_s}(\tau > t - s)$$

Taking expected values and changing variables $u = t - s$, we can write

$$g(x) = \int_0^\infty \int_s^\infty \frac{(t-s)^{k-1}}{(k-1)!} P_x(\tau > t > s) \, dt \, ds$$

Interchanging the order of integration:

$$= \int_0^\infty \int_0^t \frac{(t-s)^{k-1}}{(k-1)!} P_x(\tau > t) \, ds \, dt$$

$$= \int_0^\infty \frac{t^k}{k!} P_x(\tau > t) \, dt = E_x \tau^{k+1} / (k+1)!$$

which completes the proof.   □

7.7 Green's functions

For a discrete-time Markov chain on $\{0, 1, \ldots 2N\}$, we define the *Green's function* $G(i, j)$ to be the solution of $(P - I)g = -1_j$ with $g(0) = g(2N) = 0$, where 1_j is the function that is 1 at j and 0 otherwise. It follows from Theorem 7.10 that $G(i, j)$ is the expected number of visits to j starting from i before $\tau = T_0 \wedge T_{2N}$ and

$$E_i \sum_{0 \le m < \tau} f(X_m) = \sum_j G(i, j) f(j).$$

Theorem 7.14. *Suppose that it is possible to reach 0 and $2N$ from each $0 < i < 2N$. If we let $T_j^+ = \min\{n \ge 1 : X_n = j\}$ then*

$$G(i, j) = \frac{P_i(T_j < \tau)}{P_j(T_j^+ > \tau)} \tag{7.32}$$

Proof. The first factor is the probability we visit j at least once. If this occurs, then the number of visits to j has a geometric distribution with mean $1/P_j(T_j^+ > \tau)$.   □

In continuous time, we define the Green's function, $G(i, j)$, to be the solution of $Qg = -1_j$. It follows from Theorem 7.11 that $G(i, j)$ is the expected occupation time of j starting from i and

$$E_i \int_0^\tau f(X_s) \, ds = \sum_j G(i, j) f(j).$$

Theorem 7.15. *Suppose that it is possible to reach 0 and 2N from each $0 < i < 2N$. Let $R_j = \min\{t : X_t = j$ and $X_s \neq j$ for some $s < t\}$ be the first time the process returns to j.*

$$G(i, j) = \frac{P_i(T_j < \tau)}{q_j P_j(R_j > \tau)} \tag{7.33}$$

where $q_j = -Q(j, j)$ is the rate at which the process jumps out of j.

Proof. Again the first factor is the probability we visit j at least once. If this occurs, the number of visits to j has a geometric distribution with mean $1/P_j(R_j > \tau)$, and each visit lasts for an average amount of time $1/q_j$.   □

When space is continuous, we could, by analogy with the two previous cases, define the *Green's function* $G(x, y)$ to be the solution of $Lg = -\delta_y$, where δ_y is a point mass at y. However, as we explain in (7.44), solving this equation requires the use of calculus for "generalized functions." To keep things simple, we will instead define the Green's function $G(x, y)$ for the interval $[u, v]$ by the property that

$$g(x) = \int G(x, y) f(y) \, dy \quad \text{satisfies} \quad Lg = -f$$

for $u < x < v$ with $g(u) = g(v) = 0$.

Theorem 7.16. *Suppose that it is possible to reach u and v from each $u < x < v$. The Green's function $G(x, y)$ for the interval $[u, v]$ is*

$$\begin{aligned} &2\frac{(\phi(v) - \phi(x))(\phi(y) - \phi(u))}{\phi(v) - \phi(u)} \cdot m(y) &&y \leq x \\ &2\frac{(\phi(x) - \phi(u))(\phi(v) - \phi(y))}{\phi(v) - \phi(u)} \cdot m(y) &&x \leq y \end{aligned} \tag{7.34}$$

where $\phi(x)$ is the natural scale and $m(x) = 1/\phi'(x)a(x)$ is the speed measure.

Proof. To solve equation $Lg = -f$ now, we use (7.25) to write

$$\frac{d}{dx}\left(\frac{1}{\psi(x)}\frac{dg}{dx}\right) = -2m(x)f(x)$$

and integrate to conclude that for some constant C

$$\frac{1}{\psi(y)}\frac{dg}{dy} = C - 2\int_u^y dz\, m(z)f(z)$$

Multiplying by $\psi(y)$ on each side, integrating y from u to x, and recalling that $g(u) = 0$ and $\psi = \phi'$ we have

$$g(x) = C(\phi(x) - \phi(u)) - 2\int_u^x dy\, \psi(y) \int_u^y dz\, m(z)f(z) \tag{7.35}$$

In order to have $g(v) = 0$, we must have

$$C = \frac{2}{\phi(v) - \phi(u)} \int_u^v dy\, \psi(y) \int_u^y dz\, m(z) f(z)$$

Plugging the formula for C into (7.35) and writing

$$h_1(x) = \frac{\phi(x) - \phi(u)}{\phi(v) - \phi(u)} = P_x(T_v < T_u)$$

we have

$$f(x) = 2h_1(x) \int_u^v dy\, \psi(y) \int_u^y dz\, m(z) f(z)$$
$$- 2 \int_u^x dy\, \psi(y) \int_u^y dz\, m(z) f(z)$$

Interchanging the order of integration gives

$$f(x) = 2h_1(x) \int_u^v dz\, m(z) f(z)(\phi(v) - \phi(z))$$
$$- 2 \int_u^x dz\, m(z) f(z)(\phi(x) - \phi(z))$$

The integral over $[x, v]$ in the first term is

$$2h_1(x) \int_x^v dz\, m(z) f(z)(\phi(v) - \phi(z)) \tag{7.36}$$

Adding the integral over $[u, x]$ from the first term to the second gives

$$2 \int_u^x dz\, m(z) f(z) \left[\frac{\phi(x) - \phi(u)}{\phi(v) - \phi(u)} (\phi(v) - \phi(z)) - (\phi(x) - \phi(z)) \right]$$

A little algebra shows

$$(\phi(x) - \phi(u)) \cdot (\phi(v) - \phi(z)) - (\phi(x) - \phi(z)) \cdot (\phi(v) - \phi(u))$$
$$= -\phi(u)\phi(v) - \phi(x)\phi(z) + \phi(z)\phi(v) + \phi(x)\phi(u)$$
$$= (\phi(v) - \phi(x)) \cdot (\phi(z) - \phi(u))$$

so the second part of our formula becomes

$$2\frac{\phi(v) - \phi(x)}{\phi(v) - \phi(u)} \int_u^x dz\, m(z) f(z)[\phi(z) - \phi(u)]$$

Adding this to (7.36) gives the desired result.   □

An important consequence of (7.34) is:

Corollary. If $\tau_{u,v} = T_u \wedge T_v$ is the exit time from (u, v) then

$$E_x \tau_{u,v} = \int_u^v G(x, y)\, dy \qquad (7.37)$$

Speed (?) measure

Suppose first that the diffusion is on its natural scale, i.e., $\phi(x) = x$. In this case, if we take $\alpha = x - h$ and $\beta = x + h$ then the Green's function becomes

$$(x + h - y)m(y) \qquad x \leq y \leq x + h$$
$$(y - x + h)m(y) \qquad x - h \leq y \leq x$$

so (7.37) implies

$$E_x \tau_{x-h,x+h} = \int_x^{x+h} (x + h - y)m(y)\, dy + \int_{x-h}^x (y - x + h)m(y)\, dy \quad (7.38)$$

When h is small, $m(y) \approx m(x)$ for $y \in [x - h, x + h]$ so the above is

$$\approx m(x) \left(\int_x^{x+h} (x + h - y)\, dy + \int_{x-h}^x (y - x + h)\, dy \right) = m(x)h^2$$

Thus, $m(x)$ gives the *time* that X_t takes to exit a small interval centered at x, or to be precise, the ratio of the time for X_t to the time for a standard Brownian motion, which is h^2. Since speed is inversely proportional to the exit time, the term speed measure is a misnomer, but it is too late to change its name.

To treat a general diffusion, we have to transform it to its natural scale. Writing $\psi = \phi'$ and noting $L\phi = 0$, Theorem 7.1 implies that $Y_t = \phi(X_t)$ is a diffusion with coefficients

$$\bar{a}(y) = (a\psi^2)(\phi^{-1}(y)) \qquad \bar{b}(y) = 0 \qquad (7.39)$$

Using the previous calculation for Y, if $\bar{m}(y) = 1/(a\psi^2)(\phi^{-1}(y))$ is the speed measure for Y then $E_{\phi(x)} \tau_{\phi(x-h),\phi(x+h)}$ is

$$\approx \bar{m}(\phi(x)) \left(\int_{\phi(x)}^{\phi(x+h)} (\phi(x + h) - z)\, dz + \int_{\phi(x-h)}^{\phi(x)} (z - \phi(x - h))\, dz \right)$$

Changing variables $z = \phi(w)$, $dz = \psi(w)dw$, we see that this is

$$= \bar{m}(\phi(x))\psi(x) \left(\int_x^{x+h} (x + h - w)\, dw + \int_{x-h}^x (w - x + h)\, dw \right)$$

$$= \frac{1}{a(x)\psi(x)}h^2 = m(x)h^2$$

so the interpretation of $m(x)$ given above holds in general. As a final check that the speed measure is indeed the opposite of what the name suggests, recall that the stationary measure is $m(x)$, and the long run occupation time of a region is inversely proportional to the speed at which the process leaves it.

7.8 Examples

In all of the genetics examples in this section, there is no mutation. The calculations of the Green's functions require a fair amount of algebra, but they inform us about where we can expect the process to spend its time before reaching a boundary point at time τ and they allow us to compute $E\tau$. We begin by considering what happens when there is

No selection

Example 7.21. Symmetric simple random walk. Suppose that up jumps and down jumps each occur with probability $1/2$. X_n is a martingale, and it follows that if $a < x < b$

$$P_x(T_b < T_a) = \frac{x - a}{b - a} \qquad P_x(T_a < T_b) = \frac{b - x}{b - a}$$

Using the second formula and then the first, the numerator in (7.32) is

$$P_i(T_j < T_{2N}) = \frac{2N - i}{2N - j} \qquad j \le i$$

$$P_i(T_j < T_0) = \frac{i}{j} \qquad i \le j$$

To compute the denominator of (7.32), we note that if $\tau = T_0 \wedge T_{2N}$ then

$$P_j(T_j^+ > \tau) = \frac{1}{2}P_{j+1}(T_{2N} < T_j) + \frac{1}{2}P_{j-1}(T_0 < T_j)$$

$$= \frac{1}{2} \cdot \frac{1}{2N - j} + \frac{1}{2} \cdot \frac{1}{j} = \frac{1}{2} \cdot \frac{2N}{j(2N - j)}$$

Combining the results, we can write $G(i, j)$ as

$$2\frac{(2N - i)j}{2N} \qquad j \le i$$

$$2\frac{i(2N - j)}{2N} \qquad i \le j \qquad\qquad (7.40)$$

Summing over j and letting $k = 2N - j$, we have

$$E_i \tau = 2 \frac{(2N-i)}{2N} \sum_{j=1}^{i} j + 2 \cdot \frac{i}{2N} \sum_{k=1}^{2N-i-1} k$$

$$= 2 \left[\frac{(2N-i)}{2N} \cdot \frac{i(i+1)}{2} + \frac{i}{2N} \cdot \frac{(2N-i)(2N-i-1)}{2} \right]$$

$$= \frac{i(2N-i)}{2N} [(i+1) + (2N-i-1)] = i(2N-i)$$

in agreement with the result in Example 7.18.

Example 7.22. Moran model with no selection. In this case $P_i(T_j < \tau)$ and $P_j(R_j > \tau) = P_j(T_j^+ > \tau)$ are the same as for the symmetric simple random walk considered above, while $q_j = 2j(2N-j)/2N$. It follows from the calculation in the previous example that $q_j P_j(R_j > \tau)$ in the denominator of (7.33) is 1 and we can write $G(i,j)$ as

$$\begin{cases} \dfrac{2N-i}{2N-j} & j \le i \\[2mm] i/j & i \le j \end{cases} \tag{7.41}$$

Summing over j we have

$$E_i \tau = (2N-i) \sum_{j=1}^{i} \frac{1}{2N-j} + i \sum_{j=i+1}^{2N} \frac{1}{j}$$

If $i = 2Nx$ with $0 < x < 1$ then

$$\sum_{j=i+1}^{2N} \frac{1}{j} = \sum_{j=i+1}^{2N} \frac{1}{j/2N} \cdot \frac{1}{2N} = \int_{x}^{1} \frac{dx}{x} = -\log x$$

so we have

$$\frac{1}{N} E_{2Nx} - 2x \log x - 2(1-x) \log(1-x) \tag{7.42}$$

Example 7.23. Wright-Fisher diffusion with no selection. $\phi(x) = x$, $\psi(x) = \phi'(x) = 1$, and $a(x) = x(1-x)$ so using (7.34), $G(x,y) =$

$$\frac{2(1-x)y}{y(1-y)} = \frac{2(1-x)}{1-y} \qquad y \le x$$

$$\frac{2x(1-y)}{y(1-y)} = \frac{2x}{y} \qquad x \le y \tag{7.43}$$

If we set $i = 2Nx$ and $j = 2Nx$ in the Moran model formula, we get $(1-x)/(1-y)$ and x/y. The missing factor of 2 comes from the fact that i corresponds to $[x, x+1/2N]$, so the occupation time density is multiplied by $2N$, but time is run at rate N, so it is divided by N.

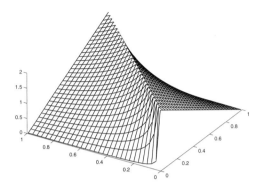

Fig. 7.5. Green's function for Wright-Fisher diffusion with no selection.

Integrating $G(x, y)$ we have

$$E_x \tau = \int_x^1 \frac{2x}{y} \, dy + \int_0^x \frac{2(1-x)}{1-y} \, dy$$
$$= -2x \log x - 2(1-x) \log(1-x)$$

which agrees with (7.42) and our computation in Example 7.19.

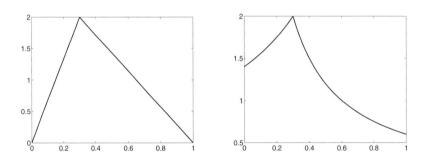

Fig. 7.6. Slices $x \to G(x, 0.3)$ and $y \to G(0.3, y)$ of the previous graph.

For fixed y, $x \to G(x, y)$ is linear on $[0, y]$ and $[y, 1]$ and vanishes at 0 and 1. To explain the form of the answer we return to our remark that for fixed y, $g(x) = G(x, y)$ is a solution of

$$\frac{1}{2} x(1-x) \frac{d^2}{dx^2} g = -\delta_y \tag{7.44}$$

When $x \ne y$, $g''(x) = 0$ so $g(x)$ is linear on $[0, y]$ and $[y, 1]$. The integral of $-\delta_y$ is 0 for $x < y$ and -1 for $x > y$, so with a little thought we realize that (7.44) can be written as $g'(y+) - g'(y-) = 2/y(1-y)$, which is correct since

$$g'(y+) - g'(y-) = \frac{2}{1-y} - \frac{2}{y} = \frac{2}{y(1-y)}$$

Selection

Example 7.24. Asymmetric simple random walk. Suppose that up jumps occur with probability $p = 1/(2-s)$ and down jumps with probability $1 - p = (1-s)/(2-s)$. We have chosen these values of p so that this is the embedded discrete-time jump chain for the Moran model with selection. The ratio $(1-p)/p = 1 - s$, so if we let $h(x) = 1 - (1-s)^x$ then by calculations in Section 6.1, $h(X_n)$ is a martingale and it follows that if $a < x < b$

$$P_x(T_b < T_a) = \frac{h(a) - h(x)}{h(a) - h(b)} \qquad P_x(T_a < T_b) = \frac{h(x) - h(b)}{h(a) - h(b)}$$

We have reversed the usual order of the numerator and denominator to make the next few calculations easier to see. Using the second formula and then the first, the numerator in (7.32) is

$$P_i(T_j < T_{2N}) = \frac{(1-s)^i - (1-s)^{2N}}{(1-s)^j - (1-s)^{2N}} \qquad j \le i$$

$$P_i(T_j < T_0) = \frac{1 - (1-s)^i}{1 - (1-s)^j} \qquad i \le j$$

To compute the denominator of (7.32), we note that

$$P_j(T_j^+ > \tau) = \frac{1}{2-s} P_{j+1}(T_{2N} < T_j) + \frac{1-s}{2-s} P_{j-1}(T_0 < T_j)$$

$$= \frac{1}{2-s} \frac{(1-s)^j - (1-s)^{j+1}}{(1-s)^j - (1-s)^{2N}} + \frac{1-s}{2-s} \frac{(1-s)^{j-1} - (1-s)^j}{1 - (1-s)^j}$$

The two numerators are $(1-s)^j - (1-s)^{j+1} = s(1-s)^j$, so the above is

$$= \frac{s(1-s)^j}{2-s} \cdot \frac{[1 - (1-s)^{2N}]}{[1 - (1-s)^j][(1-s)^j - (1-s)^{2N}]} \qquad (7.45)$$

Reintroducing $h(x) = 1 - (1-s)^x$, we can write $G(i,j)$ as

$$\frac{(h(2N) - h(i)) \cdot (h(j) - h(0))}{h(2N) - h(0)} \cdot \frac{2-s}{s(1-s)^j} \qquad j \le i$$

$$\frac{(h(i) - h(0)) \cdot (h(2N) - h(j))}{h(2N) - h(0)} \cdot \frac{2-s}{s(1-s)^j} \qquad i \le j \qquad (7.46)$$

Example 7.25. Moran model with selection. In this case, $P_i(T_j < \tau)$ and $P_j(R_j > \tau) = P_j(T_j^+ > \tau)$ are the same as for the asymmetric simple random walk considered above, while $q_j = (2-s)j(2N-j)/2N$. It follows from (7.45) that the denominator of (7.33) is

$$q_j P_j(R_j > \tau) = \frac{j(2N-j)}{2N} \frac{s(1-s)^j \cdot [1-(1-s)^{2N}]}{[1-(1-s)^j][(1-s)^j - (1-s)^{2N}]}$$

and the Green's function becomes

$$\frac{(h(2N)-h(i)) \cdot (h(j)-h(0))}{h(2N)-h(0)} \cdot \frac{2N}{s(1-s)^j j(2N-j)} \quad 0 < j < i$$

$$\frac{(h(i)-h(0)) \cdot (h(2N)-h(j))}{h(2N)-h(0)} \cdot \frac{2N}{s(1-s)^j j(2N-j)} \quad i < j < 2N \quad (7.47)$$

Example 7.26. Wright-Fisher diffusion with additive selection. To make it easier to relate the results for this case to the Moran model, we will define the natural scale to be $\phi(y) = 1 - \exp(-2\gamma y)$ which makes

$$\psi(y) = 2\gamma \exp(-2\gamma y)$$

$$m(y) = \frac{1}{a(y)\psi(y)} = \frac{1}{2\gamma \exp(-2\gamma)y(1-y)}$$

Recalling the formula for $G(x,y)$

$$2\frac{(\phi(1)-\phi(x))(\phi(y)-\phi(0))}{\phi(1)-\phi(0)} \cdot m(y) \qquad y \le x$$

$$2\frac{(\phi(x)-\phi(0))(\phi(1)-\phi(y))}{\phi(1)-\phi(0)} \cdot m(y) \qquad x \le y$$

we see that $G(x,y)$ is given by

$$2(e^{-2\gamma x} - e^{-2\gamma})\frac{1-e^{-2\gamma y}}{1-e^{-2\gamma}} \cdot \frac{1}{2\gamma e^{-2\gamma y} y(1-y)} \qquad y \le x$$

$$2(1 - e^{-2\gamma x})\frac{e^{-2\gamma y} - e^{-2\gamma}}{1-e^{-2\gamma}} \cdot \frac{1}{2\gamma e^{-2\gamma y} y(1-y)} \qquad x \le y \qquad (7.48)$$

To connect with the Moran model, note that if $x = i/2N$, $y = j/2N$, and $s = 2\gamma/2N$ then

$$h(i) = 1 - (1-\gamma/2N)^{2Nx} \to 1 - e^{-2\gamma x} = \phi(x)$$

$$\frac{2N \cdot 2N}{2Ns(1-s)^j j(2N-j)} \to \frac{1}{2\gamma e^{-2\gamma y} y(1-y)} = m(y)$$

As in the case of no selection, the missing factor of 2 comes from the fact that i corresponds to $[x, x+1/2N]$, so the density is multiplied by $2N$, but time is run at rate N to get the diffusion limit, so it is divided by N.

To help understand the Green's function, it is useful to look at slices through the graph. If y is fixed then for $x < y$ we have $G(x,y) = A(y)(1 -$

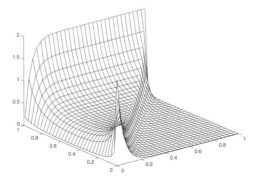

Fig. 7.7. Green's function for Wright-Fisher diffusion with additive selection $\gamma = 10$

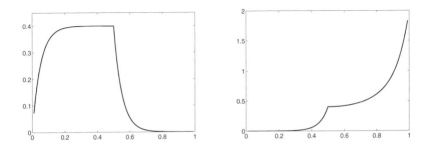

Fig. 7.8. Slices $x \to G(x, 1/2)$ and $y \to G(1/2, y)$ of the previous graph.

$e^{-2\gamma x}$) so if $x \gg 1/2\gamma$, then $G(x, y)$ is roughly constant because x will hit y with probability close to 1. For $x > y$, $G(x, y) = B(y)(e^{-2\gamma x} - e^{-2\gamma})$ since the probability of hitting y decays exponentially fast. Let $g(x) = G(x, y)$. For $x \neq y$ we have $Lg = 0$ where

$$Lg = \frac{1}{2}x(1-x)\left[\frac{d^2g}{dx^2} + 2\gamma\frac{dg}{dx}\right]$$

A little calculus shows that again we have

$$g'(y+) - g'(y-) = -2\gamma e^{-2\gamma y}\frac{2}{m(y)} = \frac{2}{y(1-y)}$$

To understand the behavior for x fixed, it is useful to multiply top and bottom of (7.48) by $e^{2\gamma y}$ to rewrite $G(x, y)$ as

$$
\begin{cases}
\dfrac{e^{-2\gamma x} - e^{-2\gamma}}{1 - e^{-2\gamma}} \cdot \dfrac{e^{2\gamma y} - 1}{\gamma y(1-y)} & y \leq x \\[3ex]
\dfrac{1 - e^{-2\gamma x}}{1 - e^{-2\gamma}} \cdot \dfrac{1 - e^{-2\gamma(1-y)}}{2\gamma y(1-y)} & x \leq y
\end{cases}
\tag{7.49}
$$

If x is fixed and γ is large, then $G(x, y)$ is approximately

$$\frac{e^{-2\gamma(x-y)}}{\gamma y(1-y)} \quad 0 \le y \le x$$

$$\frac{1}{\gamma y(1-y)} \quad x \le y \quad \text{and} \quad \gamma(1-y) \gg 1$$

$$\frac{1-e^{-2c}}{c} \quad x \le y = 1 - c/\gamma$$

This shows that the process spends a negligible amount of time $< x$ and moves through values $y < 1$ at the rate predicted by the logistic differential equation until $1 - y = O(\gamma^{-1})$. Note that as $y \to 1$, $c \to 0$ and $(1 - e^{-2c})/c \to 2$.

Example 7.27. Symmetric balancing selection. In this case $\psi(x) = e^{-2\gamma x(1-x)}$, $m(x) = e^{2\gamma x(1-x)}/x(1-x)$, and $\phi(x) = \int_0^x \psi(y)\, dy$, so $G(x, y)$ is

$$2\frac{(\phi(1) - \phi(x))\phi(y)}{\phi(1)} \cdot \frac{e^{2\gamma y(1-y)}}{y(1-y)} \quad y \le x$$

$$2\frac{(\phi(1) - \phi(y))\phi(x)}{\phi(1)} \cdot \frac{e^{2\gamma y(1-y)}}{y(1-y)} \quad x \le y \qquad (7.50)$$

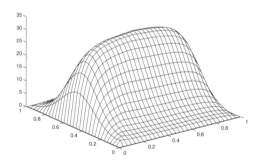

Fig. 7.9. Green's function for Wright-Fisher diffusion with symmetric balancing selection $\gamma = 10$

To help understand the Green's function, it is useful to look at slices through the graph. As we computed in Example 7.15 at the end of Section 7.4, when γ is large and x is away from the boundaries at 0 and 1,

$$\phi(1) \approx 2 \int_0^\infty e^{-2\gamma y}\, dy = \frac{1}{\gamma}$$

$$\phi(x), 1 - \phi(x) \approx \int_0^\infty e^{-2\gamma y}\, dy = \frac{1}{2\gamma}$$

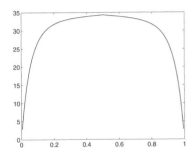 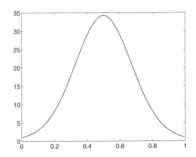

Fig. 7.10. Slices $x \to G(x, 1/2)$ and $y \to G(1/2, y)$ of the previous graph.

so the first factor in (7.50) is $\approx 1/4\gamma$ and the two cases collapse to

$$\frac{e^{2\gamma y(1-y)}}{2\gamma y(1-y)}$$

The approximation does not depend on the starting point because no matter where the diffusion starts, the frequency quickly moves to $1/2$.

Changing variables $y = 1/2 + z$, which makes $y(1-y) = 1/4 - z^2$, the above becomes

$$\frac{2e^{\gamma/2}}{\gamma} \frac{e^{-2\gamma z^2}}{1 - 4z^2}$$

Realizing that most of the contribution will come from values of z of order $O(1/\sqrt{\gamma})$,

$$E_x \tau = \int_0^1 G(x, y)\, dy \approx \frac{2e^{\gamma/2}}{\gamma} \int e^{-2\gamma z^2}\, dz$$

The integrand resembles the normal density with mean 0 and variance $1/4\gamma$, so its value is $\sqrt{\pi/2\gamma}$ and we have

$$E_x \tau \approx e^{\gamma/2} \gamma^{-3/2} \sqrt{2\pi}$$

7.9 Conditioned processes

In many situations we are interested in conditioning that the current mutation fixes or dies out. If $h(x) = P_x(T_1 < T_0)$ and we condition on fixation then the new transition probability

$$\bar{p}_t(x, y) = p_t(x, y) h(y)/h(x).$$

The same result holds for conditioning on loss, with $h(x) = P_x(T_0 < T_1)$. Integrating, we have that the conditioned Green's function

$$\bar{G}(x, y) = G(x, y)h(y)/h(x). \tag{7.51}$$

To compute the generator of the conditioned process, we note that

$$\bar{L}f(x) = \lim_{t \to 0} \frac{1}{t} \left(\int \frac{p_t(x, y)h(y)}{h(x)} f(y) \, dy - f(x) \right)$$

$$= \frac{1}{h(x)} \lim_{t \to 0} \frac{1}{t} \left(\int p_t(x, y)h(y)f(y) \, dy - h(x)f(x) \right) = \frac{1}{h(x)} L(hf)$$

Working out the derivatives

$$\frac{1}{h(x)} L(hf) = \frac{1}{h(x)} \left(b(x)(h'f + hf') + \frac{1}{2}a(x)(h''f + 2h'f' + hf'') \right)$$

Using $Lh = 0$ and simplifying

$$\bar{L}f = \frac{1}{2}a(x)f'' + \left(b(x) + a(x)\frac{h'(x)}{h(x)} \right) f'(x) \tag{7.52}$$

in agreement with (32) in Ewens (1973).

Example 7.28. Wright-Fisher diffusion with no selection. $h(x) = x$ is the probability of fixation, so using (7.43) and (7.51), $\bar{G}(x, y) =$

$$\frac{2(1-x)}{1-y} \cdot \frac{y}{x} \qquad y \le x$$

$$\frac{2x}{y} \cdot \frac{y}{x} = 2 \qquad x \le y \tag{7.53}$$

and using (7.52) we have

$$\bar{L}v_1 = \frac{1}{2}x(1-x)f''(x) + (1-x)f'(x) \tag{7.54}$$

Theorem 7.17. *For the Wright-Fisher model conditioned on fixation,* $\tau = T_0 \wedge T_1$ *has*

$$E_x(\tau|T_1 < T_0) = -2\frac{(1-x)}{x} \log(1-x) \tag{7.55}$$

$$E_x(\tau^2|T_1 < T_0) = 8 \left(\frac{(1-x)\log(1-x)}{x} - \int_x^1 \frac{\log(1-y)}{y} \, dy \right) \tag{7.56}$$

The first formula is (14) in Kimura and Ohta (1969a). This second can be obtained from (A7) of Kimura and Ohta (1969b), which is for conditioning on extinction. Both are on page 29 of Ewens (1973), but in the second case he has an erroneous minus sign. In the two references cited, formulas are given on the original time scale, so the first formula is multiplied by $2N$ and the second by $(2N)^2$.

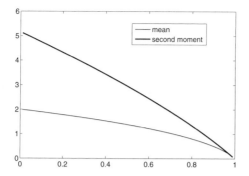

Fig. 7.11. First and second moments of τ conditioned on $T_1 < T_0$, as a function of the starting point.

Proof. We can derive the first formula from (7.53). Integrating using $y/(1 - y) = 1/(1 - y) - 1$, we have

$$E_x(\tau | T_1 < T_0) = \int_x^1 2\,dy + \frac{2(1-x)}{x}\int_0^x \frac{y}{1-y}\,dy$$

$$= 2\left((1-x) + \frac{(1-x)}{x}(-\log(1-x) - x)\right)$$

and after a little arithmetic, we have (7.55).

A second approach is to use Theorem 7.12. Let $v_1(x)$ be our formula for $E_x(\tau | T_1 < T_0)$. The first step is to check that $Lv_1 = -1$. To do this, we note that writing $-(1-x)/x = -1/x + 1$

$$v_1'(x) = \frac{2}{x^2}\log(1-x) - 2\frac{1-x}{x}\cdot\frac{-1}{1-x} = \frac{2}{x^2}\log(1-x) + \frac{2}{x}$$

$$v_1''(x) = -\frac{4}{x^3}\log(1-x) + \frac{2}{x^2}\cdot\frac{-1}{1-x} - \frac{2}{x^2}$$

so we have

$$\bar{L}v_1 = \frac{1}{2}x(1-x)v_1''(x) + (1-x)v_1'(x)$$

$$= -\frac{2(1-x)}{x^2}\log(1-x) - \frac{1}{x} - \frac{(1-x)}{x}$$

$$+ \frac{2(1-x)}{x^2}\log(1-x) + \frac{2(1-x)}{x}$$

$$= -\frac{1}{x} + \frac{1-x}{x} = -1$$

To examine the boundary conditions, we note that $y\log(y) \to 0$ as $y \to 0$ so $v_1(1) = 0$. We do not have $v_1(0) = 0$. The easiest way to see this is to note that $\log(1 - x) \le -x$, so $v'_1(x) \le 0$, i.e., $v_1(x)$ is decreasing, which is what one should expect for $v_1(x) = E_x(\tau | T_1 < T_0)$. As $x \to 0$, $\log(1 - x) \sim -x$ so $v_1(0) = 2$. As we will see in Section 7.10, the apparent inconsistency with Theorem 7.12 comes because it is impossible for the conditioned process to get to 0, so one of the assumptions of that theorem does not hold.

Let $v_2(x)$ be our formula for $E_x(\tau^2 | T_1 < T_0)$. To verify the second formula using Theorem 7.13, we want to show $Lv_2 = -2v_1$. To check this guess, we note that

$$\frac{d}{dx}\frac{v_2}{8} = -\frac{1}{x^2}\log(1 - x) + \frac{(1 - x)}{x}\cdot\frac{-1}{1 - x} + \frac{\log(1 - x)}{x}$$

$$= \left(-\frac{1}{x^2} + \frac{1}{x}\right)\log(1 - x) - \frac{1}{x}$$

$$\frac{d^2}{dx^2}\frac{v_2}{8} = \left(\frac{2}{x^3} - \frac{1}{x^2}\right)\log(1 - x) - \frac{1 - x}{x^2}\cdot\frac{-1}{1 - x} + \frac{1}{x^2}$$

$$= \frac{2 - x}{x^3}\log(1 - x) + \frac{2}{x^2}$$

Combining the last two results, we see that

$$\frac{\bar{L}v_2}{8} = \frac{1}{2}\cdot\frac{2 - x}{x^2}(1 - x)\log(1 - x) + \frac{1 - x}{x}$$

$$+ \frac{(x - 1)(1 - x)}{x^2}\log(1 - x) - \frac{(1 - x)}{x}$$

$$= \frac{(1 - x)}{x}\log(1 - x)\left[\frac{1}{x} - \frac{1}{2} + 1 - \frac{1}{x}\right]$$

and it follows that

$$\bar{L}v_2 = 4\frac{(1 - x)}{x}\log(1 - x) = -2v_1$$

Clearly $v_2(1) = 0$, so the only relevant boundary condition is satisfied.   □

Example 7.29. Age of alleles. By symmetry

$$E_x(\tau | T_0 < T_1) = E_{1-x}(\tau | T_1 < T_0) = -2\frac{x}{1 - x}\log(x)$$

As we will now show, this gives the average age of an allele A observed to be at frequency x, a classic result of Kimura and Ohta (1973). To argue this, we note that the density of the age of A given that it has frequency x is

$$f_x(t) = \lim_{\epsilon \to 0}\frac{p_t(\epsilon, x)}{\int_0^\infty p_s(\epsilon, x)\,ds}$$

Multiplying top and bottom by the speed measure $m(\epsilon)$, then using reversibility, (7.30), and noting the factors of m cancel, the above

$$= \lim_{\epsilon \to 0} \frac{m(\epsilon) p_t(\epsilon, x)}{\int_0^\infty m(\epsilon) p_s(\epsilon, x) \, ds}$$

$$= \lim_{\epsilon \to 0} \frac{p_t(x, \epsilon)}{\int_0^\infty p_s(x, \epsilon) \, ds} = g_x(t)$$

the density for the hitting time of 0 starting from x. Therefore,

$$\int t f_x(t) \, dt = \int t g_x(t) \, dt = E_x(\tau | T_0 < T_1)$$

I learned this argument from Griffiths (2003). See his paper for an account of the history and results for the expected values of the ages of alleles observed to occur k times in a sample of size n.

Example 7.30. Wright-Fisher diffusion with additive selection. In this case the probability of fixation is $h(x) = (1 - e^{-2\gamma x})/(1 - e^{-2\gamma})$ when the initial frequency is x, so

$$\bar{G}(x, y) = G(x, y) \frac{1 - e^{-2\gamma y}}{1 - e^{-2\gamma x}}$$

Using (7.49) now, we have that $\bar{G}(x, y)$ is

$$\frac{1 - e^{-2\gamma y}}{1 - e^{-2\gamma x}} \cdot \frac{e^{-2\gamma x} - e^{-2\gamma}}{1 - e^{-2\gamma}} \cdot \frac{e^{2\gamma y} - 1}{\gamma y (1 - y)} \qquad y \le x$$

$$\frac{1 - e^{-2\gamma y}}{1 - e^{-2\gamma}} \cdot \frac{1 - e^{-2\gamma(1-y)}}{\gamma y (1 - y)} \qquad x \le y$$

Note that the second formula does not depend on x (except through the condition $x \le y$). As Figure 7.12 shows, the conditioning does not change the picture very much except near $x = 0$, where we no longer have $G(x, y) \to 0$ as $x \to 0$.

If we integrate this with respect to y, then we get a result first derived by Kimura and Ohta (1969a), see their (17).

$$\bar{E}_x \tau \approx \int_x^1 \frac{[1 - e^{-2\gamma y}] \cdot [1 - e^{-2\gamma(1-y)}]}{[1 - e^{-2\gamma}] \cdot \gamma y (1 - y)} \, dy$$

$$+ \frac{e^{-2\gamma x} - e^{-2\gamma}}{1 - e^{-2\gamma x}} \int_0^x \frac{[1 - e^{-2\gamma y}] \cdot [e^{\gamma y} - 1]}{[1 - e^{-2\gamma}] \cdot 2\gamma y (1 - y)} \, dy$$

Since $1 - e^{-2\gamma a} \le 2\gamma a$, the two integrals are finite. However, they must be evaluated numerically. As a check on the last formula, we note $(e^{a\gamma} - 1)/\gamma \to a$ as $\gamma \to 0$ so

$$\frac{1 - e^{-2\gamma y}}{\gamma y} \cdot \frac{1 - e^{-2\gamma(1-y)}}{\gamma (1 - y)} \cdot \frac{\gamma}{1 - e^{-2\gamma}} \to 2$$

Using similar reasoning on the other terms,

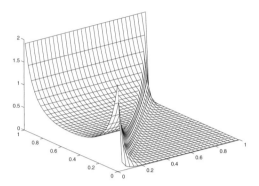

Fig. 7.12. Green's function for conditioned Wright-Fisher with selection $\gamma = 10$

$$\bar{E}_x \tau \rightarrow \int_x^1 2\,dy + 2N\left(\frac{1-x}{x}\right)\int_0^x \frac{2y}{1-y}\,dy$$

$$= -2 \cdot \frac{(1-x)}{x}\log(1-x)$$

by the calculation in Example 7.28.

Example 7.31. General diffusion. Maruyama and Kimura (1974) observed that for Wright-Fisher diffusions with general diploid selection

$$\lim_{x\to 0} E_x(\tau|T_1 < T_0) = \lim_{x\to 1} E_x(\tau|T_0 < T_1) \qquad (7.57)$$

As we will now show, and presumably the authors also realized, this is a general property of one dimensional diffusions.

Proof. We begin by recalling the formula for the Green's function given in (7.34), which we simplify by supposing $\phi(0) = 0$ and $\phi(1) = 1$.

$$2(1 - \phi(x))\phi(y)m(y) \qquad y \le x$$

$$2\phi(x)(1 - \phi(y))m(y) \qquad x \le y$$

The Green's function $G_1(x, y)$ for the process starting from x and conditioned on $T_1 < T_0$ is

$$2(1 - \phi(x))\phi(y)m(y) \cdot \frac{\phi(y)}{\phi(x)} \qquad y \le x$$

$$2\phi(x)(1 - \phi(y))m(y) \cdot \frac{\phi(y)}{\phi(x)} \qquad x \le y$$

in agreement with (12) of Maruyama and Kimura (1971). The Green's function $G_0(x, y)$ for the process starting from x and conditioned on $T_0 < T_1$ is

$$2(1 - \phi(x))\phi(y)m(y) \cdot \frac{1 - \phi(y)}{1 - \phi(x)} \qquad y \leq x$$

$$2\phi(x)(1 - \phi(y))m(y) \cdot \frac{1 - \phi(y)}{1 - \phi(x)} \qquad x \leq y$$

In each case, after cancellation we have for $x \leq y$

$$G_0(x, y) = G_1(x, y) = 2(\phi(1) - \phi(y))\phi(y) \cdot m(y)$$

and integrating gives the desired result. $\qquad \square$

Taylor, Iwasa, and Nowak (2006) have shown that this result holds for reversible Markov chains on $\{0, 1, \ldots N\}$ in which 0 and N are absorbing and these two states can only be reached directly from 1 and $N - 1$ respectively.

7.10 Boundary behavior

The consideration of diffusion processes leads to two questions that have no analogues for discrete models: "Can the process get to the boundary?" and "Once it gets to the boundary can it return to the interior of the state space?" To build some suspense, we invite the reader to guess what happens for the Wright-Fisher diffusion with mutation

$$Lf = \frac{1}{2}\frac{d^2}{dx^2}f + (\beta_1(1 - x) - \beta_2 x)\frac{d}{dx}f$$

It should not be surprising that if $\beta_1 = \beta_2 = 0$ then the diffusion stops the first time it his 0 or 1, but what if one or both of the $\beta_i > 0$?

It is enough to consider the boundary at 0. Consider a diffusion on $(0, r)$ where $r \leq \infty$, let $q \in (0, r)$, and let

$$I = \int_0^q (\phi(z) - \phi(0))\, m(z)\, dz$$

$$J = \int_0^q (M(z) - M(0))\, \psi(z)\, dz$$

where M is an antiderivative of m. Writing iff as short for "if and only if," we have the following results for a diffusion process X_t.

Theorem 7.18. X_t can get IN to the boundary point 0 iff $I < \infty$. X_t can get OUT from the boundary point 0 iff $J < \infty$.

Note that $\phi(0) = -\infty$ implies $I = \infty$ and $M(0) = -\infty$ implies $J = \infty$.

Proof. To start to prove the first result, we will show

Theorem 7.19. *Let $1/2 < b < 1$. The following are equivalent:*

(i) $\phi(0) > -\infty$ *and* $\int_0^{1/2} (\phi(z) - \phi(0))\, m(z)\, dz < \infty$

(ii) $\inf_{0<a<1/2} P_{1/2}(T_a < T_b) > 0$ *and* $\sup_{0<a<1/2} E_{1/2}(T_a \wedge T_b) < \infty$

(iii) $P_{1/2}(T_0 < T_b) > 0$

Proof. We first show that (i) and (ii) are equivalent.

$$P_{1/2}(T_a < T_b) = \frac{\phi(b) - \phi(1/2)}{\phi(b) - \phi(a)}$$

so $\inf_{\alpha < a < 0} P_0(T_a < T_b) > 0$ if and only if $\phi(0) > -\infty$. Using (7.37) and (7.34)

$$\begin{aligned}
E_x \tau_{a,b} = \ &2\,\frac{\phi(x) - \phi(a)}{\phi(b) - \phi(a)} \int_x^b (\phi(b) - \phi(z))\, m(z)\, dz \\
&+ 2\,\frac{\phi(b) - \phi(x)}{\phi(b) - \phi(a)} \int_a^x (\phi(z) - \phi(a))\, m(z)\, dz
\end{aligned}$$

The first integral always stays bounded as $a \downarrow 0$. So $E_0 \tau_{(a,b)}$ stays bounded as $a \to 0$ if and only if $\phi(0) > -\infty$ and

$$\int_0^{1/2} (\phi(z) - \phi(0)) m(z)\, dz < \infty$$

which completes the proof of the equivalence of (i) and (ii).

It is easy to see that (ii) implies (iii). For the other direction, note that Theorem 7.6 implies that if $P_{1/2}(T_0 < T_b) > 0$ then $E_{1/2}\tau_{0,b} < \infty$. □

To try to start the process X_t from 0, let ϕ be its natural scale. As (7.39) shows, $Y_t = \phi(X_t)$ has coefficients $\bar{b}(y) = 0$ and

$$\bar{a}(y) = (a\psi^2)(\phi^{-1}(y))$$

To see if we can start the process Y_t at 0, we extend $\bar{a}$ to the negative half-line by setting $\bar{a}(-y) = \bar{a}(y)$ and let Z_t be the associated diffusion. If we let $\bar{m}(y) = 1/\bar{a}(|y|)$ be the speed measure for Z_t, which is on its natural scale, we can use (7.39) and the symmetry $\bar{m}(-y) = \bar{m}(y)$ to conclude

$$\frac{1}{2} E_0 \tau_{-\epsilon,\epsilon} = \int_0^\epsilon (\epsilon - y)\bar{m}(y)\, dy$$

Changing variables $y = \phi(x)$, $dy = \psi(x)\, dx$, $\epsilon = \phi(\delta)$, the above

$$= \int_0^\delta (\phi(\delta) - \phi(x)) \frac{1}{\psi(x)a(x)}\, dx$$

$$= \int_0^\delta \left(\int_x^\delta \psi(z)\, dz \right) m(x)\, dx$$

Interchanging the order of integration, the above

$$= \int_0^\delta \int_0^z m(x)\, dx\, \psi(z)\, dz$$

At this point we have shown

$$E_0 \tau_{-\epsilon,\epsilon} = 2 \int_0^\delta (M(z) - M(0))\, \psi(z)\, dz \tag{7.58}$$

To see that $E_0 \tau_{-\epsilon,\epsilon} = \infty$ means that the process cannot escape from 0, we note that Theorem 7.6 implies that if $P_0(\tau_{-\epsilon,\epsilon} < \infty) > 0$ then $E_0 \tau_{-\epsilon,\epsilon} < \infty$. This completes the proof of the second result and of the theorem.   □

There are four possible combinations of I and J being finite or infinite, which were named by Feller as follows

$$
\begin{array}{llll}
I < \infty & J < \infty & \text{regular} \\
I < \infty & J = \infty & \text{absorbing} \\
I = \infty & J < \infty & \text{entrance} \\
I = \infty & J = \infty & \text{natural}
\end{array}
$$

The second case is called absorbing because we can get in to 0 but cannot get out. The third is called an entrance boundary because we cannot get to 0 but we can start the process there. Finally, in the fourth case, the process can neither get to nor start at 0, so it is reasonable to exclude 0 from the state space. We will now give examples of the various possibilities.

Example 7.32. Reflecting Brownian motion. Suppose $X_t = |B_t|$. In this case $\phi(x) = x$ and $m(x) = 1$ so

$$I = \int_0^{1/2} (\phi(z) - \phi(0)) m(z)\, dz = \int_0^{1/2} z\, dz < \infty$$

$$J = \int_0^{1/2} (M(z) - M(0)) \psi(z)\, dz = \int_0^{1/2} z\, dz < \infty$$

and 0 is a regular boundary point. $\phi(\infty) = \infty$ and $M(\infty) = \infty$, so ∞ is a natural boundary.

Example 7.33. Wright-Fisher diffusion. We begin with the case of no selection. From (7.26), we have

$$\psi(x) = x^{-2\beta_1}(1 - x)^{-2\beta_2}$$
$$m(x) = x^{2\beta_1 - 1}(1 - x)^{2\beta_2 - 1}$$

As $x \to 0$, $\psi(x) \sim x^{-2\beta_1}$, so if $\beta_1 \geq 1/2$, $\phi(0) = -\infty$ and the boundary cannot be reached. If $\beta_1 < 1/2$ then $\phi(z) - \phi(0) \sim C x^{-2\beta_1 + 1}$ so $I < \infty$. If $\beta_1 = 0$ then $M(0) = -\infty$. If $\beta_1 > 0$ then $M(z) - M(0) \sim C z^{2\beta_1}$, so $J < \infty$. Combining our calculations we see that

if	I	J	0 is
$\beta_1 = 0$	$< \infty$	$= \infty$	absorbing
$\beta_1 \in (0, 1/2)$	$< \infty$	$< \infty$	regular
$\beta_1 \geq 1/2$	$= \infty$	$< \infty$	entrance

Personally, I find it a little surprising that the accessibility of the boundary depends on size of the mutation rate.

If we consider a general selection scheme, then the function ψ is multiplied by $e^{-(2\delta x + \eta x^2)}$, and m by $e^{2\delta x + \eta x^2}$, which are bounded on $[0, 1]$, so the results of the tests do not change.

Example 7.34. Conditioned processes. If $p_t(x, y)$ is the transition probability of one of our diffusions X_t and $h(x) = P_x(T_1 < T_0)$, then, as we observed in Section 7.9, the process conditioned to hit 1 before 0, $\bar{X}_t$ has transition probability $\bar{p}_t(x, y) = p_t(x, y)h(y)/h(x)$, and generator

$$\bar{L}f = Lf + a(x)\frac{h'(x)}{h(x)}f'(x)$$

In the absence of mutation and selection, $h(x) = x$ and $h'(x) = 1$, so

$$\bar{L}f = \frac{1}{2}x(1-x)\frac{d^2 f}{dx^2} + (1-x)\frac{df}{dx}$$

Dropping the bars for the rest of the computation, $-2b(x)/a(x) = -2/x$, so

$$\psi(x) = e^{-2\log x} = x^{-2} \quad \text{and} \quad \phi(x) = -x^{-1}$$

$\phi(0) = -\infty$ so $I = \infty$. The speed measure

$$m(x) = \frac{1}{x(1-x)x^{-2}} = \frac{x}{1-x}$$

so $M(z) - M(0) \sim z^2$ as $z \to 0$. Since $\psi(z) = z^{-2}$, $J < \infty$. Thus, as we should have expected from the beginning, the conditioning makes 0 an entrance boundary. The process started at 0 will immediately become positive and never to return to 0.

In the next two examples, we will examine the influence of the drift and diffusion coefficients on the boundary behavior.

Example 7.35. Bessel processes. Suppose that $a(x) = 1$ and $b(x) = \gamma/2x$ for a diffusion on $[0, \infty)$. The natural scale is

$$\phi(x) = \int_1^x \exp\left(-\int_1^y \gamma/z \, dz\right) dy$$

$$= \int_1^x y^{-\gamma} \, dy = \begin{cases} \ln x & \text{if } \gamma = 1 \\ (x^{1-\gamma} - 1)/(1 - \gamma) & \text{if } \gamma \neq 1 \end{cases}$$

From the last computation, we see that if $\gamma \geq 1$ then $\phi(0) = -\infty$ and $I = \infty$. To handle $\gamma < 1$, we observe that the speed measure

$$m(z) = \frac{1}{\phi'(z)a(z)} = z^\gamma$$

So taking $q = 1$ in the definition of I,

$$I = \int_0^q (\phi(z) - \phi(0)) \, m(z) \, dz = \int_0^1 \frac{z^{1-\gamma}}{1-\gamma} z^\gamma \, dz < \infty$$

To compute J, we observe that for $\gamma \leq -1$, $M(0) = -\infty$ while for $\gamma > -1$, $M(z) = z^{\gamma+1}/(\gamma+1)$ and

$$J = \int_0^q (M(z) - M(0)) \, \psi(z) \, dz = \int_0^1 \frac{z^{\gamma+1}}{\gamma+1} z^{-\gamma} \, dz < \infty$$

Combining the two conclusions about I and J, we see that

if	I	J	0 is
$\gamma \geq 1$	$= \infty$	$< \infty$	entrance
$\gamma \in (-1, 1)$	$< \infty$	$< \infty$	regular
$\gamma \leq -1$	$< \infty$	$= \infty$	absorbing

which makes sense because as γ gets larger, the push away from 0 increases.

Example 7.36. Power law fluctuations. Suppose $a(x) = x^\delta$ and $b(x) = 0$. The natural scale is $\phi(x) = x$ and the speed measure is $m(x) = 1/(\phi'(x)a(x)) = x^{-\delta}$, so

$$I = \int_0^1 x^{1-\delta} \, dx = \begin{cases} < \infty & \text{if } \delta < 2 \\ = \infty & \text{if } \delta \geq 2 \end{cases}$$

When $\delta \geq 1$, $M(0) = -\infty$ and hence $J = \infty$. When $\delta < 1$

$$J = \int_0^1 \frac{z^{1-\delta}}{1-\delta} \, dz < \infty$$

Combining the last two conclusions, we see that

if	I	J	0 is
$\delta \in [2, \infty)$	$= \infty$	$= \infty$	natural
$\delta \in [1, 2)$	$< \infty$	$= \infty$	absorbing
$\delta < 1$	$< \infty$	$< \infty$	regular

which makes sense, because as δ gets larger the fluctuations near the boundary are smaller.

7.11 Site frequency spectrum

In this section, we will calculate the site frequency spectrum for our diffusion processes, extending the result for the Moran model in Section 1.5. Special cases of the formula, as well as pictures similar to ones given in this section can be found in Wright's (1942) paper based on his Gibbs lecture to the American Mathematical Society. The general result can be found in formula (9.27) of Kimura's (1964) paper, which appeared in the first volume of the Journal of Applied Probability.

Theorem 7.20. *Under the infinite sites model if mutations occur at rate μ and $\theta = 4N\mu$ then the site frequency spectrum in the diffusion process is*

$$\theta f(y) \quad where \quad f(y) = \frac{\psi(0)}{m(y)} \cdot \frac{\phi(1) - \phi(y)}{\phi(1) - \phi(0)} \tag{7.59}$$

Proof. We begin by recalling the result for the Moran model. Suppose a mutation occurs at time $-t$ introducing a new allele and no further mutation occurs at that locus, which is the case in the infinite sites model. The probability that there are k copies at time 0 is given by the transition probability $p_t(1, k)$. If mutations occur at times of a Poisson process with rate λ and each mutation occurs at a different site then the number of mutants with k copies at time 0 is Poisson with mean

$$\lambda \int_0^\infty p_t(1, k)\, dt = \lambda G(1, k)$$

Turning to the diffusion process, suppose without loss of generality that $\phi(0) = 0$. We cannot introduce mutants at frequency 0, so we introduce them at frequency δ at rate $(\theta/2)(\psi(0)/\phi(\delta))$. Here $\theta/2 = 2N\mu$ is the rate at which mutations occur in the population, and the factor $\psi(0)/\phi(\delta)$ is chosen so that if $\delta < \epsilon$ then the mutations that reach frequency ϵ is a Poisson process with rate

$$\frac{\theta\psi(0)}{2\phi(\delta)} \cdot \frac{\phi(\delta)}{\phi(\epsilon)} = \frac{\theta\psi(0)}{2\phi(\epsilon)}$$

The factor $\psi(0)$ is included because the natural scale with $\phi(0) = 0$ is only specified up to a constant multiple.

Using the Green's function formula (7.34) now the number of mutants with frequency in $(y, y + dy)$ with $y > \delta$ is Poisson with mean

$$\frac{\theta\psi(0)}{2\phi(\delta)} \cdot 2\frac{\phi(\delta)}{\phi(1)} \cdot (\phi(1) - \phi(y))m(y)\, dy \tag{7.60}$$

Letting $\delta \to 0$ gives the desired formula.

Remark. Most derivations of this result introduce mutations at frequency $1/2N$. In this approach, which as Sawyer and Hartl (1992) observe at the top of page 1165, the rigorous justification that they give on pages 1172–1174 is somewhat painful. In the example of additive selection which they were considering, this can be done using our remark in Section 7.8 that the Green's function for the Moran model converges to that of the Wright-Fisher diffusion as the population size $N \to \infty$. It would not be much fun to do this for every model, so we have taken the approach of introducing mutations at frequency δ and then letting $\delta \to 0$, which easily gives the result for any diffusion.

Examples

In all of the diffusions we will consider, $a(y) = y(1 - y)$.

Example 7.37. No selection. In the neutral case, $\phi(x) = x$ and (7.59) becomes

$$f(y) = \frac{(1 - y)}{y(1 - y)} = \frac{1}{y}$$

which agrees with the result derived in Section 1.5.

Example 7.38. Additive selection. In this case $\psi(x) = e^{-2\gamma x}$, $\phi(x) = [1 - \exp(-2\gamma x)]/2\gamma$, and $m(x) = e^{2\gamma x}/x(1 - x)$ so (7.59) becomes

$$f(y) = \frac{e^{2\gamma y}}{y(1 - y)} \frac{e^{-2\gamma y} - e^{-2\gamma}}{1 - e^{-2\gamma}} = \frac{1}{y(1 - y)} \frac{1 - e^{-2\gamma(1-y)}}{1 - e^{-2\gamma}} \qquad (7.61)$$

This formula can be found in slightly different notation on page 92 of Fisher (1930) and as formula (39) in Wright (1938). The next figure shows the site frequency spectrum for four values of γ. When $y \to 0$, $f(y) \sim 1/y$, while for $y \to 1$, we have $(1 - e^{-2\gamma(1-y)})/(1-y) \to 2\gamma$ and hence $f(y) \to 2\gamma/(1-e^{-2\gamma}) > 1$, so there is an excess of high frequency mutations.

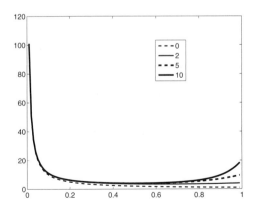

Fig. 7.13. Site frequency spectrum under directional selection.

Example 7.39. Symmetric balancing selection. In this case $\psi(x) = e^{-2\gamma x(1-x)}$ and $m(x) = e^{2\gamma x(1-x)}/x(1-x)$ so (7.59) becomes

$$f(y) = \frac{e^{2\gamma x(1-x)} \int_x^1 e^{-2\gamma y(1-y)}\, dy}{x(1-x) \int_0^1 e^{-2\gamma y(1-y)}\, dy} \tag{7.62}$$

As $x \to 1$, $\int_x^1 e^{-2\gamma y(1-y)}\, dy \sim 1-x$ so the density does not blow up there. If γ is large and x is away from the boundary then the ratio of the two integrals is close to $1/2$, and the curve is $\approx e^{2\gamma x(1-x)}/x(1-x)$, which reaches a maximum at $x = 1/2$.

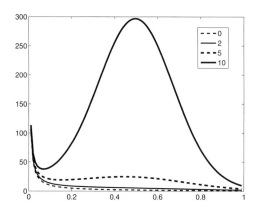

Fig. 7.14. Site frequency spectrum under balancing selection.

Fixation rate

By reasoning similar to that for (7.60), we see that if ϕ is chosen with $\phi(0) = 0$ the rate at which new mutations become fixed (when time is scaled by $2N$ generations) is

$$\frac{\theta\psi(0)}{2\phi(\delta)} \frac{\phi(\delta)}{\phi(1)} = \frac{\theta\psi(0)}{2\phi(1)}$$

In the case of additive selection $\psi(x) = e^{-2\gamma x}$ and $\phi(x) = [1 - \exp(-2\gamma x)]/2\gamma$ so this is

$$\frac{\theta}{2} \cdot \frac{2\gamma}{1 - e^{-2\gamma}} \tag{7.63}$$

7.11.1 Poisson random field model

To set up the problem, we quote from page 1166 of Sawyer and Hartl (1992): "Suppose that two species diverged $t_{div} N_e$ generations ago, and that both have the same haploid effective population size N_e. Assume that the mutation rate for silent sites in the coding region of a particular gene is μ_s per gene per generation, and that the mutation rate for nonlethal replacement mutations is μ_r per gene per generation. Assume further that (i) all new replacement mutations bestow equal fitness $w = 1 + \gamma/N_e$, (ii) each new mutation since the divergence of species occurred at a different site (in particular, the gene has not been saturated by mutations), and (iii) different sites remain in linkage equilibrium."

In this case (7.63) and (7.61) give us Table 1 of Sawyer and Hartl (1992)

	Fixation rate	Mutant frequency spectrum
Neutral	μ_s	$2\mu_s \frac{dx}{x}$
$\gamma \neq 0$	$\mu_r \frac{2\gamma}{1-e^{-2\gamma}}$	$\frac{2\mu_r}{y(1-y)} \frac{1-e^{-2\gamma(1-y)}}{1-e^{-2\gamma}}$

To make the connection note that their μ_r and μ_s are our $2N\mu$. To make it easier to compare with their paper we will keep their notation. The expected number of fixed differences between the two species are

$$2\mu_s t_{div} \quad \text{and} \quad 2\mu_r \frac{2\gamma}{1-e^{-2\gamma}}$$

for silent and replacement sites, respectively.

McDonald-Kreitman tables

Now suppose we have aligned DNA sequences from m chromosomes from the first species and n chromosomes from the second species. An allele with frequency x will be polymorphic in a sample of size m with probability $1 - x^m - (1-x)^m$, so the expected number of silent polymorphic sites in a sample of size m is

$$2\mu_s \int_0^1 \frac{1-x^m-(1-x)^m}{x} \, dx = 2\mu_s \sum_{k=1}^{m-1} \frac{1}{k}$$

Writing $L(m) = \sum_{k=1}^{m-1} 1/k \approx \log m$, the number of silent polymorphic sites in both samples together is then

$$2\mu_s(L(m) + L(n)) \tag{7.64}$$

A silent site will look like a fixed difference in species 1 in the comparison of the two samples if it is fixed in the population or if it by chance occurs in all m sampled individuals, so the expected value is

$$\mu_s t_{div} + \int_0^1 2\mu_s x^m \frac{dx}{x} = \mu_s \left(t_{div} + \frac{2}{m} \right)$$

Thus the expected number of silent fixed differences is

$$2\mu_s \left(t_{div} + \frac{1}{m} + \frac{1}{n} \right) \tag{7.65}$$

By the same reasoning the number of polymorphic replacement sites in a sample is

$$2\mu_r(H(m) + H(n)) \tag{7.66}$$

where

$$H(m) = \int_0^1 \frac{1 - x^m - (1-x)^m}{x(1-x)} \frac{1 - e^{-2\gamma(1-x)}}{1 - e^{-2\gamma}} \, dx$$

and the number of replacement fixed differences has mean

$$2\mu_r \frac{2\gamma}{1 - e^{-2\gamma}}(t_{div} + G(m) + G(n)) \tag{7.67}$$

where

$$G(m) = \int_0^1 x^{m-1} \frac{1 - e^{-2\gamma(1-x)}}{2\gamma(1-x)} \, dx$$

Since $e^{-y} \ge 1 - y$ and hence $(1 - e^{-y})/y \le 1$ for $y > 0$, $G(m) \le 1/m$ for $\gamma > 0$.

Our formulas give the expected value of the four entries in the McDonald-Krietman table.

	Divergence	Polymorphism
Silent	(7.65)	(7.64)
Replacement	(7.67)	(7.66)

Bayesian estimation

The number of mutations in one locus typically does not give us enough information to get good estimates of the parameters, so it is natural to combine the information from many loci. To do this we will follow the approach of Bustamante et al. (2002) and Sawyer et al. (2003). Changing to their notation we let $\theta_s = 4N\mu_s$ and $\theta_a = 4N\mu_r$ and denote the entries in the DPRS table by

	Divergence	Polymorphism
Silent	K_s	S_s
Replacement	K_a	S_a

The theoretical expectations for any single DPRS table include four parameters, θ_s, θ_a, γ, and the divergence time t, and contain four observations: K_s, S_s, K_a, and S_a hence there is no meaningful opportunity for model fitting. However, the divergence time is a shared parameter among all sequences. The basic idea of Bayesian analysis is to treat the parameters in a model as random variables with some underlying prior distribution. In Bustamante et al. (2002), it was assumed that for each coding sequence γ was a fixed constant but that across loci the distribution of γ was given by a normal with mean μ and standard deviation σ. The other prior distributions are $q(t)$ is uniform, $p(\theta)$ is gamma, $h(\sigma)$ is such that $1/\sigma^2$ is gamma, and $g(\mu|\sigma)$ is normal.

The posterior distribution $\pi(\gamma, t, \theta, \mu, \sigma)$ is analytically intractable but can be computed by Markov chain Monte Carlo. That is, by simulation of a Markov chain defined in such a way that the stationary distribution is precisely π. One simple method for doing this is the Metropolis algorithm in which a trial value for the new parameter is used to replace the old if the ratio of the posterior probabilities for the trial and the present values is greater than a uniform random number in $[0, 1]$. For more details about the MCMC method see page 533 of Bustamante et al. (2002).

Sawyer et al. (2003) modified the "fixed-effects" model described above to be a random-effects model so that for the ith coding sequence, the selection coefficient for a new mutation is normal with mean γ_i and standard deviation σ_w. Here σ_w is a global parameter that applies to all loci and has a uniform prior distribution.

Sawyer et al. (2003) studied a set of 72 $D.$ $simulans$ coding sequences from GenBank, which had sample sizes ranging from 4 to 70 with an average of 10.5. Nucleotide divergence between $D.$ $simulans$ and $D.$ $melanogaster$ was inferred from the reference sequence for $D.$ $melanogaster,$ see Adams (2000). In applying the random-effects model to the DPRS data, they initially found that the Markov chain did not converge, or did so excessively slowly. The output for various runs suggested that the main reason for poor convergence was that values of θ_r could be balanced off by γ. That is, an excess of replacement mutations can be caused either by a stronger intensity of positive selection or a higher mutation rate.

From runs of the fixed-effects model, they noticed that about 80% of the coding sequences had values of $\theta_r/2\theta_s$ near $1/4$, or more precisely about 0.28, so they modified the model to include a new parameter $Q = \theta_r/2\theta_s$ with a gamma prior distribution. Among the 72 genes, 14 were excluded because $\theta_r/2\theta_s > 0.28$ and two additional genes were excluded because they appeared to be spurious for other reasons. The list of genes omitted include eight male accessory gland proteins.

For the random effects model they found that the distribution of the γ_i had mean -7.3 and standard deviation $\sigma_b = 5.69$, while the within locus standard deviation was $\sigma_w = 6.79$. Most of the mean selection intensities for the 56 genes were negative but many had 95% credible intervals that overlapped 0. The fraction of beneficial new mutations ranged from 1% for Pgm to 62% for Rel with an outlier at 90% for mei-218. The average for all loci was 19.4%. Among the replacement polymorphisms in the data, an average of 46.9% were estimated to be beneficial. For the Y-linked gene kl-5 the estimated average selection intensity was -0.38. All others were positive and ranged from 2.1 for $vermillion$ to 9.4 for Rel, with an overall mean, excluding kl-5 of 5.1.

For an application of these methods to a large number of genes in the human genome, see Bustamante et al. (2005).

7.12 Fluctuating selection

Two mechanisms by which evolution can occur are the adaptive processes of natural selection and the neutral processes of genetic drift. Which of these is the principal force in the evolution of a population has been one of the central issues in evolutionary biology. An early exchange in this debate was over the changes in the frequencies of a color polymorphism in a population of the scarlet tiger moth *Callimorpha (Panaxia) dominula* near Oxford, England. Fisher and Ford (1947) argued that the population size was too large for the changes in frequencies to be due to random drift, and were caused by fluctuating selection. Wright (1948) replied by arguing that multiple factors could affect a population, and that the effective population might be much smaller than the census population size. A publicized debate ensued, see Fisher and Ford (1950), and Wright (1951).

Kimura (1954, 1962) and Ohta (1972) studied the question mathematically, but did not find the correct diffusion approximation. A little later Gillespie (1973) and Jensen (1973) did. For more on the early history see Felsenstein (1976). We will follow Karlin and Levikson (1974) and consider a model in which the fitness of A in generation n is $1+\sigma_n$ and the fitness of a is $1+\tau_n$ where σ_n, τ_n are independent and identically distributed. Dropping the subscripts to simplify the formulas, we let

$$\alpha = 2N[E(\sigma - \tau) - E(\sigma^2 - \tau^2)/2 + E(\sigma - \tau)^2/2]$$
$$\beta = 2NE(\sigma - \tau)^2$$

Theorem 7.21. *The diffusion approximation for the Karlin-Levikson model has coefficients:*

$$b(x) = x(1-x)(\alpha - \beta x) \qquad a(x) = x(1-x)[1 + \beta x(1-x)] \qquad (7.68)$$

The drift looks like balancing selection, but the variance has an additional term.

Proof. To derive the diffusion approximation, note that reasoning as in Section 6.2, the change in frequency in one generation is

$$\frac{x(1+\sigma)}{x(1+\sigma) + (1-x)(1+\tau)} - x = \frac{x(1+\sigma) - x - \sigma x^2 - \tau x(1-x)}{1 + \sigma x + \tau(1-x)}$$

$$= \frac{(\sigma - \tau)x(1-x)}{1 + \sigma x + \tau(1-x)} \approx (\sigma - \tau)x(1-x)[1 - \sigma x - \tau(1-x)]$$

Writing $x = 1/2 - (1/2 - x)$ and $1 - x = 1/2 + 1/2 - x$, the above is

$$= (\sigma - \tau)x(1-x)[1 - (\sigma + \tau)/2 + (\sigma - \tau)(1/2 - x)]$$

$$= x(1-x)[(\sigma - \tau) - (\sigma^2 - \tau^2)/2 + (\sigma - \tau)^2(1/2 - x)]$$

Taking expected value and speeding up time by a factor of $2N$ the drift coefficient is

$$b(x) = x(1-x)(2N)[E(\sigma - \tau) - E(\sigma^2 - \tau^2)/2 + E(\sigma - \tau)^2(1/2 - x)]$$

To compute the variance, let ΔX be the change in frequency and Y be the environment.

$$\text{var}\,(\Delta X) = E\,\text{var}\,(\Delta X|Y) + \text{var}\,(E(\Delta X|Y))$$

To evaluate the first term, recall that the variance of Binomial$(2N, p)$ is $2Np(1-p)$ and the allele frequencies by $O(1/N)$ in one generation so

$$\text{var}\,(\Delta X|Y) \approx \frac{x(1-x)}{2N}$$

As we computed above

$$E(\Delta X|Y) = \frac{(\sigma - \tau)x(1-x)}{1 + \sigma x + \tau(1-x)}$$

Since $\sigma x, \tau(1-x) \ll 1$, we can drop these terms from the denominator:

$$\text{var}\,(E(\Delta X|Y)) = x^2(1-x)^2 E(\sigma - \tau)^2$$

Adding the two results and speeding up time by a factor of $2N$ gives

$$a(x) = x(1-x) + x^2(1-x)^2 2N E(\sigma - \tau)^2$$

and completes the proof.   □

Remark. Takahata, Ishii, Matsuda (1975) considered a Wright-Fisher diffusion with varying selection

$$\frac{1}{4N}x(1-x)\frac{d^2}{dx^2} + s(t)x(1-x)\frac{d}{dx}$$

They let $\bar{s} = Es(t)$ and $V = \int_0^\infty E([s(t) - \bar{s}][s(0) - \bar{s}])\,dt$, and found that in the diffusion approximation

$$a(x) = \frac{1}{2N}x(1-x) + 2Vx^2(1-x)^2$$

$$b(x) = \bar{s}x(1-x) + Vx(1-x)(1-2x)$$

To connect with the Karlin-Levikson result, suppose $E(\sigma^2 - \tau^2) = 0$, let $\bar{s} = E(\sigma - \tau)$, and note that in discrete time $V = E(\sigma - \tau)^2/2$. This suggests that if $\bar{s} = E(\sigma - \tau) = 0$ and we have (σ_n, τ_n) that are correlated in time all we do is replace $E(\sigma - \tau)^2/2$ by

$$\sum_{n=0}^{\infty} E[(\sigma_0 - \tau_0)(\sigma_n - \tau_n)].$$

For the rest of the section we will only consider the special case that is closely related to the model of Takahata, Ishii, and Matsuda (1975).

Theorem 7.22. *Consider the Karlin-Levikson model with $E(\sigma - \tau) = 0$ and $E(\sigma^2 - \tau^2) = 0$. The derivative of the natural scale*

$$\psi(y) = \frac{1}{y(1-y) + 1/\beta}$$

The speed measure is exactly the same as for the neutral case

$$m(y) = \frac{1}{y(1-y)}$$

Proof. Since $a(y) = y(1-y)(1 + \beta y(1-y))$, the second formula follows from the first. To compute ψ, we begin by noting

$$\frac{-2b(x)}{a(x)} = \frac{-2[\alpha - \beta x]}{1 + x(1-x)\beta}$$

To find the roots of the quadratic in the denominator, we write it as $x^2 - x - 1/\beta = 0$ and solve to find roots $r_1 < 0 < 1 < r_2$ given by

$$r_i = \frac{1 \pm \sqrt{1 + 4/\beta}}{2}$$

Note that the two roots are symmetric about $1/2$. To evaluate the integral we write

$$\frac{-2b(x)}{a(x)} = \frac{-2[\alpha - \beta x]}{1 + x(1-x)\beta} = \frac{-2[\alpha/\beta - x]}{1/\beta + x(1-x)} = \frac{C}{x - r_1} + \frac{D}{r_2 - x}$$

To find the constants we solve $-C + D = 2$ and $Cr_2 - Dr_1 = -2\alpha/\beta$ to find

$$C = \frac{2r_1 - 2\alpha/\beta}{r_2 - r_1} \qquad D = \frac{2r_2 - 2\alpha/\beta}{r_2 - r_1}$$

which, as the reader can easily check, satisfies the two equations. Integrating

$$\int^y \frac{C}{x - r_1} + \frac{D}{r_2 - x} \, dx = C\log(y - r_1) - D\log(r_2 - y)$$

so we have

$$\psi(y) = \exp\left(\int^y \frac{-2b(x)}{a(x)} \right) = (y - r_1)^C (r_2 - y)^{-D}$$

Consider now the special case in which σ and τ have the same distribution so $E(\sigma - \tau) = 0$, $E(\sigma^2 - \tau^2) = 0$, and hence $\alpha = \beta/2$.

$$C = \frac{2r_1 - 2\alpha/\beta}{r_2 - r_1} = \frac{2(r_1 - 1/2)}{r_2 - r_1} = -1$$

$$D = \frac{2r_2 - 2\alpha/\beta}{r_2 - r_1} = \frac{2(r_2 - 1/2)}{r_2 - r_1} = 1$$

and we have the very nice formula

$$\psi(y) = (y - r_1)^{-1}(r_2 - y)^{-1} = \frac{1}{y(1 - y) + 1/\beta} \qquad \square$$

Karlin and Levikson (1974) find $\psi(y) = [1 + \beta y(1 - y)]^{-1}$ on their page 402, but this agrees with our computation since the solution of $\psi'(y) = -2b(y)\psi(y)/a(y)$ is only determined up to a constant multiple. To make it easier to compare with their formulas, for the rest of the section we will use

$$\psi(y) = \frac{1}{\beta y(1 - y) + 1} = \beta^{-1}(y - r_1)^{-1}(r_2 - y)^{-1} \qquad (7.69)$$

Theorem 7.23. *Let $r_1 < r_2$ be the roots $(1 \pm \sqrt{1 + 4/\beta})/2$. Under the assumptions of Theorem 7.22, the probability of fixation starting from frequency x is*

$$\frac{1}{2} + \frac{\log\left[\frac{x - r_1}{r_2 - x}\right]}{2\log[r_2/(-r_1)]}$$

This is (8) in Jensen (1973). As $\beta \to \infty$, $r_1 \to 0$ and $r_2 \to 1$ so $\phi(x) \to 1/2$. The next graph shows the hitting probabilities for $\beta = 0, 10, 40$.

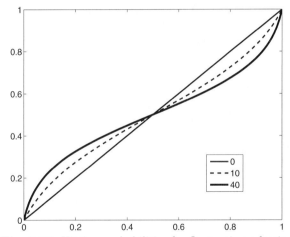

Fig. 7.15. Hitting probabilities for fluctuating selection

Proof. To compute the natural scale ϕ, we integrate to find

$$\phi(x) = \beta^{-1} \int_0^x (y - r_1)^{-1}(r_2 - y)^{-1} \, dy$$

$$= \frac{1}{\beta(r_2 - r_1)} \int_0^x \frac{1}{y - r_1} + \frac{1}{r_2 - y} \, dy \qquad (7.70)$$

$$= \frac{1}{\sqrt{\beta^2 + 4\beta}} [\log(x - r_1) - \log(-r_1) - \log(r_2 - x) + \log(r_2)]$$

This is close to but not exactly the same as Karlin and Levikson (1974). Their roots are $\lambda_2 = r_1$ and $\lambda_1 = r_2$, and they write $w = \beta/2$, so their constant has 2β instead of 4β under the square root.

Since $\phi(0) = 0$, the probability of fixation starting from frequency x is

$$\phi(x)/\phi(1) = \log\left[\frac{x - r_1}{-r_1} \cdot \frac{r_2}{r_2 - x}\right] \bigg/ \log\left[\frac{1 - r_1}{-r_1} \cdot \frac{r_2}{r_2 - 1}\right] \quad (7.71)$$

$r_2 - 1/2 = 1/2 - r_1$ and $r_2 - 1 = -r_1$ so $\phi(1/2)/\phi(1) = 1/2$ and we can write above as

$$\frac{1}{2} + \frac{\phi(x) - \phi(1/2)}{\phi(1)} = \frac{1}{2} + \frac{\log\left[\frac{x-r_1}{r_2-x}\right]}{2\log[r_2/(-r_1)]} \qquad \qquad \square$$

Theorem 7.24. *Let $\tau = T_0 \wedge T_1$ be the time until one allele is lost. Under the assumptions of Theorem 7.22*

$$E_x\tau = \begin{cases} \int_0^x \frac{2}{1+\beta y(1-y)} \log\left(\frac{1-y}{y}\right) dy & \text{when } x \leq 1/2 \\ \int_x^1 \frac{2}{1+\beta y(1-y)} \log\left(\frac{y}{1-y}\right) dy & \text{when } x \geq 1/2 \end{cases}$$

Note that in each case the log is nonnegative throughout the range of integration, so $E_x\tau$ is a decreasing function of β. This result, which is on page 402 of Karlin and Levikson (1974) is somewhat surprising since (7.68) shows that the diffusion has a drift toward $1/2$, which will encourage it to spend more time at intermediate values. However, this effect is counteracted by the increase in $a(x)$.

Proof. Since $m(y) = 1/y(1 - y)$ and $\phi(0) = 0$, the Green's function $G(x, y)$ from (7.34) is

$$2\frac{\phi(x)}{\phi(1)} \cdot \frac{\phi(1) - \phi(y)}{y(1 - y)} \qquad x \leq y$$

$$2\frac{\phi(1) - \phi(x)}{\phi(1)} \cdot \frac{\phi(y)}{y(1 - y)} \qquad y \leq x$$

The expected time to fixation is

$$E_x\tau = 2\frac{\phi(x)}{\phi(1)} \int_x^1 \frac{\phi(1) - \phi(y)}{y(1 - y)} dy + 2\frac{\phi(1) - \phi(x)}{\phi(1)} \int_0^x \frac{\phi(y)}{y(1 - y)} dy$$

Since $1/(1 - y)y = 1/(1 - y) + 1/y$ has antiderivative $-\log(1 - y) + \log(y)$, integrating by parts gives

$$= 2\frac{\phi(x)}{\phi(1)} \left(\phi(1) - \phi(y)\right) \cdot \left(-\log(1-y) + \log(y)\right)\big|_x^1$$

$$+ 2\frac{\phi(x)}{\phi(1)} \int_x^1 \psi(y) \log\left(\frac{y}{1-y}\right) dy$$

$$+ 2\frac{\phi(1) - \phi(x)}{\phi(1)} \phi(y) \cdot \left(-\log(1-y) + \log(y)\right)\big|_0^x$$

$$- 2\frac{\phi(1) - \phi(x)}{\phi(1)} \int_0^x \psi(y) \log\left(\frac{y}{1-y}\right) dy$$

Since $\phi(1) - \phi(y) \sim \phi'(1)(1-y)$ as $y \to 1$ and $(1-y)\log(1-y) \to 0$ as $y \to 1$, evaluating the first term at 1 gives 0. Similarly $\phi(y) \sim \phi'(0)y$ as $y \to 0$ and $y\log(y) \to 0$ as $y \to 0$, so evaluating the third term at 0 gives 0. Evaluating the first term at x cancels with evaluating the third at x, so the above

$$= 2\frac{\phi(x)}{\phi(1)} \int_x^1 \psi(y) \log\left(\frac{y}{1-y}\right) dy - 2\frac{\phi(1) - \phi(x)}{\phi(1)} \int_0^x \psi(y) \log\left(\frac{y}{1-y}\right) dy$$

Adding and subtracting $2(\phi(x)/\phi(1)) \int_0^x$, then flipping the fraction inside the log to get rid of the minus sign, the above

$$= 2\frac{\phi(x)}{\phi(1)} \int_0^1 \psi(y) \log\left(\frac{y}{1-y}\right) dy + \int_0^x 2\psi(y) \log\left(\frac{1-y}{y}\right) dy$$

$\psi(y)$ is symmetric about $1/2$ and $\log(y/(1-y)) = \log(y) - \log(1-y)$ is antisymmetric about $1/2$, so the first integral vanishes, and

$$E_x \tau = \int_0^x \frac{2}{1 + \beta y(1-y)} \log\left(\frac{1-y}{y}\right) dy$$

When $x \geq 1/2$ we can use the fact that the total integral is 0 to write

$$E_x \tau = \int_x^1 \frac{2}{1 + \beta y(1-y)} \log\left(\frac{y}{1-y}\right) dy \qquad \square$$

Using Kimura's formula (7.59), we have

Theorem 7.25. *Under the assumptions of Theorem 7.22, the site frequency spectrum is*

$$\frac{\theta}{y(1-y)} \cdot \log\left(\frac{1-r_1}{y-r_1} \cdot \frac{r_2 - y}{r_2 - 1}\right) \Big/ \log\left(\frac{1-r_1}{-r_1} \cdot \frac{r_2}{r_2 - 1}\right)$$

Proof. Using either formula for $\psi(y)$

$$\frac{\psi(0)}{\psi(y)a(y)} = \frac{1}{y(1-y)}$$

Using (7.71), shows that $(\phi(1) - \phi(y))/(\phi(1) - \phi(0)) =$ the second factor.

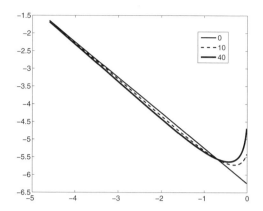

Fig. 7.16. Log-log plot of the site frequency spectrum for fluctuating selection

Figure 7.16 shows the site frequency spectrum for $\beta = 0, 10, 40$. To make the differences more visible we have done a log-log plot. In the presence of fluctuating selection high frequency derived alleles (y near 1) are overrepresented, and intermediate frequency alleles are underrepresented with respect to the neutral case. Somewhat remarkably,

Theorem 7.26. *The integral of the site frequency spectrum does not depend on β.*

Proof. It follows from Theorem 7.23 that

$$g(\beta, y) = \frac{\partial}{\partial \beta} \frac{\phi(1) - \phi(y)}{\phi(1) - \phi(0)}$$

has $g(\beta, y) = -g(\beta, 1 - y)$. From this it follows that

$$\frac{\partial}{\partial \beta} \int_0^1 \frac{\theta}{y(1-y)} \frac{\phi(1) - \phi(y)}{\phi(1) - \phi(0)} \, dy = \int_0^1 \frac{\theta}{y(1-y)} g(\beta, y) \, dy = 0 \qquad \square$$

Parameter estimation

Most studies of fluctuating selection have based their inferences on time series data for allele frequencies. See Mueller et al (1985), Lynch (1987), Cook and Jones (1996), and O'Hara (2005). Recently, Huerta-Sanchez, Durrett, and Bustamante (2007) have used the Poisson random field framework to develop an alternative approach that uses DNA polymorphism data from a sample of individuals collected at a single point in time. To do this they used methods described in Bustamante et al. (2001), which we will begin by describing in general. Let $f(y, \beta)$ be the site frequency spectrum. Since this represents the distribution of mutation frequencies at any time, the probability of finding i mutant alleles in a sample of size n

$$F_n(k, \beta) = \int_0^1 \binom{n}{k} y^k (1-y)^{n-k} f(y, \beta) \, dy$$

Our definition of $f(y, \beta)$ is $1/2$ the usual one so there is no factor of 2, as in (2) of Bustamante et al. (2001).

Let y_i be the number of sites at which there are i mutant alleles. In the Poisson random field framework, different sites are independent so the likelihood is given by

$$L(\theta, \beta) = \prod_{i=1}^{n-1} \exp(-\theta F_n(i, \beta)) \frac{(\theta F_n(i, \beta))^{y_i}}{y_i!}$$

Therefore, the log likelihood function (dropping the term $\log(y_i!)$ which is independent of the parameters) is

$$\ell(\theta, \beta) = \sum_{i=1}^{n-1} -\theta F_n(i, \beta) + y_i \log(\theta F_n(i, \beta))$$

Differentiating with respect to θ we see that

$$\frac{\partial}{\partial \beta} \log L(y, \beta) = -\sum_{i=1}^{n} F_n(i, \beta) + \frac{y_i}{\theta}$$

so for fixed β the maximum likelihood estimate of θ is

$$\hat{\theta}(\beta) = S_n / \sum_{i=1}^{n} F_n(i, \beta)$$

which is a generalization of Watterson's estimate.

Given the last result we can work with the profile likelihood

$$\ell^*(\beta) = L(\hat{\theta}(\beta), \beta)$$

which can be maximized numerically using standard optimization techniques such as Newton-Raphson iteration. In the current example, that task simplifies because Theorem 7.26 implies

Theorem 7.27. $ES_n = \sum_{i=1}^{n} F_n(i, \beta)$ *does not depend on* β.

Proof. Let $h(k, y) = \binom{n}{k} y^k (1-y)^{n-k} + \binom{n}{n-k} y^{n-k} (1-y)^k$. Since $h_n(k, y)$ is symmetric about $1/2$,

$$\frac{\partial}{\partial \beta} [F_n(k, \beta) + F_n(n-k, \beta)] = 0$$

Summing from $k = 1$ to $n-1$ now gives the desired result.   □

To perform the optimization, we have to compute the first and second derivatives of the log likelihood with respect to its parameters. We have nothing insightful to say about these details so we refer the reader to Bustamante et al. (2001) or Huerta-Sanchez, Durrett, and Bustamante (2007) for details, and simulation results which show the performance of the estimators. For another approach to fitting fluctuating selection models to data see Mutsonen and Lässig (2007). Using data for 271 loci in 12 *Droxsophila melanogaster* and a *D. simulans* sequence, they find strong support ($p < 10^{-17}$) for time dependent selection.

Before leaving the topic of fluctuating selection, we must mention the work of Gillespie. To quote the preface of his 1991 book *The Causes of Molecular Evolution*: "If we are to propose that molecular evolution is due to the action of natural selection, we need a mathematical theory to demonstrate that the dynamics of selection are compatible with the observations of molecular variation. It is my conviction that the only viable model of selection is one based on temporal and spatial fluctuations in the environment. The mathematics of selection in a random environment have never been systematically developed or brought to a point where they serve as a model of molecular evolution. Both situations will be remedied in Chapter 4. Unfortunately, the mathematics are very difficult. Yet, if molecular evolution is in response to a changing environment, then this is the sort of mathematical challenge we must be willing to face. Chapter 4 is littered with unresolved problems that should prove of interest to those with a mathematical bent." In addition to the source just cited the reader should consult his more recent papers on the SAS-CFF model (stochastic additive scale-concave fitness function).

8

Multidimensional Diffusions

"As long as algebra is taught in school, there will be prayer in school."
Cokie Roberts

It is straightforward to extend our one dimensional definition to higher dimensions. To make formulas smaller, let

$$D_i = \frac{\partial}{\partial x_i} \qquad D_{ij} = \frac{\partial^2}{\partial x_i \partial x_j}$$

The infinitesimal generator can then be written as

$$Lf = \frac{1}{2} \sum_{i,j} a_{ij}(x) D_{ij} f + \sum_i b_i(x) D_i f$$

where $b(x)$ is the infinitesimal drift vector, and $a(x)$ is the infinitesimal covariance matrix. In one dimension $a(x) \geq 0$. Here, since a is a covariance matrix, a must be nonnegative definite, i.e., $\sum_{i,j} y_i a_{ij} y_j \geq 0$ for all vectors y, since this gives the infinitesimal variance of $\sum_i y_i X_i(t)$.

While the definition of a diffusion, and the derivation of the coefficients in examples, are not much different in higher dimensions, the theory changes drastically. In one dimension, differential equations can be solved by calculus, so we can find formulas for most quantities of interest. In higher dimensions, there is no systematic method for solving the equations, so there are far fewer explicit formulas.

8.1 K allele model

This section is devoted to the study of the K allele Moran model with mutation and selection. If $1 - s_i$ is the fitness of type i, μ_{ij} is the rate of mutations from i to j, and $n = (n_1, n_2, \ldots, n_K)$ gives the number of individuals with the K alleles then

R. Durrett, *Probability Models for DNA Sequence Evolution*,
DOI: 10.1007/978-0-387-78168-6_8, © Springer Science+Business Media, LLC 2008

$$n \to n + e_i - e_j \quad \text{at rate} \quad n_j \left(\frac{n_i}{2N}(1 - s_i) + \mu_{ji} \right)$$

where e_i is the vector that is 1 in the ith position and 0 otherwise. In words, the transition $n \to n + e_i - e_j$ says we lose an individual of type j and add one of type i. The first term arises because each of the n_j individuals with the jth allele is replaced at rate 1. The randomly chosen individual to do the replacement will be of type i with probability $n_i/2N$, and the replacement will occur with probability $1 - s_i$. The second term comes from mutations from type j to type i.

To find the drift in the diffusion approximation

$$\frac{d}{dt} E X_i(t) = \frac{1}{2N} \left[\sum_{j \neq i} n_j \left(\frac{n_i}{2N}(1 - s_i) + \mu_{ji} \right) - \sum_{j \neq i} n_i \left(\frac{n_j}{2N}(1 - s_j) + \mu_{ij} \right) \right]$$

$$= -x_i \mu_{i\cdot} + \sum_j x_j \mu_{ji} + x_i \sum_j x_j (s_j - s_i)$$

where $x_i = n_i/2N$ and $\mu_{i\cdot} = \sum_j \mu_{ij}$. We do not need $j \neq i$ on the second line because $\mu_{ii} = 0$ and $s_i - s_i = 0$.

Letting $\beta_{ij} = N\mu_{ij}$, $\beta_{i\cdot} = N\mu_{i\cdot}$, and $\gamma_i = Ns_i$ we see that the drift coefficient for the process run at rate N is

$$b_i(x) = -x_i \beta_{i\cdot} + \sum_j x_j \beta_{ji} + x_i \sum_j x_j (\gamma_j - \gamma_i)$$

To compute the second order terms, we note

$$\frac{d}{dt} E(X_i(t) - x_i)^2 = \frac{1}{(2N)^2} \left[\sum_{j \neq i} n_j \left(\frac{n_i}{2N}(1 - s_i) + \mu_{ji} \right) \right.$$

$$\left. + \sum_{j \neq i} n_i \left(\frac{n_j}{2N}(1 - s_j) + \mu_{ij} \right) \right]$$

$$= \frac{1}{2N} \cdot 2x_i(1 - x_i) + o(1/N)$$

since $s_i, \mu_{ij} \to 0$. For $i \neq j$ we have

$$\frac{d}{dt} E((X_i(t) - x_i)(X_j(t) - x_j)) = \frac{-1}{(2N)^2} \left[n_j \left(\frac{n_i}{2N}(1 - s_i) + \mu_{ji} \right) \right.$$

$$\left. + n_i \left(\frac{n_j}{2N}(1 - s_j) + \mu_{ij} \right) \right]$$

$$= -\frac{1}{2N} (2x_i x_j + O(1/N))$$

Letting $\delta_{ij} = 1$ if $i = j$ and 0 otherwise, we have shown that the infinitesimal covariance for the process run at rate N is

$$a_{ij}(x) = x_i(\delta_{ij} - x_j)$$

which is the covariance matrix for multinomial sampling. We get the same limit for the Wright-Fisher model if we let $\beta_{ij} = 2N\mu_{ij}$ and $\gamma_i = 2Ns_i$. Since the sum of the coordinates is 0, we drop the Kth coordinate to have a $K-1$ dimensional process (just as we did when $K = 2$).

8.1.1 Fixation probabilities and time

When there is no mutation or selection,

$$Lf = \frac{1}{2} \sum_{i,j} x_i(\delta_{ij} - x_j)D_{ij}f$$

In this case the coordinate functions $f_i(x) = x_i$ have $Lf_i = 0$, so X_t^i is a martingale, and if we start from initial frequencies given by x, the probability the ith allele is the one that takes over the population is x_i.

Consider now the case of $K = 3$ alleles, which is a two dimensional process with generator

$$\frac{1}{2}x_1(1 - x_1)D_{11} - x_1x_2D_{12} + \frac{1}{2}x_2(1 - x_2)D_{22}$$

The middle term has no $1/2$, since for nice functions $D_{12}f = D_{21}f$. The state space is $\{(x_1, x_2) : x_1 \geq 0, \ x_2 \geq 0, \ x_1 + x_2 \leq 1\}$ which is a triangle.

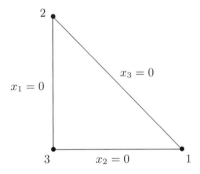

Fig. 8.1. State space for the three allele diffusion.

When the process begins in the interior of the triangle, it runs until it hits one of the sides, which represents the loss of one allele. Then it moves along that side until it becomes absorbed in one the corners. The numbers in the corners indicate the allele that has frequency 1 there.

Theorem 8.1. *Let $E_3 = \{$allele 3 is first to get lost$\}$.*

$$P_x(E_3) = h(x) = x_1 x_2[(1 - x_1)^{-1} + (1 - x_2)^{-1}] \qquad (8.1)$$

Proof. To check this, we note that if $x_1 = 0$ or $x_2 = 0$ then $h = 0$, while if $x_3 = 0$, $x_1 = 1 - x_2$ and $x_2 = 1 - x_1$ so

$$h(x) = \frac{x_1(1 - x_1)}{1 - x_1} + \frac{x_2(1 - x_2)}{1 - x_2} = 1$$

assuming that $0 < x_1, x_2 < 1$. The function h is not defined at the corners where $x_1 = 1$ or $x_2 = 1$ but this is not a problem since two alleles will not be lost simultaneously.

To prove (8.1) now, it is enough to show that $Lh = 0$, since this implies $h(X_t)$ is a martingale, and if T_2 is the first time there are only two alleles, it follows that

$$h(x) = \lim_{t \to \infty} E_x h(X(t \wedge T_2)) = P_x(E_3)$$

because of the boundary conditions.

To check that $Lh = 0$ now, we note that

$$D_1 h = x_2[(1 - x_1)^{-1} + (1 - x_2)^{-1}] + x_1 x_2(1 - x_1)^{-2}$$
$$D_{11} h = 2x_2(1 - x_1)^{-2} + 2x_1 x_2(1 - x_1)^{-3}$$
$$D_{12} h = (1 - x_1)^{-1} + (1 - x_2)^{-1} + x_2(1 - x_2)^{-2} + x_1(1 - x_1)^{-2}$$

By symmetry, we can interchange the roles of x_1 and x_2 to conclude

$$D_{22} h = 2x_1(1 - x_2)^{-2} + 2x_1 x_2(1 - x_2)^{-3}$$

Using these equations, we have

$$2Lh = x_1[2x_2(1 - x_1)^{-1} + 2x_1 x_2(1 - x_1)^{-2}]$$
$$- 2x_1 x_2[(1 - x_1)^{-1} + (1 - x_2)^{-1} + x_2(1 - x_2)^{-2} + x_1(1 - x_1)^{-2}]$$
$$+ x_2[2x_1(1 - x_2)^{-1} + 2x_1 x_2(1 - x_2)^{-2}]$$

Numbering the terms on the right-hand side in the order they appear, we have $1 + 3 = 0$, $7 + 4 = 0$, $2 + 5 = 0$, and $8 + 6 = 0$, showing that $Lh = 0$. $\qquad \square$

Keeping with the case of 3 alleles, we have

Theorem 8.2. *Let T_1 be the time at which there is only one allele.*

$$E_x T_1 = g_1(x) = -2 \sum_i (1 - x_i) \log(1 - x_i) \qquad (8.2)$$

where $0 \log 0 = \lim_{h \to 0} h \log h = 0$.

Proof. The first step is to note that $g_1(x) = 0$ at the corners of the triangles where some $x_i = 1$ and the other coordinates are 0. To verify the answer it is enough to show that $Lg_1 = -1$. Writing

$$g_1(x) = -2(1 - x_1)\log(1 - x_1) - 2(1 - x_2)\log(1 - x_2) - 2(x_1 + x_2)\log(x_1 + x_2)$$

and differentiating we have

$$D_1 g_1 = 2\log(1 - x_1) + 2 - 2\log(x_1 + x_2) - 2$$

$$D_{11} g_1 = -\frac{2}{1 - x_1} - \frac{2}{x_1 + x_2} \qquad D_{12} g_1 = -\frac{2}{x_1 + x_2}$$

Using symmetry to evaluate $D_{22} g_1$, we have

$$Lg_1 = \frac{1}{2}x_1(1 - x_1)\left[-\frac{2}{1 - x_1} - \frac{2}{x_1 + x_2}\right] - x_1 x_2 \cdot \frac{-2}{x_1 + x_2}$$
$$+ \frac{1}{2}x_2(1 - x_2)\left[-\frac{2}{1 - x_2} - \frac{2}{x_1 + x_2}\right]$$

A little algebra converts this into

$$= -x_1 - x_2 - \frac{1}{x_1 + x_2}[x_1 - x_1^2 - 2x_1 x_2 + x_2 - x_2^2]$$

$$= -x_1 - x_2 - 1 - \frac{(x_1 + x_2)^2}{x_1 + x_2} = -1$$

which completes the proof. □

Theorem 8.3. *Let T_2 be the first time at which there are only two alleles.*

$$E_x T_1 = -2\left[\sum_i x_i \log x_i - \sum_i (1 - x_i)\log(1 - x_i)\right] \qquad (8.3)$$

Proof. To verify the boundary condition, we note that if $x_3 = 0$ then the quantity in brackets is

$$x_1 \log x_1 + x_2 \log x_2 - x_2 \log x_2 - x_1 \log x_1 = 0$$

By symmetry, we also have $g_1 = 0$ when $x_1 = 0$ or $x_2 = 0$. To check the differential equation, let $g_2(x) = \sum_i x_i \log x_i$. The proposed formula is $-2g_2(x) - g_1(x)$, where $g_1(x)$ is the function in (8.2). Thus, it suffices to show that $Lg_2 = 1$ since this will imply $L(-2g_2 - g_1) = -2 + 1 = -1$. Writing $x_3 = 1 - x_1 - x_2$, we have

$$g_2(x) = x_1 \log x_1 + x_2 \log x_2 + (1 - x_1 - x_2)\log(1 - x_1 - x_2)$$

Differentiating gives

$$D_1 g_2 = \log x_1 + 1 - \log(1 - x_1 - x_2) - 1$$

$$D_{11} g_2 = \frac{1}{x_1} + \frac{1}{1 - x_1 - x_2} \qquad D_{12} g_2 = \frac{1}{1 - x_1 - x_2}$$

Using symmetry to evaluate $D_{22} g_1$, we have

$$Lg_2 = \frac{1}{2}(1 - x_1) + \frac{1}{2}(1 - x_2)$$

$$+ \frac{1}{1 - x_1 - x_2} \left[\frac{1}{2} x_1 (1 - x_1) - x_1 x_2 + \frac{1}{2} x_2 (1 - x_2) \right]$$

Two times the term in square brackets is $x_1 + x_2 - x_1^2 - 2x_1 x_2 - x_2^2$, so we have

$$Lg_2 = \frac{1}{2} \left[(1 - x_1) + (1 - x_2) + \frac{(x_1 + x_2)(1 - x_1 - x_2)}{(1 - x_1 - x_2)} \right] = 1$$

which completes the proof.   □

Littler (1975) found in the K allele case that if T_i is the first time there are only i alleles

$$ET_i = -2 \sum_{k=1}^{i} (-1)^{i-k} \binom{K - 1 - k}{i - k} \sum \left(1 - \sum_{j=1}^{k} p_{i_j} \right) \log \left(1 - \sum_{j=1}^{k} p_{i_j} \right)$$

where the inner sum is over all possible values $1 \leq i_1 < i_2 \ldots < i_k \leq K$. This reduces to (8.2) when $i = 1$ and to (8.3) when $i = 2$ showing that these formulas are not restricted to the case $K = 3$.

8.1.2 Stationary distributions

To derive the stationary distribution for the K allele diffusion, we begin by finding the stationary distribution for the Moran model.

Theorem 8.4. *Suppose that $\mu_{ij} = \mu_j$ does not depend on i and let $\beta_j = N\mu_j > 0$. Then the stationary distribution is*

$$\pi(n) = c \prod_{j} \prod_{k=1}^{n_j} \left(1 - s_j + \frac{2\beta_j - (1 - s_j)}{k} \right) \qquad (8.4)$$

The assumption that $\mu_{ij} = \mu_j$ does not depend on i is called *parent independent mutation*. This assumption excludes the interesting case of two loci with two alleles, because in that case if we let $0 = ab$, $1 = aB$, $2 = Ab$, and $3 = AB$ then mutations from $0 \to 3$ of $1 \to 2$ are not possible.

To see why this assumption is useful, we extend the reasoning in Section 1.3 to conclude that if we are following a sample backwards in time, then when there are k lineages coalescence occurs at rate $k(k-1)/2$, and lineages will be killed at rate $k\sum_j \mu_j$ because if a mutation is encountered, we know the genetic state of that individual and all of its descendants in the sample. Of course, for this recipe to work the mutation rate must be independent of the current state of the individual.

Proof. By definition, the stationary distribution will be reversible if and only if for all n, and $i \neq j$

$$\pi(n)q(n, n + e_i - e_j) = \pi(n + e_i - e_j)q(n + e_i - e_j, n)$$

Filling in the rates, this means

$$\pi(n)n_j \left(\frac{n_i}{2N}(1 - s_i) + \mu_i\right)$$
$$= \pi(n + e_i - e_j)(n_i + 1)\left(\frac{n_j - 1}{2N}(1 - s_j) + \mu_j\right)$$

Multiplying on both sides by $2N/(n_j(n_i + 1))$, this becomes

$$\pi(n)\frac{n_i(1 - s_i) + 2N\mu_i}{n_i + 1}$$
$$= \pi(n + e_i - e_j)\frac{(n_j - 1)(1 - s_j) + 2N\mu_j}{n_j}$$

For our choice of π this is true because both sides are $= \pi(n + e_i)$. $\square$

To find the stationary distribution for the limiting diffusion, we will now take the limit as $N \to \infty$ in (8.4), assuming $Ns_j \to \gamma_j$. This assumption implies $s_j \to 0$, so we can drop that term from the numerator.

$$\pi(n) \approx c \prod_j \prod_{k=1}^{n_j} \left(1 - s_j + \frac{2\beta_j - 1}{k}\right)$$
$$\approx c \prod_j \exp(-s_j n_j + (2\beta_j - 1)\log n_j)$$

Changing variables $x_j = n_j/2N$, and altering the constant the above is

$$C \prod_j x_j^{2\beta_j - 1} \exp(-2\gamma_j x_j) \tag{8.5}$$

If $K = 2$ and $\gamma_1 = 0$ this reduces to the formula for the additive selection case given in Section 7.5.

Having found a stationary distribution by taking the limit of the Moran model, we will now verify that it works for the diffusion. The details are somewhat involved and the proof relies on several last minute algebraic manipulations, so before we embark on this task, we note that it has to work. The stationary distribution for the Moran model has $\pi_N Q_N f = 0$ for all f and as $N \to \infty$, $\pi_N \to \pi$ and $NQ_N \to L$. Thus, if nothing bad happens, we will have $\int \pi L f = 0$ for all f.

Theorem 8.5. *Suppose that in the K allele Wright-Fisher diffusion $\beta_{ij} = \beta_j$ does not depend on i. Then the stationary distribution is given by (8.5).*

Proof. As in one dimension, the stationary distribution satisfies $L^*\pi = 0$. The computation of $L^*\pi$ will involve a lot of terms so we begin with a special case.

Proof with no selection. Motivated by knowing that the stationary distribution is reversible, we isolate the terms that come from transitions $n \to n + e_i - e_j$ and write the Moran model generator as

$$2Lf = \sum_{i \neq j} [x_i x_j (D_{ii} - D_{ij}) + (-2\beta_j x_i + 2\beta_i x_j + (2\gamma_j - 2\gamma_i)x_i x_j)D_i] f$$

$$2L^*\pi = \sum_{i \neq j} D_{ii}(x_i x_j \pi) - D_{ij}(x_i x_j \pi)$$
$$- D_i([(-2\beta_j x_i + 2\beta_i x_j) + (2\gamma_j - 2\gamma_i)x_i x_j]\pi)$$

Writing L_{ij} and L^*_{ij} for the i, j term in the sum, reversibility suggests that we will find $(L^*_{ij} + L^*_{ji})\pi = 0$. In the special case, we will have $L^*_{ij}\pi = 0$.
When the $\gamma_j = 0$, (8.5) becomes

$$\pi = x_1^{2\beta_1 - 1} \cdots x_{K-1}^{2\beta_{K-1} - 1} z^{2\beta_K - 1} \quad \text{where} \quad z = 1 - \sum_{i=1}^{K-1} x_i \qquad (8.6)$$

Each x_i appears twice, once by itself and once in z. Writing 1 and 2 for the two locations, when we differentiate twice with respect to i we can do 11, 22, or 12 (and 21). To try to keep the formulas simple, we will keep π in the answer. For example, $x_i x_j \pi$ has $x_i^{2\beta_i}$ and $x_j^{2\beta_j}$, so in term 11 below, the power of x_i is reduced twice and we end up with a constant times $x_i^{-1}x_j\pi$.

$$D_{ii}(x_i x_j \pi) = 2\beta_i(2\beta_i - 1)x_i^{-1}x_j\pi$$
$$+ (2\beta_K - 1)(2\beta_K - 2)x_i x_j z^{-2}\pi(-1)^2$$
$$+ 2 \cdot 2\beta_i(2\beta_K - 1)x_j z^{-1}\pi(-1)$$
$$= y_1 + y_2 + 2y_3$$

Using a similar approach to the mixed derivative term

$$-D_{ij}(x_i x_j \pi) = -2\beta_i(2\beta_j)\pi - (2\beta_K - 1)(2\beta_K - 2)x_i x_j z^{-2}\pi(-1)^2$$
$$- 2\beta_i(2\beta_K - 1)x_j z^{-1}\pi(-1) - 2\beta_j(2\beta_K - 1)x_i z^{-1}\pi(-1)$$
$$= y_4 + y_5 + y_6 + y_7$$

while the first order term produces

$$-D_i[(-2\beta_j x_i + 2\beta_i x_j)\pi] = 2\beta_j(2\beta_i)\pi + 2\beta_j(2\beta_K - 1)x_i z^{-1}\pi(-1)$$
$$- 2\beta_i(2\beta_i - 1)x_i^{-1}x_j\pi - 2\beta_i(2\beta_K - 1)x_j z^{-1}\pi(-1)$$
$$= y_8 + y_9 + y_{10} + y_{11}$$

To get the result now we note that $y_1 = -y_{10}$, $y_2 = -y_5$, $y_3 = -y_6 = -y_{11}$, $y_4 = -y_8$, and $y_7 = -y_9$.

Proof with selection. To suppress terms that arise from differentiating the x_K in the exponential, we can subtract γ_K from all of the γ_j, which does not change the generator or the proposed stationary measure (because $\sum_i x_i = 1$) and suppose without loss of generality that $\gamma_K = 0$. Let

$$\rho = x_1^{2\beta_1 - 1} \cdots x_{K-1}^{2\beta_{K-1} - 1} z^{2\beta_K - 1} \quad \text{and} \quad h = \exp\left(-\sum_j 2\gamma_j x_j\right)$$

Computing the second order term as before

$$D_{ii}(x_i x_j \rho h) = D_{ii}(x_i x_j \rho) \cdot h + x_i x_j \rho (2\gamma_i)^2 h + 2D_i(x_i x_j \rho)(-2\gamma_i h)$$

Using a similar approach to the mixed derivative term

$$-D_{ij}(x_i x_j \pi h) = -D_{ij}(x_i x_j \rho) \cdot h - x_i x_j \rho (2\gamma_i)(2\gamma_j) h$$
$$-D_i(x_i x_j \rho)(-2\gamma_j h) - D_j(x_i x_j \rho)(-2\gamma_i h)$$

while the first order term produces

$$-D_i[(-2\beta_j x_i + 2\beta_i x_j)\rho] \cdot h + (2\beta_j x_i - 2\beta_i x_j)\rho(-2\gamma_i h)$$
$$-(2\gamma_j - 2\gamma_i)D_i(x_i x_j \rho)h - (2\gamma_j - 2\gamma_i)x_i x_j \rho(-2\gamma_i h)$$

By the result for no selection

$$\sum_{i \neq j} D_{ii}(x_i x_j \rho) - D_{ij}(x_i x_j \rho) - D_i[(-2\beta_j x_i + 2\beta_i x_j)\rho] = 0$$

In what remains of our three formulas, the terms that contain $x_i x_j \rho h$ are

$$x_i x_j \rho h[(2\gamma_i)^2 - 2\gamma_i(2\gamma_j) + (2\gamma_j - 2\gamma_i)2\gamma_i] = 0$$

At this point, there are $(3 + 4 + 4) - (3 + 3) = 5$ terms left

$$2D_i(x_i x_j \rho)(-2\gamma_i h)$$
$$-D_i(x_i x_j \rho)(-2\gamma_j h) - D_j(x_i x_j \rho)(-2\gamma_i h)$$
$$+(2\beta_j x_i - 2\beta_i x_j)\rho(-2\gamma_i h) - (2\gamma_j - 2\gamma_i)D_i(x_i x_j \rho)h$$

To make the fourth term match the others, we write

$$(2\beta_j x_i - 2\beta_i x_j)\rho(-2\gamma_i h) = [D_j(x_i x_j \rho) + D_i(x_i x_j \rho)](-2\gamma_i h)$$

The sum over i and j is not affected if we replace $D_j(x_i x_j \rho)(-2\gamma_i h)$ by $D_i(x_i x_j \rho)(-2\gamma_j h)$. If we do this then the terms with $D_i(x_i x_j \rho)$ are

$$\sum_{i \neq j} D_i(x_i x_j \rho)[-2(2\gamma_i) + 2(2\gamma_j) - 2(2\gamma_j - 2\gamma_i)] = 0$$

which completes the proof.   □

8.2 Recombination

Linkage disequilibrium can be built up between two segregating loci in a population due to the action of selection, but is also produced due to random frequency drift in a finite population. In this section, we will return to a topic considered in Section 3.4: How much linkage disequilibrium is produced by random drift? To answer this question, we consider two loci, each with two alleles that are separated by recombination with a probability r per generation. We will use a Wright-Fisher model which assumes that the diploid individuals in the population are a random union of gametes, and the following notation:

$$\begin{array}{ccccc} \text{genotype} & AB & Ab & aB & ab \\ \text{frequency} & Z_1(t) & Z_2(t) & Z_3(t) & Z_4(t) \end{array}$$

Let $D(t) = Z_4(t)Z_1(t) - Z_2(t)Z_3(t)$. The D is for linkage disequilibrium. To see the reason for the name note that

$$\begin{aligned} p_{AB} - p_A p_B &= z_1 - (z_1 + z_2)(z_1 + z_3) \\ &= z_1(1 - z_1 - z_2 - z_3) - z_2 z_3 = z_1 z_4 - z_2 z_3 \end{aligned}$$

The first step is to study the evolution of the means.

Lemma 8.1. *If we write $z_i = Z_i(t)$ and $z_i' = EZ_i(t+1)$, then*

$$z_i' = z_i + \eta_i r(z_2 z_3 - z_1 z_4)$$

where $\eta_1 = \eta_4 = 1$ and $\eta_2 = \eta_3 = -1$.

Proof. Considering the ways we can end up with AB gives

$$z_1' = (1 - r)z_1 + r z_1^2 + \frac{r}{2}(2z_1 z_2 + 2z_1 z_3 + 2z_2 z_3)$$

The first term says that if there is no recombination then the ancestor must be AB. If both ancestors are AB and a recombination occurs then an AB is always produced. When $i \neq j$ the probability the ancestors are i and j is $2z_i z_j$, but in the three cases indicated, when there is a recombination we only end up with a new AB half of the time. AB offspring are impossible for the other combinations. The identity for $i = 1$ now follows from the fact that $z_1 + z_2 + z_3 = 1 - z_4$, so combining terms on the right-hand side

$$\begin{aligned} z_1' &= (1 - r)z_1 + r(z_1(1 - z_4) + z_2 z_3) \\ &= z_i + \eta_i r(z_2 z_3 - z_1 z_4) \end{aligned}$$

The other three equations are similar.   □

In the previous result and throughout this section, capital letters are random variables, while lowercase letters are possible values or related functions.

Theorem 8.6. *If we run time at rate $2N$ and $2Nr \to R$ then (Z_1, Z_2, Z_3, Z_4) converges to a diffusion with drift vector $-\eta_i RD$ and covariance matrix*

$$
\begin{matrix}
z_1(1 - z_1) & -z_1 z_2 & -z_1 z_3 & -z_1 z_4 \\
-z_1 z_2 & z_2(1 - z_2) & -z_2 z_3 & -z_2 z_4 \\
-z_1 z_3 & -z_2 z_3 & z_3(1 - z_3) & -z_3 z_4 \\
-z_1 z_4 & -z_2 z_4 & -z_3 z_4 & z_4(1 - z_4)
\end{matrix}
$$

Proof. The drift follows from Lemma 8.1. To compute the covariance note that the result of the $2N$ draws in the Wright-Fisher model is multinomial with probabilities z_i'. If N_i are the number of type i drawn then $\text{var}(N_i/2N) = z_i'(1 - z_i')/2N$ and $\text{cov}(N_i/2N, N_j/2N) = -z_i' z_j'/2N$. Since $z_i - z_i' = O(1/N)$ the desired result follows. $\square$

8.2.1 A clever change of variables

Following Ohta and Kimura (1969a) we will now change variables to

$$
\begin{aligned}
X(t) &= Z_1(t) + Z_2(t) \\
Y(t) &= Z_1(t) + Z_3(t) \\
D(t) &= Z_1(t) Z_4(t) - Z_2(t) Z_3(t)
\end{aligned}
$$

To do this we need a multivariate extension of the change of variables formula in Theorem 7.1.

Theorem 8.7. Change of variables. *Suppose $X(t) = (X_1(t), \ldots, X_m(t))$ is an m-dimensional diffusion. If $f : \mathbf{R}^m \to \mathbf{R}$ has continuous partial derivatives of order ≤ 2, then the infinitesimal mean of $f(X_t)$ is*

$$
\sum_i D_i f(x) b_i(x) + \frac{1}{2} \sum_{i,j} D_{ij} f(x) a_{ij}(x) \tag{8.7}
$$

If g is another such function then the infinitesimal covariance of $f(X_t)$ and $g(X_t)$ is

$$
\sum_{i,j} D_i f(x) D_j g(x) a_{ij}(x) \tag{8.8}
$$

Proof. Let L be the infinitesimal generator of $X(t)$. If $h : \mathbf{R} \to \mathbf{R}$ is smooth

$$
Lh(f(x)) = \sum_i b_i(x) h'(f(x)) D_i f(x)
$$

$$
+ \frac{1}{2} \sum_{i,j} a_{i,j}(x) [h''(f(x)) D_i f(x) D_j f(x) + h'(f(x)) D_{ij} f(x)]
$$

Rearranging the right-hand side becomes

$$= h'(f(x)) \left[\sum_i b_i(x) D_i f(x) + \frac{1}{2} \sum_{ij} a_{ij}(x) D_{ij} f(x) \right]$$
$$+ \frac{1}{2} h''(f(x)) \sum_{i,j} a_{i,j}(x) D_i f(x) D_j f(x)$$

This identifies the infinitesimal mean and variance of $f(X_t)$. To get the co-variance of $Y_t = f(X_t)$ and $Z_t = g(X_t)$ we use

$$2 \operatorname{cov}(Y_t, Z_t) = \operatorname{var}(Y_t + Z_t) - \operatorname{var}(Y_t) - \operatorname{var}(Z_t)$$

and apply the previous result to $f + g$, f and g. $\square$

Theorem 8.8. *If we run time at rate $2N$ and $2Nr \to R$ then (X, Y, D) converges to a diffusion with drift vector $(0, 0, -(1+R)D)$ and covariance matrix*

$$\begin{array}{ccc} x(1-x) & D & D(1-2x) \\ D & y(1-y) & D(1-2y) \\ D(1-2x) & D(1-2y) & F \end{array}$$

where $F = xy(1-x)(1-y) + D(1-2x)(1-2y) - D^2$.

This is 2 times the diffusion defined in formula (2) of Ohta and Kimura (1969), since they run time at rate N.

Proof. Taking $f(z_1, z_2, z_3, z_4) = z_1 + z_2$ and using (8.7) with Theorem 8.6 we see that the infinitesimal mean of $X(t)$ is $b_1(Z(t)) + b_2(Z(t)) = 0$. Similarly the infinitesimal mean of $Y(t)$ is 0.

When $f(z_1, z_2, z_3, z_4) = z_1 z_4 - z_2 z_3$,

$$D_1 f = z_4, \quad D_2 f = -z_3, \quad D_3 f = -z_2, \quad D_4 f = z_1,$$
$$D_{1,4} f = D_{4,1} f = 1, \quad D_{2,3} f = D_{3,2} f = -1$$

and the other second order partials are 0. So using (8.7) with Theorem 8.6 the infinitesimal mean of $D(t)$ is

$$z_4 b_1(Z(t)) - z_3 b_2(Z(t)) - z_2 b_3(Z(t)) + z_1 b_4(Z(t))$$
$$+ \frac{1}{2} \cdot 2(a_{1,4}(Z(t)) - a_{2,3}(Z(t)))$$
$$= -RD(z_4 + z_3 + z_2 + z_1) + (-z_1 z_4 + z_2 z_3) = -(1+R)D$$

Using (8.8) with $f = g = z_1 + z_2$ we see that the infinitesimal variance of $X(t)$ is

$$z_1(1 - z_1) - 2z_1 z_2 + z_2(1 - z_2) = (z_1 + z_2)(1 - z_1 - z_2) = x(1-x)$$

Similarly the infinitesimal variance of $Y(t)$ is $y(1-y)$ Using (8.8) with $f = z_1 + z_2$ and $g = z_1 + z_3$ we see that the infinitesimal covariance of $X(t)$ and $Y(t)$ is

$$z_1(1 - z_1) - z_1 z_2 - z_1 z_3 - z_2 z_3 = z_1 z_4 - z_2 z_3 = D$$

Using (8.8) with $f = z_1 + z_2$ and $g = z_1 z_4 - z_2 z_3$ we see that the infinitesimal covariance of $X(t)$ and $D(t)$ is

$$z_4 z_1(1 - z_1) - z_3(-z_1 z_2) - z_2(-z_1 z_3) + z_1(-z_1 z_4)$$
$$+ z_4(-z_1 z_2) - z_3 z_2(1 - z_2) - z_2(-z_2 z_3) + z_1(-z_2 z_4)$$
$$= z_4 z_1(1 - 2z_1) + 2z_1 z_2 z_3 - z_3 z_2(1 - 2z_2) - 2z_1 z_2 z_4$$
$$= z_4 z_1(1 - 2z_1 - 2z_2) - z_3 z_2(1 - 2z_1 - 2z_2) = D(1 - 2x)$$

Similarly the infinitesimal covariance of $Y(t)$ and $D(t)$ is $D(1 - 2y)$.

The last, and most complicated step, is to compute the infinitesimal variance of $D(t)$, and show that it is equal to F. Using (8.8) with $f = g = z_1 z_4 - z_2 z_3$ which has partial derivatives $\nabla f = (z_4, -z_3, -z_2, z_1)$, we see that the variance $\sigma_D^2 = \nabla f \cdot a \nabla f$. To simplify the computation, we write the covariance matrix as

$$
\begin{array}{cccc}
z_1(z_2 + z_3 + z_4) & -z_1 z_2 & -z_1 z_3 & -z_1 z_4 \\
-z_1 z_2 & z_2(z_1 + z_3 + z_4) & -z_2 z_3 & -z_2 z_4 \\
-z_1 z_3 & -z_2 z_3 & z_3(z_1 + z_2 + z_4) & -z_3 z_4 \\
-z_1 z_4 & -z_2 z_4 & -z_3 z_4 & z_4(z_1 + z_2 + z_3)
\end{array}
$$

We do this so that all of the 24 terms in σ_D^2 will be of order 4 in the z_i. This is useful, because it will turn out that all of the terms in F are of order 4. Sorting the 24 terms in σ_D^2 according to the powers of z:

powers		number
(3,1)	$z_i^3 z_{5-i}$	4
(2,2)	$-2z_1^2 z_4^2 - 2z_2^2 z_3^2$	4
(2,1,1)	$-z_i^2 z_{5-i} z_j$	8
(1,1,1,1)	$8z_1 z_2 z_3 z_4$	8

Here, and in what follows, all the indices shown are supposed to be different, and we find the number of the different types by considering how they can be generated. For example, the (3,1)'s and (2,1,1)'s come from the diagonal a_{ii}. The (2,2)'s come from the anti-diagonal $a_{i,5-i}$. The (1,1,1,1)'s come from the other eight entries in a_{ij}.

In computing F, it is convenient to introduce now some notation for its three parts that will be the subject of the next theorem. Writing $u = xy(1 - x)(1 - y) = (z_1 + z_2)(z_1 + z_3)(z_3 + z_4)(z_2 + z_4)$, we generate 16 terms:

powers		number
(2,2)	$z_1^2 z_4^2 + z_2^2 z_3^2$	2
(2,1,1)	$z_i^2 z_j z_k$	12
(1,1,1,1)	$2z_1 z_2 z_3 z_4$	2

Writing $v = D(1 - 2x)(1 - 2y) = (z_1 z_4 - z_2 z_3)(z_3 + z_4 - z_1 - z_2)(z_1 + z_3 - z_2 - z_4)$, we generate 32 terms:

$$
\begin{array}{ccc}
\text{powers} & & \text{number} \\
(3,1) & z_i^3 z_{5-i} & 4 \\
(2,2) & -2z_1^2 z_4^2 - 2z_2^2 z_3^2 & 4 \\
(2,1,1) & -2z_i^2 z_{5-i} z_j - z_i z_{5-i} z_j^2 & 16 + 4 \\
(1,1,1,1) & 4z_1 z_2 z_3 z_4 & 4
\end{array}
$$

Finally $-w = -D^2 = -(z_1 z_4 - z_2 z_3)^2$ generates only four terms:

$$
\begin{array}{ccc}
\text{powers} & & \text{number} \\
(2,2) & -z_1^2 z_4^2 - z_2^2 z_3^2 & 2 \\
(1,1,1,1) & 2z_1 z_2 z_3 z_4 & 2
\end{array}
$$

To prove that $\sigma_D^2 = F$, we only have to compare the four tables. The only subtle part of this to note that the 20 $(2,1,1)$'s in v cancel with the 12 in u to leave the 8 in σ_D^2. □

8.2.2 Time-dependent behavior

The two new functions that made their appearance in the definition of F in Theorem 8.8 turn out to be special. The miracle here is that Lu, Lv, and Lw are linear combinations of u, v, and w.

Theorem 8.9. If we let $u = xy(1-x)(1-y)$, $v = D(1-2x)(1-2y)$ and $w = D^2$ then

$$
\begin{aligned}
Lu &= -2u + v \\
Lv &= -(5+R)v + 4w \\
Lw &= u + v - (3+2R)w
\end{aligned}
$$

Again this agrees with (9) in Ohta and Kimura (1969a) once one recalls the factor of 2 from Theorem 8.8, and another factor of 2 that comes from the fact that their $R = Nr$. Thus, to get from our equation to theirs we divide by 2 and then change $R \to 2R$.

Proof. Given the definition of the infinitesimal generator in Theorem 8.8 it suffices to compute the relevant partial derivatives. $u = (x - x^2)(y - y^2)$ has

$$
\begin{array}{lll}
D_x u = (1-2x)(y-y^2) & D_y u = (1-2y)(x-x^2) & \\
D_{xx} u = -2y(1-y) & D_{xy} u = (1-2x)(1-2y) & D_{yy} u = -2x(1-x)
\end{array}
$$

so we have

$$
\begin{aligned}
Lu &= \frac{1}{2}[-2y(1-y) \cdot x(1-x) + 2(1-2x)(1-2y) \cdot D - 2x(1-x) \cdot y(1-y)] \\
&= -2u + v
\end{aligned}
$$

The function $v = D(1-2x)(1-2y)$ has

$$D_x v = -2(1-2y)D \qquad D_y v = -2(1-2x)D \qquad D_D v = (1-2x)(1-2y)$$
$$D_{xx} v = D_{yy} v = D_{DD} v = 0 \qquad D_{xy} v = 4D$$
$$D_{xD} v = -2(1-2y) \qquad D_{yD} = -2(1-2x)$$

so we have

$$Lv = -(1+R)D \cdot (1-2x)(1-2y) + 4D \cdot D - 4D(1-2x)(1-2y)$$
$$= -(5+R)v + 4w$$

The function $w = D^2$ has $D_D w = 2D$ and $D_{DD} w = 2$, so using $F = u+v-w$

$$Lw = -2(1+R)w + (u+v-w) = u+v-(3+2R)$$

which completes the proof. $\qquad\qquad\qquad\qquad\qquad\qquad\qquad\square$

Lemma 8.9 shows that the expected values of u, v, w satisfy a 3 dimensional linear system of ODE's. If we consider linear combinations $au+bv+cw$ where a, b, c is a left eigenvalue of the matrix

$$\begin{matrix} -2 & 1 & 0 \\ 0 & -(5+R) & 4 \\ 1 & 1 & -(3+2R) \end{matrix} \qquad (8.9)$$

with eigenvalue λ_i then $E(au + bv + cw)(t) = e^{\lambda_i t} E(au + bv + cw)(0)$. Once the eigenvectors associated with the eigenvalues are found one can express u, v, and w as linear combinations of them and find exact formulas for their expected values. The results are somewhat messy, so we refer the interested reader to Ohta and Kimura (1969a) for details. Without going through this work, we can see that if $0 > \lambda_1 > \lambda_2 > \lambda_3$ then all three quantities decay asymptotically like $e^{\lambda_1 t}$.

The eigenvalues λ_i of the matrix satisfy a cubic equation

$$\det \begin{pmatrix} -(2+\lambda) & 1 & 0 \\ 0 & -(5R+\lambda) & 4 \\ 1 & 1 & -(3+2R+\lambda) \end{pmatrix} = 0$$

Calculating the determinant by expanding it in the first row, and then multiplying the equation by -1 we have

$$(2+\lambda)(5R+\lambda)(3+2R+\lambda) - (2+\lambda) \cdot 4 - 4 = 0$$

which after some algebra becomes

$$\lambda^3 + \lambda^2(10+3R) + \lambda(27+19R+2R^2) + (18+26R+4R^2) = 0$$

The first step in solving a cubic equation $\lambda^3 + b\lambda^2 + c\lambda + d = 0$ is to let $\lambda = x - (b/3)$, which eliminates the x^2 term and gives

$$x^3 + x\left(c - \frac{b^2}{3}\right) + \left(d - \frac{cb}{3} + \frac{2b^3}{27}\right)$$

In our case, this is

$$x^3 - x\left(\frac{19 + 3R + 3R^2}{3}\right) + \left(\frac{56 + 63R - 45R^2}{27}\right) \tag{8.10}$$

To solve the equation, we use the following trick

Theorem 8.10. *The solutions of $y^3 - (a/12)y + b/108 = 0$ are given by setting $k = 0, 1, 2$ in*

$$y_k = \frac{a^{1/2}}{3}\cos\left(\frac{\theta}{3} + k\frac{2\pi}{3}\right) \quad where \quad \theta = \cos^{-1}(-b/a^{3/2})$$

Proof. Using the trig identities

$$\sin(2\theta) = 2\sin\theta\cos\theta$$
$$\cos(\alpha + \beta) = \cos\alpha\cos\beta - \sin\alpha\sin\beta$$
$$\cos^2\theta = 1 - \sin^2\theta$$

we have

$$\cos 3\theta = \cos(2\theta)\cos\theta - \sin(2\theta)\sin\theta$$
$$= \cos^3\theta - \sin^2\theta\cos\theta - 2\sin^2\theta\cos\theta$$
$$= 4\cos^3\theta - 3\cos\theta$$

From this, it follows that

$$y_k^3 - \frac{a}{12}y_k = \frac{a^{3/2}}{27}\cos^3\left(\frac{\theta}{3} + k\frac{2\pi}{3}\right) - \frac{a^{3/2}}{36}\cos\left(\frac{\theta}{3} + k\frac{2\pi}{3}\right)$$
$$= \frac{a^{3/2}}{108}\cos(\theta) = -b/108$$

when $\theta = \cos^{-1}(-b/a^{3/2})$. $\qquad\square$

Using this on (8.10), we find the solutions given in formula (14) of Ohta and Kimura (1969a), modulo the usual conversion that to get their formula from ours we have to let $R \to 2R$ and the eigenvalues will be twice as large. The formulas are ugly, but they can be easily evaluated numerically. In Figure 8.2, we graph $-\lambda_i$, where the eigenvalues in increasing order of magnitude are: $\lambda_1 = y_0 - (10 + 3R)/2$, $\lambda_2 = y_2 - (10 + 3R)/2$, and $\lambda_3 = y_1 - (10 + 3R)/2$. When $R = 0$ the cubic equation is

$$\lambda^3 + 10\lambda^2 + 27\lambda + 18 = (\lambda + 1)(\lambda + 3)(\lambda + 6)$$

so the eigenvalues are -1, -3, and -6. At first it may be surprising that disequilibrium decays when $R = 0$, but recall that $D = z_4z_1 - z_2z_3$, so D will become 0 when one of the alleles becomes fixed in the population. It is interesting to note that as $R \to \infty$, $-\lambda_1 \to 2$ while $-\lambda_2$ and $-\lambda_3$ grow linearly.

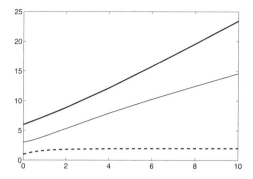

Fig. 8.2. -1 times the eigenvalues of (8.9).

8.2.3 Equilibrium when there is mutation

If we introduce mutation $A \to a$ with probability u_1, $a \to A$ with probability v_1, $B \to b$ with probability u_2, and $b \to B$ with probability v_2 then the mean frequencies x and y of the A and B alleles change according to

$$x' - x = -u_1 x + v_1(1 - x) = v_1 - (u_1 + v_1)x$$
$$y' - y = -u_2 y + v_2(1 - y) = v_2 - (u_2 + v_2)y$$

Ignoring the possibility of two changes at once:

$$p'_{AB} - p_{AB} = -(u_1 + u_2)p_{AB} + v_1 p_{aB} + v_2 p_{Ab}$$
$$p'_A p'_B - p_A p_B = -(u_1 + u_2)p_A p_B + v_1 p_a p_B + v_2 p_A p_b$$

so the change in D due to mutation is

$$D' - D = -(u_1 + u_2 + v_1 + v_2)D$$

Let $\mu_i = 2Nu_i$, $\nu_i = 2Nv_i$, and $\kappa = \mu_1 + \mu_2 + \nu_1 + \nu_2$. If we incorporate mutation, then the covariance matrix of the (X, Y, Z) diffusion remains the same:

$$\begin{matrix} x(1-x) & D & D(1-2x) \\ D & y(1-y) & D(1-2y) \\ D(1-2x) & D(1-2y) & F \end{matrix}$$

but the drift vector is now

$$\nu_1 - (\mu_1 + \nu_1)x, \quad \nu_2 - (\mu_2 + \nu_2)y, \quad -(1 + R + \kappa)D$$

Using $0 = (d/dt)Ef(X, Y, D) = E(Lf)$ for different choices of f gives information about the equilibirum state of the (X, Y, D) diffusion with mutation.

Theorem 8.11. *In equilibrium, $ED = 0$, $E(XD) = 0$, $EX = \nu_1/(\mu_1 + \nu_1)$,*

$$E(XY) = \frac{\nu_1}{\mu_1 + \nu_1} \cdot \frac{\nu_2}{\mu_2 + \nu_2}$$

$$E(X(1-X)) = \frac{\nu_1}{\mu_1 + \nu_1} \cdot \frac{2\mu_1}{2\mu_1 + 2\nu_1 + 1}$$

$$E(X(1-X)Y) = \frac{\nu_1}{\mu_1 + \nu_1} \cdot \frac{2\mu_1}{2\mu_1 + 2\nu_1 + 1} \cdot \frac{\nu_2}{\mu_2 + \nu_2}$$

In Section 3.4, we showed $ED = 0$. The second conclusion shows that the allele frequencies are uncorrelated with D. The third and fifth formulas were obtained in Section 7.5, with $\mu_1 = \beta_1$ and $\nu_1 = \beta_2$. The fourth formula follows from the first, third, and $ED = E(XY) - EX \cdot EY$. We can obtain other formulas from these using two rules: (i) to replace Y by $1 - Y$ subtract two formulas:

$$E(X(1-Y)) = EX - EXY = \frac{\nu_1}{\mu_1 + \nu_1} \cdot \frac{\mu_2}{\mu_2 + \nu_2}$$

$$E(X(1-X)(1-Y)) = \frac{\nu_1}{\mu_1 + \nu_1} \cdot \frac{2\mu_1}{2\mu_1 + 2\nu_1 + 1} \cdot \frac{\mu_2}{\mu_2 + \nu_2}$$

and (ii) to replace X by Y exchange the roles of the two subscripts,

$$E(Y(1-Y)X) = \frac{\nu_2}{\mu_2 + \nu_2} \cdot \frac{2\mu_2}{2\mu_2 + 2\nu_2 + 2} \cdot \frac{\nu_1}{\mu_1 + \nu_1}$$

$$E(Y(1-Y)(1-X)) = \frac{\nu_2}{\mu_2 + \nu_2} \cdot \frac{2\mu_2}{2\mu_2 + 2\nu_2 + 1} \cdot \frac{\mu_1}{\mu_1 + \nu_1}$$

Proof. It follows from the infinitesimal generator, that

$$LD = -(1 + R + \kappa)D$$
$$L(xD) = D[\nu_1 - (\mu_1 + \nu_1)x] - x(1 + R + \kappa)D + D(1 - 2x)$$

In equilibrium, the two right-hand sides have 0 expected value, so we conclude from the first that $ED = 0$, and then from the second that $E(XD) = 0$.

$$Lx = [\nu_1 - (\mu_1 + \nu_1)x] \quad \text{implies} \quad EX = \frac{\nu_1}{\mu_1 + \nu_1}.$$

Continuing with the pattern of using Lf to compute Ef,

$$L(xy) = y[\nu_1 - (\mu_1 + \nu_1)x] + x[\nu_2 - (\mu_2 + \nu_2)x] + D$$

so using $ED = 0$ we have

$$(\mu_1 + \nu_1 + \mu_2 + \mu_2)EXY = \nu_1 \frac{\nu_2}{\mu_2 + \nu_2} + \nu_2 \frac{\nu_1}{\mu_1 + \nu_1}$$
$$= \frac{\nu_1}{\mu_1 + \nu_1} \frac{\nu_2}{\mu_2 + \nu_2}(\mu_1 + \nu_1 + \mu_2 + \mu_2)$$

Using the identity $\nu_1 - (\mu_1 + \nu_1)x = -\mu_1 + (\mu_1 + \nu_1)(1 - x)$ in the third line

$$L(x(1 - x)) = [(1 - x) - x](\nu_1 - (\mu_1 + \nu_1)x) + \frac{1}{2}(-2)x(1 - x)$$
$$= -(\mu_1 + \nu_1)x(1 - x) + \nu_1(1 - x)$$
$$\quad -(\mu_1 + \nu_1)x(1 - x) + \mu_1 x - x(1 - x)$$
$$= -(2\mu_1 + 2\nu_1 + 1)x(1 - x) + \nu_1(1 - x) + \mu_1 x$$

so we have

$$EX(1 - X) = \left(\nu_1 \frac{\mu_1}{\mu_1 + \nu_1} + \mu_1 \frac{\nu_1}{\mu_1 + \nu_1}\right) \cdot \frac{1}{2\mu_1 + 2\nu_1 + 1}$$
$$= \frac{\nu_1}{\mu_1 + \nu_1} \cdot \frac{2\mu_1}{2\mu_1 + 2\nu_1 + 1}$$

Using the formula for the generator, we have

$$L(x(1 - x)y) = [(1 - x)y - xy](\nu_1 - (\mu_1 + \nu_1)x)$$
$$\quad + x(1 - x)(\nu_2 - (\mu_2 + \nu_2)y)$$
$$\quad + \frac{1}{2}(-2y)x(1 - x) + (1 - 2x)D$$

Using the identity from the previous calculation on the $-xy$ term in the first line, the right-hand side is

$$= -(\mu_1 + \nu_1)x(1 - x)y + \nu_1(1 - x)y - (\mu_1 + \nu_1)x(1 - x)y + \mu_1 xy$$
$$\quad - (\mu_2 + \nu_2)x(1 - x)y + \nu_2 x(1 - x) - x(1 - x)y + (1 - 2x)D$$

Taking expected value and using $E(1 - 2X)D = 0$, we have

$$(2\mu_1 + 2\nu_1 + \mu_2 + \nu_2 + 1)EX(1 - X)Y$$
$$= \nu_1 \cdot \frac{\mu_1}{\mu_1 + \nu_1} \cdot \frac{\nu_2}{\mu_2 + \nu_2} + \mu_1 \cdot \frac{\nu_1}{\mu_1 + \nu_1} \cdot \frac{\nu_2}{\mu_2 + \nu_2}$$
$$\quad + \nu_2 \cdot \frac{\nu_1}{\mu_1 + \nu_1} \cdot \frac{2\mu_1}{2\mu_1 + 2\nu_1 + 1}$$
$$= \mu_1 \cdot \frac{\nu_1}{\mu_1 + \nu_1} \cdot \nu_2 \left(\frac{2}{\mu_2 + \nu_2} + \frac{2}{2\mu_1 + 2\nu_1 + 1}\right)$$

Combining the two fractions over a common denominator, we have

$$EX(1 - X)Y = \frac{\nu_1}{\mu_1 + \nu_1} \cdot \frac{\nu_2}{\mu_2 + \nu_2} \cdot \frac{2\mu_1}{2\mu_1 + 2\nu_1 + 1}$$

which completes the proof. $\qquad\square$

Theorem 8.12. *If we let $U = XY(1 - X)(1 - Y)$, $V = D(1 - 2X)(1 - 2Y)$ and $W = D^2$ then in equilibrium the (X, Y, D) diffusion with mutation has*

$$(2 + 2\kappa)EU - EV = \alpha$$
$$(5 + R + 2\kappa)EV - 4EW = 0$$
$$EU + EV - (3 + 2R + 2\kappa)EW = 0$$

where $\kappa = \mu_1 + \nu_1 + \mu_2 + \nu_2$ and

$$\alpha = E[\mu_2 X(1 - X)Y + \nu_2 X(1 - X)(1 - Y)$$
$$+ \mu_1 XY(1 - Y) + \nu_1(1 - X)Y(1 - Y)]$$

After the factors of 2 and the differences in notation are taken into account, this is (16) of Ohta and Kimura (1969b), but with a more symmetric definition of α.

Proof. In equilibirium $ELu = ELv = ELw = 0$. Given the formulas for derivatives in the proof of Theorem 8.9, it is enough to find the terms that result from the new drifts.

$$Lu = -2u + v + (1 - 2x)y(1 - y)[\nu_1 - (\mu_1 + \nu_1)x]$$
$$+ (1 - 2y)x(1 - x)[\nu_2 - (\mu_2 + \nu_2)y]$$

Using $1 - 2x = 2(1 - x) - 1$, we can rewrite $(1 - 2x)[\nu_1 - (\mu_1 + \nu_1)x]$ as

$$= -2(\mu_1 + \nu_1)x(1 - x) + 2\nu_1(1 - x) - \nu_1 + (\mu_1 + \nu_1)x$$
$$= -2(\mu_1 + \nu_1)x(1 - x) + \nu_1(1 - x) + \mu_1 x$$

so multiplying by $y(1 - y)$, we see that the first extra term is

$$-2(\mu_1 + \nu_1)u + \nu_1(1 - x)y(1 - y) + \mu_1 xy(1 - y)$$

By symmetry, the second extra term is

$$-2(\mu_2 + \nu_2)u + \nu_2(1 - y)x(1 - x) + \mu_2 yx(1 - x)$$

so recalling the definition of α, gives the first equation:

$$Lv = -(5 + R)v + 4w - 2(1 - 2y)D[\nu_1 - (\mu_1 + \nu_1)x]$$
$$-2(1 - 2x)D[\nu_2 - (\mu_2 + \nu_2)y] - (1 - 2x)(1 - 2y)\kappa D$$

The final term is $-\kappa v$. Writing

$$\nu_1 - (\mu_1 + \nu_1)x = \frac{1}{2}[(\mu_1 + \nu_1)(1 - 2x) + (\nu_1 - \mu_1)]$$

we see that the first extra term is $-(\mu_1 + \nu_1)v - (\nu_1 - \mu_1)(1 - 2y)D$. Theorem 8.11 implies that $E(1 - 2Y)D = 0$. Performing a similar manipulation on the second extra term gives the second equation.

The last equation is the simplest:

$$Lw = u + v - (3 + 2R)w + 2D \cdot (-\kappa D) = u + v - (3 + 2R + 2\kappa)w \quad \square$$

Solving the three equations in three unknowns in Theorem 8.12, we have (17) in Ohta and Kimura (1969b), a result that was found earlier by Hill and Robertson (1966).

Theorem 8.13. *In equilibrium, the (X, Y, D) diffusion with mutation has*

$$EU = \frac{\alpha}{2 + 2\kappa}\left(1 + \frac{1}{h}\right)$$

$$EV = \frac{\alpha}{h}$$

$$EW = \frac{5 + R + 2\kappa}{4} \cdot \frac{\alpha}{h}$$

where α and κ are given in Theorem 8.12 and

$$h = (1 + \kappa)(3 + 2R + 2\kappa)(2.5 + 0.5R + \kappa) - (2 + 2\kappa) - 1.$$

From this it follows that

$$\sigma_d^2 \equiv \frac{EW}{EU} = \frac{5 + R + 2\kappa}{(3 + 2R + 2\kappa)(2.5 + 0.5R + \kappa) - 4}$$

If κ is $2N$ times the mutation rate at a nucleotide, while R is $2N$ times the recombination rate between two nucleotides separated by hundreds or thousands of nucleotides, then $\kappa \ll R$ and the last result simplifies to

$$\sigma_d^2 = \frac{5 + R}{11 + 13R + 2R^2}$$

Replacing R by $2R$ gives the formula on page 577 of Ohta and Kimura (1971). Replacing R by $\rho/2$ gives Theorem 3.9.

Proof. To solve the equations in Theorem 8.12, we note that the first two equations imply

$$EU = \frac{\alpha + EV}{2 + 2\kappa} \qquad EW = \frac{5 + R + 2\kappa}{4}EV$$

Using these in the third equation, we have

$$\frac{\alpha + EV}{2 + 2\kappa} + EV - (3 + 2R + 2\kappa)\frac{5 + R + 2\kappa}{4}EV = 0$$

Rearranging gives

$$\frac{\alpha}{1 + \kappa} = \left((3 + 2R + 2\kappa)\frac{5 + R + 2\kappa}{2} - 2 - \frac{1}{1 + \kappa}\right)EV = 0$$

and solving gives $EV = \alpha/h$. Substitution into the first two equations gives EU and EW, and shows

$$\frac{EW}{EU} = \frac{(5 + R + 2\kappa)/4}{(h + 1)/(2 + 2\kappa)}$$

which after a little algebra becomes the formula given. □

The quantity α cancelled out in the computation of σ_d^2. For completeness, we note that the formulas in Theorem 8.11 imply that

$$\alpha = \mu_2 \cdot \frac{\nu_1}{\mu_1 + \nu_1} \cdot \frac{2\mu_1}{2\mu_1 + 2\nu_1 + 1} \cdot \frac{\nu_2}{\mu_2 + \nu_2}$$
$$+ \nu_2 \cdot \frac{\nu_1}{\mu_1 + \nu_1} \cdot \frac{2\mu_1}{2\mu_1 + 2\nu_1 + 1} \cdot \frac{\mu_2}{\mu_2 + \nu_2}$$
$$+ \mu_1 \cdot \frac{\nu_2}{\mu_2 + \nu_2} \cdot \frac{2\mu_2}{2\mu_2 + 2\nu_2 + 2} \cdot \frac{\nu_1}{\mu_1 + \nu_1}$$
$$+ \nu_1 \cdot \frac{\nu_2}{\mu_2 + \nu_2} \cdot \frac{2\mu_2}{2\mu_2 + 2\nu_2 + 1} \cdot \frac{\mu_1}{\mu_1 + \nu_1}$$

so collecting terms, we have (10) of Ohta and Kimura (1969b)

$$\alpha = \nu_2 \cdot \frac{\nu_1}{\mu_1 + \nu_1} \cdot \frac{\nu_2}{\mu_2 + \nu_2} \cdot \frac{4\mu_1}{2\mu_1 + 2\nu_1 + 1}$$
$$+ \nu_1 \cdot \frac{\nu_2}{\mu_2 + \nu_2} \cdot \frac{\nu_1}{\mu_1 + \nu_1} \cdot \frac{4\mu_2}{2\mu_2 + 2\nu_2 + 2}$$

since our $\alpha = 2NA$.

8.3 Hill-Robertson interference

Hill and Robertson (1966) were among the first to try to quantify the observation that linked advantageous mutations interfere with each other. That is, if an advantageous allele B arises while a second one A is on its way to fixation, then the fixation probability of B is reduced. Let s_A and s_B be the fitness advantages of the A and B alleles, and assume that fitnesses are additive. There are two possibilities for the mutant chromosome (i) aB in which case the fixation probability is decreased, and (ii) AB in which case the fixation probability is increased.

To analyze this situation, we will, at the beginning, assume that there is no recombination between the two loci. This gives us a three allele model in which allele 1 is the type of the mutant chromosome, allele 2 is Ab and allele 3 is ab. If we write σ_i, $i = 1, 2, 3$ as the rescaled fitness advantages of the three alleles then the limiting diffusion process has generator:

$$L = \frac{1}{2}x_1(1 - x_1)D_{11} - x_1x_2D_{12} + \frac{1}{2}x_2(1 - x_2)D_{22}$$
$$+ x_1 \sum_j x_j(\sigma_j - \sigma_1)D_1 + x_2 \sum_j x_j(\sigma_j - \sigma_2)D_2 \qquad (8.11)$$

Suppose $s_A = s_B = s$. We want to compute $h_a(x_1, x_2)$ and $h_A(x_1, x_2)$, which are the probability of $F_1 = \{$ 1's become fixed $\}$ when the initial frequencies of alleles 1 and 2 are x_1 and x_2, and the subscript gives the background on which the B allele arose. That is, allele 1 is aB in the first case and AB in the second.

Dropping the subscript from h because the next calculation applies to either situation, we have

Theorem 8.14. $h(x) = P_x(F_1)$ *is the unique solution of* $Lh = 0$ *with boundary conditions* $h(0, x_2) = 0$ *and* $h(1, 0) = 1$.

Proof. If we let $\mathcal{F}_t$ be the information known at time t then the Markov property implies
$$P(F_1|\mathcal{F}_t) = h(X_1(t), X_2(t))$$
In words, given the behavior of the process up to time t the fixation probability only depends on the current allele frequencies. The left-hand side is a martingale, so the right-hand side is also and we conclude $Lh = 0$.

In addition to this differential equation for h, we have the boundary conditions $h(0, x_2) = 0$ and $h(1, 0) = 1$, which come from the fact that the B allele has been lost in the first case, and fixed in the second. This is enough to determine the function, because if the process hits $x_2 = 0$, then it remains on this side until it hits $(0, 0)$ or $(1, 0)$. Likewise, if the process hits $x_1 + x_2 = 1$ it will continue on that side until it hits $(0, 1)$ or $(1, 0)$. $\square$

Case 1. Suppose B arises on the a background, so $\sigma_1 = \sigma_2 = \sigma$, $\sigma_3 = 0$. In this case, if we consider the competition of the two fitter types against the 3's, we can use results for one dimensional diffusions to conclude that the probability the 3's die out is
$$\frac{1 - e^{-2\sigma(x_1+x_2)}}{1 - e^{-2\sigma}}$$
Since the competition of 1's and 2's is a fair game, we guess that

Theorem 8.15. *The fixation probability when* B *arises on the* a *background is*
$$h_a(x) = \frac{1 - e^{-2\sigma(x_1+x_2)}}{1 - e^{-2\sigma}} \cdot \frac{x_1}{x_1 + x_2} \tag{8.12}$$

Proof. Clearly $h_a(0, x_2) = 0$ and $h_a(1, 0) = 1$, so it suffices to show that $Lh_a = 0$. The drift term in (8.11) is
$$\sigma x_1(1 - x_1 - x_2)D_1 + \sigma x_2(1 - x_1 - x_2)D_2 \tag{8.13}$$
To make computations easier, write $f_i = D_i f$, $f_{ij} = D_{ij}f$, etc., and note that
$$\begin{aligned} D_i(fg) &= f_i g + f g_i \\ D_{ii}(fg) &= f_{ii}g + 2f_i g_i + f g_{ii} \\ D_{12}(fg) &= f_{12}g + f_1 g_2 + f_2 g_1 + f g_{12} \end{aligned}$$
so we have
$$\begin{aligned} L(fg) = {}&gLf + fLg \\ &+ x_1(1 - x_1)f_1 g_1 - x_1 x_2(f_1 g_2 + f_2 g_1) + x_2(1 - x_2)f_2 g_2 \end{aligned}$$

To see this, note that the first two terms contain all the situations when only one function in the product is differentiated. What remains are the second order terms with each function differentiated once.

Letting $f = 1 - e^{-2\sigma(x_1+x_2)}$ and $g = x_1/(x_1+x_2) = 1 - x_2/(x_1+x_2)$

$$f_i = 2\sigma e^{-2\sigma(x_1+x_2)} \qquad f_{ij} = -4\sigma^2 e^{-2\sigma(x_1+x_2)}$$

$$g_1 = \frac{x_2}{(x_1+x_2)^2} \qquad g_2 = \frac{-x_1}{(x_1+x_2)^2}$$

$$g_{11} = \frac{-2x_2}{(x_1+x_2)^3} \qquad g_{22} = \frac{2x_1}{(x_1+x_2)^3}$$

$$g_{12} = \frac{1}{(x_1+x_2)^2} - \frac{2x_2}{(x_1+x_2)^3} = \frac{x_1-x_2}{(x_1+x_2)^3}$$

From this we get in the special case with drift (8.13)

$$\frac{Lf}{2\sigma e^{-2\sigma(x_1+x_2)}} = \left[\frac{1}{2}x_1(1-x_1) - x_1 x_2 + \frac{1}{2}x_2(1-x_2)\right](-2\sigma)$$
$$+\sigma(x_1+x_2)(1-x_1-x_2)$$
$$= -\sigma[x_1(1-x_1-x_2) + x_2(1-x_1-x_2)]$$
$$+\sigma(x_1+x_2)(1-x_1-x_2) = 0$$

To show that $Lg = 0$, we begin by observing that

$$\sigma x_1(1-x_1-x_2)g_1 + \sigma x_2(1-x_1-x_2)g_2 = 0$$

The second order terms in Lg are

$$\frac{1}{2}x_1(1-x_1)\frac{-2x_2}{(x_1+x_2)^3} - x_1 x_2 \frac{x_1-x_2}{(x_1+x_2)^3} + \frac{1}{2}x_2(1-x_2)\frac{2x_1}{(x_1+x_2)^3}$$
$$= x_1 x_2 \frac{-(1-x_1) - (x_1-x_2) + (1-x_2)}{(x_1+x_2)^3} = 0$$

Since $f_1 = f_2$, the last term we have to evaluate is

$$x_1(1-x_1)g_1 - x_1 x_2(g_2+g_1) + x_2(1-x_2)g_2$$
$$= \frac{x_1(1-x_1)x_2 - x_1 x_2(x_2-x_1) - x_2(1-x_2)x_1}{(x_1+x_2)^2} = 0$$

which completes the proof of (8.12).   □

Case 2. Suppose B arises on the A background, so $\sigma_1 = 2\sigma$, $\sigma_2 = \sigma$, $\sigma_3 = 0$. It is easy to determine what $h_A(x) = P(\text{ 1's fix })$ does on the boundary of the triangle but there does not seem to be a formula for the values on the interior. Here, we will combine the analytical result from Case 1 with an observation based on simulations to get a very accurate approximate formula.

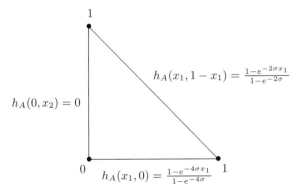

Fig. 8.3. Boundary conditions for h_A

Theorem 8.16. *Let* $g(x_2) = x_2 h_A(1/2N, x_2) + (1 - x_2) h_a(1/2N, x_2)$ *be the probability B becomes fixed when the mutation occurs on a randomly chosen chromosome*

$$g(x_2) \approx \frac{1 - e^{-2\sigma x_2}}{2\sigma x_2} (x_2[\eta_0(1 - x_2) + \eta_1 x_2] + (1 - x_2)) \qquad (8.14)$$

$$\text{where} \quad \eta_0 = \frac{2(1 - e^{-2\sigma})}{1 - e^{-4\sigma}} \quad \text{and} \quad \eta_1 = \frac{2\sigma}{1 - e^{-2\sigma}}$$

Proof. The starting point is to note that

$$h_A(1/2N, 0) \approx \frac{4\sigma/2N}{1 - e^{-4\sigma}} \qquad h_A(1/2N, 1 - 1/2N) \approx \frac{2\sigma/2N}{1 - e^{-2\sigma}}$$

From the solution in (8.12) for Case 1, we see that if $x_1 + x_2$ is small

$$h_a(x) \approx \frac{2\sigma x_1}{1 - e^{-2\sigma}}$$

while if $x_1 + x_2 = 1$, $h_a(x) = x_1$. From the last two results

$$h_a(1/2N, 0) \approx \frac{2\sigma/2N}{1 - e^{-2\sigma}} \qquad h_a(1/2N, 1 - 1/2N) = 1/2N$$

Thus, if we let $H_A(x_2) = h_A(1/2N, x_2)$ and $H_a(x_2) = h_a(1/2N, x_2)$

$$\eta_0 \equiv \lim_{x_2 \to 0} \frac{H_A(x_2)}{H_a(x_2)} = \frac{2(1 - e^{-2\sigma})}{1 - e^{-4\sigma}}$$

$$\eta_1 \equiv \lim_{x_2 \to 1} \frac{H_A(x_2)}{H_a(x_2)} = \frac{2\sigma}{1 - e^{-2\sigma}}$$

A remarkable feature of the graphs in Figure 2 in McVean and Charlesworth (2000) is that $H_A(x_2)/H_a(x_2)$ is almost linear. Combining this with our exact formula for $H_a(x_2)$ we can get an approximate formula for $H_A(x_2)$ and for the ratio of the success probability of a B mutant inserted on a random chromosome to its success in the absence of the A mutation:

$$\frac{x_2 H_A(x_2) + (1-x_2)H_a(x_2)}{H_a(0)} = \frac{H_a(x_2)}{H_a(0)}\left(x_2\frac{H_A(x_2)}{H_a(x_2)} + (1-x_2)\right)$$

The approximate linearity implies $H_A(x_2)/H_a(x_2) \approx \eta_0(1-x_2) + \eta_1 x_2$. For the first term we note

$$\frac{H_a(x_2)}{H_a(0)} = \frac{h_a(1/2N, x_2)}{h_a(1/2N, 0)} = \frac{1 - e^{-2\sigma(x_2+1/N)}}{1 - e^{-2\sigma/N}} \cdot \frac{1/N}{1/N + x}$$

Since $(1/N)/(1 - e^{-2\sigma/N}) \to 2\sigma$ as $N \to \infty$, a little algebra gives the approximation in (8.14).   □

The next figure shows the approximation when $Ns = 1, 2, 4$. The result is very close to the simulated data given in Figure 1 of McVean and Charlesworth (2000).

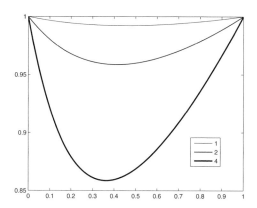

Fig. 8.4. Plot of the approximation (8.14) for $Ns = 1, 2, 4$.

Results for strong selection

Barton (1995) considered the situation in which the rescaled selection coefficients $2Ns_A$ and $2Ns_B$ are large. In this case, we can suppose that the frequency of the A locus is given by the solution to the logistic differential equation:

$$u(t) = \frac{1}{1 + \exp(-s_A t)}$$

Here time has been shifted so that $u(0) = 1/2$. The selective advantage of the B allele depends on whether it arises linked to the A allele or a allele:

$$s_A(t) = s_B + [1 - u(t)]s_A$$
$$s_a(t) = s_B - u(t)s_A$$

Theorem 8.17. *Let $p_A(t)$ and $p_a(t)$ be the probability that a B allele arising at time t fixes depending on the background on which it arises.*

$$-\frac{dp_A}{dt} = r(1 - u(t))(p_a(t) - p_A(t)) + s_A(t)p_A(t) - \frac{p_A(t)^2}{2}$$
$$-\frac{dp_a}{dt} = ru(t))(p_A(t) - p_a(t)) + s_a(t)p_a(t) - \frac{p_a(t)^2}{2} \qquad (8.15)$$

Proof. To get the same result as Barton has for Wright-Fisher dynamics, we use a Moran model with $2N$ chromosomes, time run at rate N, and selection coefficients multiplied by 2. If h is small, then we can ignore the probability of more than one jump, and we will have

$$p_A(t) = rh(1 - u(t))(p_a(t + h) - p_A(t + h))$$
$$+ \frac{h}{2}[2p_A(t + h) - p_A(t + h)^2] + [1 - (1 - s_A(t))h]p_A(t, h) + o(h)$$

The first term comes from the B allele changing background due to recombination. To explain the second and third, we note that when there is one B allele, up jumps happen at rate $1/2$ and down jumps at rate $(1 - 2s_A(t))/2$ for a total rate $1 - s_A(t)$. A down jump means that fixation occurs with probability 0. After an up jump there are two copies and at least one will start a successful branching process with probability $2p_A(t + h) - p_A(t + h)^2$ since

$$P(E_1 \cup E_2) = P(E_1) + P(E_2) - P(E_1 \cap E_2) = 2P(E_i) - P(E_i)^2$$

Rearranging, we have

$$\frac{p_A(t) - p_A(t + h)}{h} = rh(1 - u(t))(p_a(t + h) - p_A(t + h))$$
$$+ s_A(t)p_A(t + h) - \frac{p_A(t + h)^2}{2}$$

Letting $h \to 0$, we have the first equation. The second is similar.   □

To solve (8.15), it is convenient to change variables $q_c(t) = p_c(t/s_A)/s_A$ for $c = A, a$ to get

$$-\frac{dq_A}{dt} = \frac{r}{s_A}(1 - u(t))(q_a(t) - q_A(t)) + \frac{s_A(t)}{s_A}q_A(t) - \frac{q_A(t)^2}{2}$$
$$-\frac{dq_a}{dt} = \frac{r}{s_A}u(t))(q_A(t) - q_a(t)) + \frac{s_a(t)}{s_A}q_a(t) - \frac{q_a(t)^2}{2}$$

On this time scale $u(t) = 1/(1 + e^{-t})$ and there are only two parameters r/s_A and s_B/s_A.

To solve the equation, we need a boundary condition, which we can theoretically take as $q_A(+\infty) = 2s_B/s_A$. That is, after the sweep is complete, the B allele arises on a homogeneous background and has fixation probability $2s_B$. To implement this numerically we will follow Barton and set $q_A(+10) = 2s_B/s_A$. The next figure shows results for $v(t)/2s_B$ where $v(t) = u(t)p_A(t) + (1 - u(t))p_a(t)$ is the probability of fixation of a mutant that occurs on a randomly chosen individual.

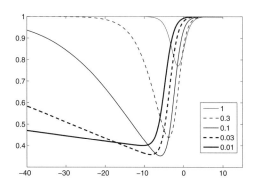

Fig. 8.5. Barton's approximation for the scaled fixation probability $v(t)/2s_B$ plotted versus the time of the mutation, for the indicated values of s_B/s_A.

The curve for $s_B/s_A = 0.03$ rises only slowly as we move to the left because on this time scale the time for the sweep to be complete is $O(s_A/s_B)$, so the value will only return to $2s_B$ if the sweep is complete before the A allele has a significant frequency.

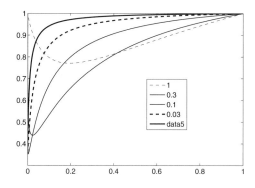

Fig. 8.6. Quantities from the previous figure plotted versus the frequency of A's.

To compare with our previous result, we will plot $(u(t), v(t))$ in Figure 8.6 to show the reduction in the fixation probability as a function of the initial frequency of A. As s_B/s_A decreases the region in which there is a change shifts to the left, because otherwise B will not rise to a significant frequency before the A sweep is over.

8.4 Gene duplication

Studies have shown that a surprisingly large number of duplicated genes are present in all sequenced genomes, revealing that there is frequent evolutionary conservation of genes that arise through local events that generate tandem duplications, larger-scale events that duplicate chromosomal regions or entire chromosomes, or genome-wide events that result in complete genome duplication (polyploidization). Analyses of the human genome by Li et al. (2001) have revealed that at least 15% of human genes are indeed duplicates, with segmental duplications covering 5.2% of the genome, see Bailey et al. (2002).

Gene duplications have been traditionally considered to be a major evolutionary source of new protein functions. The conventional view, pioneered by Ohno (1970), holds that gene duplication produces two functionally redundant copies, called *paralogous genes*, and thereby frees one of them from selective constraints. To quote Kimura and Ohta (1974) "The crucial point pertinent here is that the existence of two copies of the same gene allows one of the copies to accumulate mutations and to eventually emerge as a new gene, while another copy retains the old function required by the species for survival through the transitional period."

While this is an appealing idea, it is not supported by data. Hughes and Hughes (1993) studied 17 pairs of duplicated genes in the tetraploid frog *Xenopus laevis* and showed that both copies were subject to purifying selection. Lynch and Conrey (2000) examined duplicated genes in humans, mice, Drosophila, yeast, *C. elegans*, and Arabidopsis and concluded that the vast majority of gene pairs with a synonymous site divergence S of more than 10% exhibit a ratio of replacement to silent substitutions $R/S \ll 1$.

More recently Kondrashov et al. (2002) examined data from 26 bacterial, six archael, and seven eukaryotic genomes. They found that the ratio of non-synonymous substitutions (K_a/K_s) in most paralogous pairs is $\ll 1$ and the paralogs typically evolve at similar rates, without significant asymmetry, indicating that both paralogs produced by a duplication are subject to purifying selection.

Perhaps the most puzzling aspect of gene duplication is that duplicate copies are preserved. Virtually all models predict that the usual fate of a duplicate gene pair is the nonfunctionalization of one copy. The expected time before a gene is silenced is thought to be relatively short, on the order of the null mutation rate per locus, typically a few million years. However the rate of retention of duplicated gene preservation following ancient polyploidization

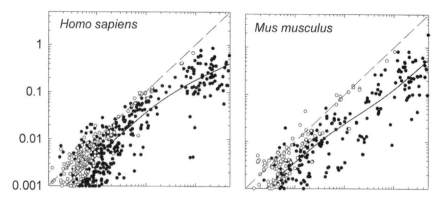

Fig. 8.7. Ratios of replacement to synonymous substitutions plotted versus age of duplicate in humans and mice.

events, e.g. in tetraploid fishes, is often suggested to be in the neighborhood of 30% over tens to hundreds of millions of years.

For more background, see Prince and Pickett (2002), Zhang (2003), and Taylor and Raes (2004). Fisher (1935) presented the first population genetic model of the fate of duplicate genes, and the 1970s saw an explosion of work on this topic. In this section and the next two, we will review some of the more recent mathematical models of gene duplication. Walsh (2003) gives a nice survey of the historical development of the topic.

Neofunctionalization

Walsh (1995) computed the probability that a duplicated gene would evolve a new function rather than be lost to a deleterious mutation. Let μ and $\rho\mu$ be the per-copy, per-generation mutation rates of null and advantageous alleles, respectively. Typically, we expect the ratio ρ of advantageous to null mutations to be much less than 1. Null mutations in the duplicated copy are assumed to be neutral while advantageous mutations are assumed to have additive fitnesses. If we let $S = 4N_e s$ where s is the selective advantage then by (6.3) the probability of fixation of the advantageous allele is $2s/(1 - e^{-S})$. Multiplying by the population size $2N$ and the relevant mutations rates, the rates of fixation of null and advantageous alleles are

$$a = \mu \quad \text{and} \quad b = \mu\rho\frac{S}{1 - e^{-S}}$$

The probability that a beneficial mutation fixates first is thus

$$u_B = \frac{b}{a + b} = \left(1 + \frac{a}{b}\right)^{-1} = \left(1 + \frac{1 - e^{-S}}{\rho S}\right)^{-1}$$

If S is very small $1 - e^{-S} \approx 2S$ while if S is very large $1 - e^{-S} \approx 1$, so

$$u_B \approx \begin{cases} \rho/(1+\rho) \approx \rho & S \ll 1 \\ S\rho/(1+S\rho) \approx S\rho & S \gg 1 \quad S\rho \ll 1 \\ 1 - (\rho S)^{-1} & S\rho \gg 1 \end{cases}$$

If for example $\rho = 10^{-4}$ then from the formula for u_B

S	0.0316	0.1	0.316	1	3.16	10	31.6	100
$100u_B$	0.102	0.105	0.117	0.158	0.333	1	3.15	9.9

or if you prefer a graph

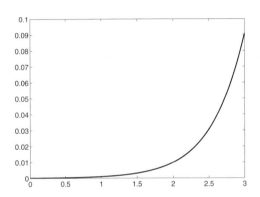

Fig. 8.8. Walsh's formula for u_B as a function of $\log_{10} S$ when $\rho = 10^{-4}$.

These probabilities are small, reaching 10% only when $S = 1000$. However, one has to remember that these are the probability of success per attempt and evolution will provide for many attempts.

8.5 Watterson's double recessive null model

In the absence of selection, how long does it take after gene duplication before one gene is lost from the population? To answer this question, Watterson considered a Wright-Fisher model with unlinked loci and diploid individuals. He assumed that the duplication had already spread and become fixed in the population, so in generation n we have $2N$ letters that are either A (working copy of gene 1) or a (nonfunctional copy) and $2N$ letters that are either B (working copy of gene 2) or b (nonfunctional copy). To build up generation $n+1$, we repeat the following procedure until we have N successes:

- Pick with replacement two copies of gene 1 and two copies of gene 2.
- An A that is picked may mutate to a with probability μ. Likewise a B that is picked may mutate to b with probability μ.

- We think of mutation as changing one of the several hundred nucleotides in the gene. Hence the reverse mutation, which corresponds to undoing a specific nucleotide substitution is assumed to have probability 0.
- We assume that all individuals with at least one working copy have fitness 1, but $aabb$ is lethal. In the model this translates into the rule: if the result of our choices after mutation is not $aabb$ then the new individual is added to the collection.

Note that after forming the new individual we do not keep track of the copies that reside in a single individual. This standard practice is referred to as the assumption of a "random union of gametes."

Letting x and y be the frequencies of a and b alleles, Kimura and King (1979) derived the diffusion limit of this model to be

$$L_1 f = \frac{1}{4N} x(1-x) \frac{\partial^2 f}{\partial x^2} + (1-x)(\mu - x^2 y^2) \frac{\partial f}{\partial x}$$
$$+ \frac{1}{4N} y(1-y) \frac{\partial^2 f}{\partial y^2} + (1-y)(\mu - x^2 y^2) \frac{\partial f}{\partial y} \quad (8.16)$$

Here we are writing the diffusion as they did with time run at rate 1. To get the usual diffusion limit time is run at rate $2N$, and the operator L_1 is multiplied by $2N$.

To explain the coefficients in (8.16), we note that if we have a one locus Wright-Fisher model in which mutation from A to a occurs at rate μ and allele a has a selective disadvantage s, then, when time is run at rate $2N$, the diffusion approximation, by (7.3), is

$$\frac{1}{2} x(1-x) \frac{\partial^2 f}{\partial x^2} + 2N[\mu(1-x) - sx(1-x)] \frac{\partial f}{\partial x}$$

In our case an A allele has fitness 1 and a allele has fitness 0 if it is paired with another a and two b's, so the selective disadvantage of an a allele is $s = xy^2$.

This diffusion limit in (8.16) is unusual since it does not assume that s and μ are of order $1/N$. Since all individuals have fitness 1 or 0, it is not sensible to make this assumption about s, but one can, as Watterson did, assume $4N\mu \to \theta$. By using arguments that were clever but not completely rigorous, Watterson concluded that the mean time until loss of A or B had mean approximately

$$N[\log(2N) - \psi(\theta/2)]$$

where ψ is the digamma function.

Sketch of proof. Watterson changes variables $\rho = N^{1/2}$ and

$$\eta = \frac{x - y}{1 - \min\{x, y\}}$$

ρ measures the distance from the curve of equilibria, while η is a function of $(1-y)/(1-x)$ and measures the position along it. Watterson's calculation is based on the heuristic that ρ evolves much faster than η, so the evolution of η can be computed by assuming that ρ is always in equilibrium. This reduces the problem to studying a one dimensional distribution and the fixation time can be computed using the methods described in Chapter 7. □

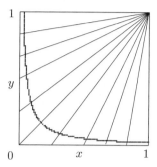

Fig. 8.9. Curve of equilibria in Watterson's model and solutions to ODE.

Here we will take a different approach to the problem following Durrett and Popovic (2007), who assume that μ is a constant. To state their results, we begin by observing that solutions of the ordinary differential equation (ODE)

$$\frac{dx}{dt} = (1 - x)(\mu - x^2 y^2)$$

$$\frac{dy}{dt} = (1 - y)(\mu - x^2 y^2) \tag{8.17}$$

have $(1 - y_t)/(1 - x_t)$ constant, so the solution moves along the line through $(1, 1)$ until it hits the equilibrium curve $xy = \sqrt{\mu}$. When time is run at rate 1, the variance in the diffusion is $O(1/N)$, so it should not be surprising that the diffusion (i) first follows the solution of the ODE to reach the set of equilibria and then (ii) wanders around near the curve $xy = \sqrt{\mu}$.

Theorem 8.18. *Let $Z_t = (X_t, Y_t)$ be the diffusion in (8.16) and Z_t^0 the solution of (8.17) starting from (X_0, Y_0). Let $0 < \epsilon < \sqrt{\mu}$. There is a constant γ, which depends on ϵ so that if N is large then for all $(X_0, Y_0) \in [\epsilon, 1 - \epsilon]^2$*

$$E\left(\sup_{0 \le t \le \gamma \log N} |Z_t - Z_t^0|^2\right) \le N^{-1/2}$$

As the reader may have noticed, this is not what the frequencies in the discrete Wright-Fisher model do on this time scale. They would move to the curve by a series of jumps. With more effort one could prove our results for the Wright-Fisher model, however, for simplicity, we will, as Watterson did, consider the diffusion process.

Since Theorem 8.18 shows that the frequencies come rapidly to the curve of equilibria, it suffices to consider what happens when the diffusion starts near the curve. To explain the assumption about the starting point in the next result, we note that the ODE converges to equilibrium exponentially fast, so at time $\gamma \log N$ the distance from the curve will be $\leq N^{-\delta}$ for some $\delta > 0$.

Given (x, y) define $\Phi(x, y) = (x^*, y^*)$ by $(1 - y^*)(1 - x^*) = (1 - y)/(1 - x)$ and $x^* y^* = \sqrt{\mu}$. In words, (x^*, y^*) is the point on the equilibrium curve that we would reach by flowing according to the ODE starting from (x, y). Our next result concerns the behavior of the process (X_t^*, Y_t^*) projected onto the curve of equilibria.

Theorem 8.19. *Consider the diffusion* (X_t, Y_t) *in (8.16). Let* $\tau = \inf\{t : X_t = 1 \text{ or } Y_t = 1\}$ *be the time to loss of A or B.*

(i) Let $0 < \delta < 1/2$*. Suppose* $|\mu - X_0^2 Y_0^2| \leq N^{-\delta}$*. Then if N is large, with high probability we have* $|\mu - X_t^2 Y_t^2| \leq N^{-\delta}$ *for all* $t \leq \tau$*.*

(ii) If the diffusion process is run at rate $2N$ *then the process* $X_t^* - Y_t^*$ *converges in distribution to a diffusion process on the interval* $[-1 + \mu, 1 - \mu]$*.*

(iii) $E_0 \tau \sim 2N c_2(\mu)$*.*

Sketch of proof. The coefficients of the limiting diffusion process can be explicitly calculated by applying the change of variables formula, Theorem 8.7 to $h(\Phi(X_t, Y_t))$ where $h(x, y) = x - y$. This is not very much fun since

$$x^* = g\left(\frac{1 - x}{1 - y}\right) \qquad \text{where} \quad g(u) = \frac{1 - u + \sqrt{(1 - u)^2 + bu\sqrt{\mu}}}{2}$$

The coefficients have very complicated formulas, but they have the important property that they are independent of N, since the function Φ is constant in the direction of the strong push of the drift. Once we show that the diffusion stays near the curve of equilibria, we can calculate the expected fixation time using the methods of Chapter 7. $\square$

Figure 8.10 shows the drift $b(z)$, the variance $a(z)$ of the limiting process and the quantity $-2b(z)/a(z)$, which appears in the derivative of its natural scale when $\mu = 10^{-4}$. From these we can compute the Green's function using (7.34). Recalling

$$E_0 \tau = 2 \int_0^{1-\mu} G(0, x)\, dx$$

and integrating numerically we see that in our concrete case $c_2(\mu) = 6.993302$, so for the rate 1 diffusion we have $E_0 \tau \approx 14N$.

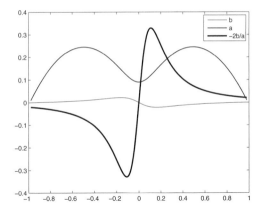

Fig. 8.10. Coefficients of the limiting diffusion.

8.6 Subfunctionalization

Subfunctionalization proposes that, after duplication, the two gene copies acquire complementary loss-of-function mutations in independent subfunctions, so that both genes are required to produce the full complement of functions of the ancestral genes. For example, a gene that is originally expressed in two tissues may, after duplication, diverge into two copies, each being expressed in one of the two tissues.

This general process has been described in detail in the *duplication-degeneration-complementation (DDC) model*. For simplicity, consider a gene with two independently mutable subfunctions, for example, because they are controlled by two different regulatory regions, which are spatially nonoverlapping with each other and with the coding region.

In the drawing, the large rectangle is the gene, typically several hundred nucleotides, while the two small rectangles are the transcription factor binding sites, typically about 10 nucleotides. It is supposed that mutations which cause loss of a regulatory site happen at rate μ_r while those which cause the gene to completely lose function happen at rate μ_c.

Consider first what happens within one haploid individual. In order to have the outcome called *subfunctionalization* in which the two genes specialize to do different functions, the first event must be a loss of a regulatory unit.

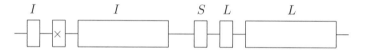

After this occurs, mutations in the indicated regions lead to inactivation of one gene copy, I, subfunctionalization, S, or are lethal L, since one of the functions is missing. It follows that the probability of subfunctionalization in one lineage is

$$P_s = \frac{4\mu_r}{4\mu_r + 2\mu_c} \cdot \frac{\mu_r}{2\mu_r + \mu_c} = 2\left(\frac{\mu_r}{2\mu_r + \mu_c}\right)^2 \qquad (8.18)$$

This is equation (2) of Lynch and Force (2000). If we make the simple assumption that $\mu_r = \mu_c$, this probability is $2/9$, but if we observe that the gene region may easily be 30 times as large as the regulatory elements and set $\mu_c = 30\mu_r$ then the probability is $1/512$.

Lynch and Force (2000) investigated the probability of subfunctionalization for diploid individuals for the cases of complete linkage which is appropriate for tandem duplicates, and for free recombination, which would occur if the duplicates are on different chromosomes. They set $\mu_c = 10^{-5}$ and considered $\mu_r/\mu_c = 3, 1, 0.3, 0.1$. The qualitative behavior for the three ratios is similar, so here we will follow Ward and Durrett (2004) and study only the case $\mu_r = \mu_c$, taking $\mu_c = 10^{-4}$ to speed up the simulations by a factor of 10. In addition to the diploid case, we will consider the haploid case, which has the same qualitative behavior but is simpler to analyze.

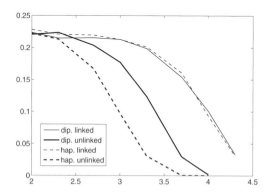

Fig. 8.11. Plot of $\log_{10} P_s$ versus n, where P_s is the subfunctionalization probability.

Lynch and Force (2000), see their Figure 3, plotted the probability of subfunctionalization, P_s, versus the logarithm of the population size, n. In Figure 8.11, following Ward and Durrett (2004), we have instead plotted the logarithm of P_s versus the population size revealing the exponential decay of the probability with increasing n.

To compute the mean of T, the time to resolution (subfunctionalization or loss of function), when $N = 1$, we note that the mean time to the first mutation is $1/(4\mu_r + 2\mu_c)$. If the first mutation is complete lost of function for one gene then the process is done. If not, an event of probability $2\mu_r/(2\mu_r + \mu_c)$, then the waiting time to the next event has mean $1/(2\mu_r + \mu_c)$. Adding the two terms we have

$$ET = \frac{1}{2(2\mu_r + \mu_c)} + \frac{2\mu_r}{2\mu_r + \mu_c} \frac{1}{2\mu_r + \mu_c} = \frac{6\mu_r + \mu_c}{2(2\mu_r + \mu_c)^2} \tag{8.19}$$

When $\mu_r = \mu_c = \mu$, this is $7/18\mu = .3888/\mu$.

Lynch and Force (2000), see their Figure 2, plotted the average time to resolution versus log population size. In Figure 8.12, following Ward and Durrett (2004), we plot time to resolution versus population size, revealing the linear increase with population size.

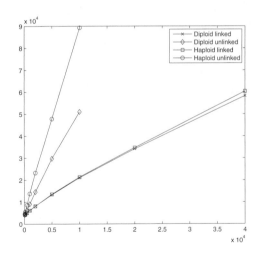

Fig. 8.12. Plot of time to resolution ET versus n.

Linked loci

A common type of gene duplication is called *tandem duplication* when a short segment of chromosome is duplicated resulting in two copies of the gene that are close to each other, and to a first approximation, the probability of a recombination between the two copies is 0. Simulations of Ward and Durrett

(2004) show that the haploid and diploid linked models have almost identical behavior, so we will consider the simpler haploid case. Using four digit binary numbers to indicate the states of the subfunctions in the two genes, there are nine possible states for viable individuals in the haploid model. To reduce the dimension we have color coded them as follows:

white	all working	1111
yellow	3 out of 4 functions	1110, 1101, 1011, 0111
green	subfunctionalization	1001, 0110
red	loss of function	1100, 0011

Simulations in Ward and Durrett (2004) show that as the population size increases the stochastic model converges to the deterministic model in which offspring are produced with the expected frequencies. Writing x_w, x_y, x_g, and x_r for the frequencies of the different colors, and using $a = \mu_c$ and $b = \mu_r$ to simplify notation the deterministic equations in discrete time are

$$x'_w = x_w(1 - 2a - 4b)/z$$
$$x'_y = [4bx_w + x_y(1 - 2a - 3b)]/z$$
$$x'_g = [bx_y + x_g(1 - 2a - 2b)]/z$$
$$x'_r = [2ax_w + (a + b)x_y + x_r(1 - a - 2b)]/z$$

where $z = x_w + (1 - a - b)x_y + (1 - 2a - 2b)x_g + (1 - a - 2b)x_0$ is a normalization that makes the x' sum to 1.

To solve these equations it is convenient to consider

$$X'_w = X_w(1 - 2a - 4b)$$
$$X'_y = 4bX_w + (1 - 2a - 3b)X_y$$
$$X'_g = bX_y + (1 - 2a - 2b)X_g$$
$$X'_r = 2aX_w + (a + b)X_y + (1 - a - 2b)X_r$$

Since the original equations are linear except for the renormalization, it follows that if $x_i(n)$ and $X_i(n)$ are the values in generation n then $x_i(n) = X_i(n)/Z(n)$ where $Z(n) = X_w(n) + X_y(n) + X_g(n) + X_r(n)$. The second set of equations is triangular, so they can be solved by starting with the first equation and working down. Noting that $X_w(0) = 1$, $X_y(0) = X_g(0) = X_r(0)$ and letting $\lambda_{i,j} = 1 - ai - bj$ we have

$$X_w(n) = \lambda_{2,4}^n$$
$$X_y(n) = 4[\lambda_{2,3}^n - \lambda_{2,4}^n]$$
$$X_g(n) = 2[\lambda_{2,2}^n - 2\lambda_{2,3}^n + \lambda_{2,4}^n]$$
$$X_r(n) = \left(2 + \frac{4b}{a + 2b}\right)\lambda_{1,2}^n - 2\lambda_{2,2}^n - 4\lambda_{2,3}^n + \frac{4(a + b)}{a + 2b}\lambda_{2,4}^n$$

$X_r(n)$ decays at the slowest rate so $Z(n) \sim X_r(n)$ and $x_r(n) \to 1$ as $n \to \infty$. Ignoring lower order terms and constants, in a population of size n we will have lost all of the first three types when

$$\frac{1}{N} \approx \frac{X_g(n)}{X_r(n)} \approx \left(1 - \frac{a}{1 - a - 2b}\right)^n$$

Approximating $1 - a - 2b$ in the denominator by 1, it follows that $n \approx (\log N)/a$.

Unlinked loci

Again simulations show that the haploid and diploid cases have similar qualitative behavior, so we will study the simpler haploid case. To introduce our model, consider first an infinitely large population for which the allele frequency dynamics will be deterministic. Let $3 = 11$, $2 = 10$, $1 = 01$, and $0 = 00$ denote the four possible states of each gene copy, where 1 and 0 indicate presence or absence of the two functions, and let x_i and y_j denote the frequencies of states i and j at the first and second copy with $x_0 = 1 - x_3 - x_2 - x_1$, and $y_0 = 1 - y_3 - y_2 - y_1$. To simplify, we will assume $\mu_r = \mu_c = b$. Let

$$w = x_3 + y_3 - x_3 y_3 + x_1 y_2 + x_2 y_1 \qquad (8.20)$$

be the mean fitness, i.e., the probability the new individual chosen to replace the old one is viable. To explain the formula for w, we note that if either gene is in state 3, an event of probability $x_3 + y_3 - x_3 y_3$, then the offspring is always viable, whereas if neither gene is in state 3, the only viable combinations are (1,2) and (2,1). We are assuming the copies are unlinked, so the events are independent.

The diffusion limit of this model is

$$L_2 f = \frac{1}{4N} \sum x_i(1 - x_i) \frac{\partial^2 f}{\partial x_i^2}$$
$$+ \sum \frac{1}{4N} y_i(1 - y_i) \frac{\partial^2 f}{\partial y_i^2} + F(x_3, x_2, x_1, y_3, y_2, y_1) \cdot \nabla f \quad (8.21)$$

where $F : \mathbf{R}^6 \to \mathbf{R}^6$ is the vector field determining the evolution of the ODE system

$$dx_3/dt = -x_3 w + x_3 - 3bx_3$$
$$dx_2/dt = -x_2 w + x_2(y_3 + y_1) + bx_3 - 2bx_2$$
$$dx_1/dt = -x_1 w + x_1(y_3 + y_2) + bx_3 - 2bx_1$$
$$dy_3/dt = -y_3 w + y_3 - 3by_3 \qquad (8.22)$$
$$dy_2/dt = -y_2 w + y_2(x_3 + x_1) + by_3 - 2by_2$$
$$dy_1/dt = -y_1 w + y_1(x_3 + x_2) + by_3 - 2by_1$$

If we let $\alpha = 1 - 3b$, then the equations for x_3 and y_3 become

$$\frac{dx_3}{dt} = x_3(\alpha - w)$$
$$\frac{dy_3}{dt} = y_3(\alpha - w) \qquad (8.23)$$

so the first and fourth equations for an equilibrium reduce to the single equation $w = \alpha$. Thus, if things are nice we will have a one dimensional curve of equilibria.

To find one set of solutions we can begin by investigating the case in which $x_2 = x_1 = x$ and $y_2 = y_1 = y$ which gives us four equations:

$$dx_3/dt = -x_3 w + x_3 - 3bx_3$$
$$dx/dt = -xw + x(y_3 + y) + bx_3 - 2bx$$
$$dy_3/dt = -y_3 w + y_3 - 3by_3$$
$$dy/dt = -yw + y(x_3 + x) + by_3 - 2by$$

which after some algebra can be explicitly solved. See Durrett and Popovic (2007) for this and other details. The next figure gives a graph of the solutions in the special case $b = 0.001$. Because of symmetry we only show the part where $x_3 \le y_3$.

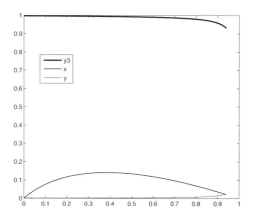

Fig. 8.13. Curve of equilibria for the haploid unliked model as a function of x_3.

Equation (8.23) implies that if $y_3(0)/x_3(0) = r$ then $y_3(t)/x_3(t) = r$ for all t. Given $z = (x_3, x_2, x_1, y_3, y_2, y_1)$, let $z^* = \Phi(z)$ be the point on the curve of equilibria with $y_3^*/x_3^* = y_3/x_3$. Numerical results show that starting from any z the ODE will converge to $\Phi(z)$, but we do not know how to prove this for the dynamical system, so we will only consider the situation when the process starts close to the curve.

Theorem 8.20. *Suppose* $b \le 0.01$. *Let* $\tau = \inf\{t : X_0(t) = 1 \text{ or } Y_0(t) = 1\}$ *be the time to loss of gene 1 or gene 2.*

(i) Suppose $|Z_0 - Z_0^*| \le 1/N^{1/4}$. *Then if N is large, with high probability we have* $|Z_t - Z_t^*| \le 2/N^{1/4}$ *for all* $t \le \tau$.

(ii) When run at rate $2N$ the process $X_3^*(t) - Y_3^*(t)$ converges in distribution to a diffusion process.

(iii) $E_0\tau \sim 2Nc_3(b)$.

The proof of this result is similar to the proof of Theorem 8.19. However, the key step of proving that the diffusion stays close to the curve is more complicated. In Watterson's model $(1 - y)/(1 - x) = r$ is a line and it is clear that the ODE will move along the line to the equilibrium. In the subfunctionalization model, $y_3/x_3 = r$ is five dimensional. Since we are only interested in the behavior near the curve, it is enough to consider the linearization of the ODE, but then we must show that all of the eigenvalues have negative real part and use this to construct a suitable Lyapunov function which will allow us to control the distance from the curve.

The coefficients of the limiting diffusion can again be computed in terms of the derivatives of $\Phi(X_3, X_2, X_1, Y_3, Y_2, Y_1)$ and $h = x_3 - y_3$. Figure 8.14 shows the drift $b(z)$, the variance $a(z)$ of the limit of $X_3^* - Y_3^*$ and the quantity $-2b(z)/a(z)$, which appears in the derivative of the natural scale, when $b = 10^{-3}$. Even though the model is quite different, the curves are similar to those for Watterson's model. The Green's function can be computed from the coefficients using (7.34). Integrating the Green's function, we see that in our concrete example $c_3(b) = 3.284906$, so for the process run at rate 1 we have $E_0\tau \approx 6.5N$. In the case of *Drosophila* who have an effective population size of $N = 10^6$ this is 6.5 million generations or 650,000 years.

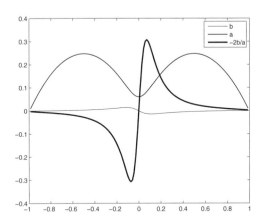

Fig. 8.14. Coefficients of the subfunctionalization diffusion.

In the last result τ is essentially the time until one of the two copies is lost, since subfunctionalization has a very small probability. To see this note that the curve of equilibria lie in the four dimensional space where $x_2 = x_1$

and $y_2 = y_1$, while subfunctionalization corresponds to $x_1 = 1$ and $y_2 = 1$ or $x_2 = 1$ and $y_1 = 1$. Since the diffusion stays close to the curve it has a very small probability of ending up at one of the subfunctionalization outcomes. Based on the theory of large deviations for random perturbations of dynamical systems we would guess that the probability of subfunctionalization decays exponentially fast with population size. However, we don't have to guess. Simulations given in Figure 8.11 show us that this is true.

9

Genome Rearrangement

"As far as the laws of mathematics refer to reality, they are not certain; and as far as they certain, they do not refer to reality." Albert Einstein

Up to this point, we have only considered the effect of small-scale processes: nucleotide substitutions, insertions, and deletions. In this chapter, we will consider a variety of large-scale processes, as well as the possibility of whole genome duplication.

9.1 Inversions

Analysis of genome rearrangements in molecular evolutions was pioneered by Dobzhansky and Sturtevant (1938), who introduced the notion of a *breakpoint* (disruption of gene order) and published a milestone paper with a rearrangement scenario for the species *D. pseuodobscura* and *D. miranda*.

Palmer and Herbon (1988) discovered that while the genes in the mitochondrial genomes of cabbage (*Brassica oleracea*) and turnip (*Brassica campestris*) are 99% identical, the gene order is quite different. If the order of the segments is 1, 2, 3, 4, 5 in turnip then the order in cabbage is 1, −5, 4, −3, 2, where the negative numbers indicate that the segment in cabbage is reversed relative to turnip. A little experimentation reveals that it is possible to turn cabbage into turnip with three inversions. Here parentheses indicate the segment that will be flipped.

$$
\begin{array}{ccccc}
1 & -5 & 4 & -3 & (2) \\
1 & -5 & (4) & -3 & -2 \\
1 & (-5 & -4 & -3 & -2) \\
1 & 2 & 3 & 4 & 5
\end{array}
$$

9.1.1 Breakpoint graph

In this simple example, it is not hard to see that three inversions are necessary, but to handle more complicated examples such as the comparison between

R. Durrett, *Probability Models for DNA Sequence Evolution*,
DOI: 10.1007/978-0-387-78168-6_9, © Springer Science+Business Media, LLC 2008

cytomegalovirus (CMV) 1, 2, 3, 4, 5, 6, 7 and Epstein-Barr virus (EBV) 1, -2, -3, -6, -5, 7, -4 studied by Hannenhalli et al. (1995), we will need to develop some theory. Our first step is to define the *breakpoint graph*. To do this, we first replace k by $2k - 1\,2k$, $-k$ by $2k\,2k - 1$, and then add 0, at the beginning and ,15 at the end to generate two strings

$$CMV \quad 0,1\,2,3\,4,5\,6,7\,8,9\,10,11\,12,13\,14,15$$
$$EBV \quad 0,1\,2,4\,3,6\,5,12\,11,10\,9,13\,14,8\,7,15$$

The vertices separated by commas in EBV are connected by "black edges," the thick lines in the picture below. Those separated by commas in CMV are connected by "gray edges," the thin lines.

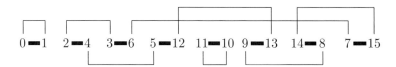

Each vertex in the breakpoint graph has degree 2 and is an endpoint of one black edge and one gray edge. Starting with a black edge and following the alternating sequence of black and gray edges, we can see that this graph has three cycles: 0-1-0, 2-4-5-12-13-9-8-14-15-7-6-3-2, 11-10-11. When EBV has been rearranged to match CMV there will be eight cycles.

$$0\,{-}\,1 \quad 2\,{-}\,3 \quad 4\,{-}\,5 \quad 6\,{-}\,7 \quad 8\,{-}\,9 \quad 10\,{-}\,11 \quad 12\,{-}\,13 \quad 14\,{-}\,15$$

A move corresponds to cutting two black edges and reversing the segment in between. Since the best thing that can happen is that one cycle is cut into two, this can at most increase the number of cycles by 1. Therefore, if we let $d(\pi)$ denote the minimum number of reversals to make the signed permutation π into the identity and let $c(\pi)$ be the number of cycles in the breakpoint graph, then we have

Theorem 9.1. *The reversal distance for a permutation π of n objects satisfies*

$$d(\pi) \geq n + 1 - c(\pi) \tag{9.1}$$

In the current example, $n = 7$ and $c(\pi) = 3$, so this says that $d(\pi) \geq 5$. Here, and in what follows, we use inversion for what happens to chromosomes and reversal for the corresponding change in permutations.

Another consequence of (9.1) is that if at each stage we can increase the number of cycles by 1, then we can find a path of length 5. A reversal is called *proper* if it increases the number of cycles by 1. A little thought shows this holds if and only if when the two black edges are eliminated, then the left end of one black edge is connected to the left edge of the other black edge by the edges that remain. An example of a proper reversal above is the one involving the black edges 2-4 and 3-6. The result is

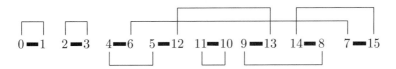

Using black edges 4-6 and 5-12 now leads to

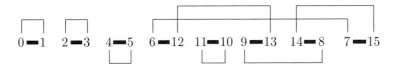

At this point, if we tried the black edges 9-13 and 14-8, then the number of cycles would not change. This happens whenever the left end of one black edge is connected to the right end of the other. 6-12 and 7-15 is a proper reversal and leads to

Using 8-14 and 12-15 now leads to

A fifth and final move using 8-12 and 9-13 puts all of the numbers in order. The construction shows $d(\pi) \leq 5$ so with (9.1) we can conclude that the distance to the identity is indeed equal to 5.

The method above has been applied by Bafna and Pevzner (1995, 1996) to a number of examples in the biology literature. In a few cases, the distance they found was smaller than the most parsimonious path biologists found by hand. The latter is not surprising. When the number of segments is 12, the number of permutations is $12! = 479{,}001{,}600$ and the number of possible reversals at each stage is $\binom{13}{2} = 78$.

Example 9.1. Human versus mouse X chromosome. All of the examples considered above have been mitochondrial, chloroplast, or viral genomes, where there is a single chromosome. As we will see in Section 9.4, the situation is much more complicated for multichromosome genomes, where nonhomologous chromosomes exchange genetic material by a process called reciprocal translocation. The sex chromosomes do not participate in this process, so the change in their gene order is due only to reversals. Taking the order of conserved blocks in mice to be 1, 2, 3, 4, 5, 6, 7, 8, the order in humans is -4, -6, 1, 7, -2, -3, 5, 8. Ignoring block 8, the breakpoint graph is as follows

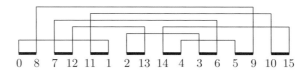

$$0 \quad 8 \quad 7 \quad 12 \quad 11 \quad 1 \quad 2 \quad 13 \quad 14 \quad 4 \quad 3 \quad 6 \quad 5 \quad 9 \quad 10 \quad 15$$

There are two cycles here: 0-8-9-5-4-14-15-10-11-1-0 and 7-12-13-2-3-6-7, so $d(\pi) \geq 8 - 2 = 6$. The following sequence shows that six moves suffice. Here we have reinstated the final segment 8, and for reasons that will become clear in a moment have introduced a 0 at the beginning. Parentheses mark the segments that are reversed in going to the next line.

$$
\begin{array}{l}
0\ (-4\ -6\ \ 1\ \ \ \ 7\ \ -2\ -3)\ 5\ 8 \\
0\ (3\ \ \ 2\ \ -7\ -1)\ \ 6\ \ \ \ 4\ \ \ 5\ 8 \\
0\ \ 1\ \ (7\ -2)\ \ 3\ \ \ \ 6\ \ \ \ 4\ \ \ 5\ 8 \\
0\ \ 1\ \ \ 2\ \ -7\ (3\ \ \ 6)\ \ \ 4\ \ \ 5\ 8 \\
0\ \ 1\ \ \ 2\ (-7\ -6\ \ 3\ \ \ 4\ \ 5)\ 8 \\
0\ \ 1\ \ \ 2\ (-5\ -4\ -3)\ \ 6\ \ \ 7\ 8 \\
0\ \ 1\ \ \ 2\ \ \ 3\ \ \ \ 4\ \ \ 5\ \ \ 6\ \ \ 7\ 8
\end{array}
$$

A remarkable aspect of the reversal distance problem is the dramatic lack of uniqueness of the minimal path. Bafna and Pevzner (1995) have computed that there are 1872 scenarios that transform the human into the mouse X chromosome with six reversals.

9.1.2 Hurdles

The point of this subsection is (i) to show that lower bound in (9.1) is not always right answer, and then (ii) to describe how the distance is computed. The graph theoretic complications are not needed in later sections and it will turn out that the simple lower bound in (9.1) is the right answer is most biological examples, so the reader should not feel compelled to go through all of the details.

The next example, which is the breakpoint graph for the permutation 3, 2, 1, shows that the lower bound is not sharp even when $n = 3$.

($\star$)

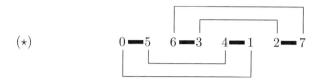

There are two cycles: 0-5-4-1-0 and 6-3-2-7-6. However, since in each case the left end of one black edge is connected to the right end of the other, there is no proper reversal. The reversal involving 0-5 and 4-1 does not decrease the number of cycles. However, after this is done, the situation becomes

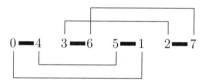

At this point, the reversal involving 3-6 and 2-7 has become proper. Performing it leads to

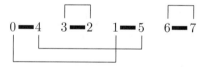

Now the reversal involving 0-4 and 1-5 is proper. Doing it puts the numbers in order. Since there are three segments in this example and two cycles in the breakpoint graph, the bound that (2.1) gives is $d(\pi) \geq 4 - 2$. However, on our first move we could not increase the number of cycles, so we ended up with $d(\pi) = 3$.

To explain the obstruction we encountered here, requires some terminology. Let π be the current permutation. At the beginning of the example above, i.e., in ($\star$), we have

$$\pi_0 = 0, \ \pi_1 = 5, \ \pi_2 = 6, \ \pi_3 = 3, \ \pi_4 = 4, \ \pi_5 = 1, \ \pi_6 = 2, \ \pi_7 = 7$$

We say that a reversal $\rho(i, j)$ acts on black edges (π_{i-1}, π_i) and (π_{j-1}, π_j). Here i and j will always be odd numbers. A gray edge is said to be *oriented* if the reversal acting on the two black edges incident to it is proper, and *unoriented* otherwise. In ($\star$) all gray edges are unoriented. A cycle is said to be *oriented* if it contains an oriented gray edge. Otherwise it is *unoriented*. In ($\star$) both cycles are unoriented.

Gray edges (π_i, π_j) and (π_k, π_ℓ) are interleaving if the intervals $[i, j]$ and $[k, \ell]$ overlap but neither contains the other. In ($\star$), (0,1) and (6,7) are interleaving but (0,1) and (5,4) are not. Two cycles, C_1 and C_2, are said to be *interleaving* if there are gray edges $g_i \in C_i$ so that g_1 and g_2 are interleaving. In ($\star$), the two cycles are interleaving. Let C_π be the set of cycles in the breakpoint graph of a permutation π. Define the interleaving graph H_π to have vertices C_π and edges connecting any two interleaving cycles. In ($\star$) the graph consists of two vertices connected by one edge.

A connected component of H_π is said to be *oriented* if it contains an oriented cycle and is unoriented otherwise. In ($\star$), there is only one unoriented cycle in the interleaving graph, so it is a *hurdle*. Hannenhalli and Pevzner (1995a) have shown that if $h(\pi)$ is the number of hurdles, then

$$d(\pi) \geq n + 1 - c(\pi) + h(\pi) \tag{9.2}$$

In ($\star$), $n = 3$, $c(\pi) = 2$ and $h(\pi) = 1$, so this new lower bound $4 - 2 + 1 = 3$ gives the right answer.

To see the reason for distinguishing oriented from unoriented cycles we need a more complicated example. Consider the permutation 3, -5, 8, -6, 4, -7, 9, 2, 1, which has the following breakpoint graph.

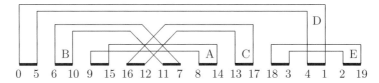

This time there are five cycles, as indicated in the figure. A is oriented but the other four are not. Cycle A is interleaving with B and C but not with D and E. Cycles B and C, and D and E, are interleaving, but the other pairs we have not mentioned are not.

The interleaving graph consists of two components: a triangle ABC that is oriented and an edge DE that is not. Again, there is only one unoriented component, so the number of hurdles is 1. Since $n = 9$ and $c(\pi) = 5$, the lower bound from (9.2) is $10 - 5 + 1 = 6$. To see that this can be achieved, note that if we start with the reversal that cuts the black edges 9-15 and 8-14, then we increase the number of cycles by 1 and the result is

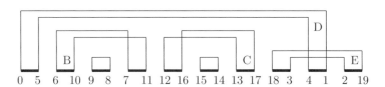

The cycles B and C are now oriented. Reversals using 6-10 and 7-11 then 12-16 and 13-17 put the numbers 6 to 17 in order. The pair DE is equivalent to $(\star)$ so it can be sorted in three moves for a total distance of 6.

To understand the final complexities in the general definition of hurdles we need another more complicated example. Consider 5, 7, 6, 8, 1, 3, 2, 4, for which the breakpoint graph is

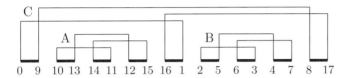

This time there are three cycles, as indicated in the figure. All three are unoriented and no two of them are interleaving, so the interleaving graph H_π consists of three connected components that are isolated points, all of which are unoriented. For a connected component U of H_π, define the leftmost and rightmost positions in U by

$$U_{min} = \min_{\pi_i \in C \in U} i \quad \text{and} \quad U_{max} = \max_{\pi_i \in C \in U} i$$

Here $A_{min} = 2$, $A_{max} = 7$, $B_{min} = 10$, $B_{max} = 15$, $C_{min} = 0$, and $C_{max} = 17$. Consider the set of unoriented components, and define the *containment partial order* $U \prec V$ if $[U_{min}, U_{max}] \subset [V_{min}, V_{max}]$. An unoriented component that is minimal with respect to this order is a *minimal hurdle*.

There are also potential hurdles at the top of the partial order. We say that a component U separates V and W if there is a gray edge (π_i, π_j) in U so that $[V_{min}, V_{max}] \subset [i, j]$ but $[W_{min}, W_{max}] \not\subset [i, j]$. In our most recent example C separates A and B by virtue of the edge $(0,1)$. If U is a maximal oriented component with respect to $\prec$ and it does not separate any two hurdles, then U is called a *maximal hurdle*. This example has two minimal hurdles and no maximal hurdle, so $h(\pi) = 2$. Since $n = 8$ and $c(\pi) = 3$, the lower bound that results from (2.2) is $9 - 3 + 2 = 8$.

To see the reason for the no separation condition in the definition of maximal hurdle, we apply the reversal that involves black edges 12-15 and 2-5:

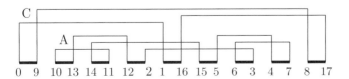

The gray edges (0,1) and (16,17) in cycle C are now oriented. The cycles A and B have merged into 1, but the gray edges (14,15) and (2,3) are now oriented. The number of cycles is now 2 but both hurdles have been destroyed so the lower bound from (9.2) is now $9 - 2 + 0 = 7$. With both cycles oriented we can compute as before to find seven moves that bring this permutation to the identity.

The bound in (9.2) is very close to the right answer. Hannenhalli and Pevzner (1995a) showed that

Theorem 9.2. *The reversal distance for a permutation π of n objects is given by*

$$d(\pi) = n + 1 - c(\pi) + h(\pi) + f(\pi) \tag{9.3}$$

where $c(\pi)$ is the number of cycles in the breakpoint graph, $h(\pi)$ is the number of hurdles, and $f(\pi) = 1$ if the permutation is a fortress and 0 otherwise.

The details of the definition of a fortress are somewhat complicated, so we refer the reader to Chapter 10 of Pevzner (2000) for it and the proofs of (9.2) and (9.3).

Computational complexity

All of the steps in the proof of (9.3) are constructive, so it leads to an algorithm for computing the reversal distance in $O(n^4)$ time, where n is the number of segments. The n^4 makes this method somewhat painful for large n. However, it was a remarkable achievement to find a polynomial time algorithm. Subsequent work has produced more efficient algorithms. Kaplan, Shamir, and Tarjan (2000) have an algorithm that runs in time $O(n^2)$. However if one wants to only know the distance without computing a sequence of reversals that achieves it, Bader, Moret, and Yan (2000) can do this in time $O(n)$. This results are restricted to the case of signed permutations. Caprara (1997, 1999a) has shown that the corresponding problem for unoriented permutations is NP-hard. Much earlier, Even and Goldreich (1981) had shown that the general problem of computing the distance to the identity for a given set of generators of the permutation group is NP-hard.

While it is nice to have an algorithm that computes the exact distance, all of the complications that come from considering hurdles are not really necessary. Caprara (1999b) has shown that for a randomly chosen permutation of n markers, the probability that the lower bound (9.1) is not the right answer is of order $1/n^5$. For this reason, we will content ourselves with computing the graph distance, which is very easy to compute. One can count the cycles in the breakpoint graph in time $O(n)$.

9.2 When is parsimony reliable?

Bourque and Pevzner (2002) approached this question by taking 100 markers in order, performing k randomly chosen reversals to get a permutation π_k, computing the minimum number of reversals needed to return to the identity, $d(\pi_k)$, and then plotting the average value of $d(\pi_k) - k \leq 0$ for 100 simulations. They concluded, based on their simulations, that the parsimony distance for n markers was a good estimate as long as the number of reversals performed was at most $0.4n$. In Figure 9.1, we have repeated their experiment for the graph distance $d_0(\pi) = n + 1 - c(\pi)$ and plotted with squares the average value of $k - d_0(\pi_k) \geq 0$ for 10,000 replications. Our curve is less random, but close to data of Bourque and Pevzner (2002) for $k - d(\pi)$, indicated in the figure by crosses.

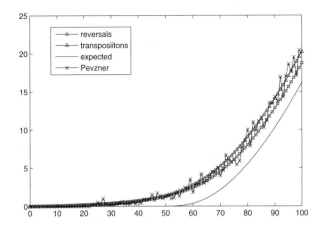

Fig. 9.1. Bourque and Pevzner's data simulation compared with other computer experiments on reversals and transpositions, and Theorem 9.5.

The biological question concerns the random reversals. However, it is also an interesting mathematical problem to consider the analogous question for random transpositions. In that case the distance from the identity can be easily computed: it is the number of markers n minus the number of cycles in the permutation. For an example, consider the following permutation of 14 objects written in its cyclic decomposition:

$$(1\,7\,4)\,(2)\,(3\,12)\,(5\,13\,9\,11\,6)\,(8\,10\,14)$$

which indicates that $1 \to 7$, $7 \to 4$, $4 \to 1$, $2 \to 2$, $3 \to 12$, $12 \to 3$, etc. There are five cycles so the distance from the identity is 9. If we perform a transposition that includes markers from two different cycles (e.g., 7 and 9) the two cycles coalesce into one, while if we pick two in the same cycle (e.g., 13 and 11) it fragments into two.

The situation is similar but slightly more complicated for reversals. There a reversal that involves edges in two different components of the breakpoint graph merges them into one, but a reversal that involves two edges of the same cycle may or may not increase the number of cycles. One can attempt to couple the components of the breakpoint graph for random reversals on $n-1$ markers and the cycles of random transposition of n markers as follows: number the edges between markers in the reversal chain (including the ends 0 and n); when markers i and j are transposed, do the inversion of edges numbered i and j. The result of the coupled simulation is given Figure 9.1. As expected, time minus distance is smaller for reversals. However, the qualitative behavior is similar. Thus, we will begin by considering the biologically less relevant case of

Random transpositions

Let $(\sigma_t, t \geq 0)$ be the continuous-time random walk on the group of permutations, starting at the identity, in which, at times of a rate 1 Poisson process, we perform a transposition of two elements chosen uniformly at random, with replacement, from $\{1, \ldots, n\}$. Choosing with replacement causes the chain to do nothing with probability $1/n$, but makes some of the calculations a little nicer.

Define the distance to the identity, D_t, to be the minimum number of transpositions one needs to perform on σ_t to go back to the identity element. It is clear that if N_t is the number of transpositions distinct from the identity performed up to time t, a Poisson random variable with mean $t(1-1/n)$, then $D_t \leq N_t$. As mentioned earlier D_t is given by $D_t = n - |\sigma_t|$, where $|\sigma_t|$ is the number of cycles in the cycle decomposition of σ_t. This formula allows us to turn any question about D_t into a question about $|\sigma_t|$.

To study the evolution of the cycles in the random permutation, we construct a random graph process G_t^*. Start with the initial graph on vertices $\{1, \ldots, n\}$ with no edge between the vertices. When a transposition of i and j occurs in the random walk, we draw an edge between the vertices i and j, even if one is already present. Elementary properties of the Poisson process imply that if we collapse multiple edges in G_t^* into one, then the resulting graph G_t is a realization of the Erdös-Renyi random graph $G(n, p)$, in which edges are independently present with probability $p = 1 - \exp(-2t/n^2)$. The probability of picking an edge twice is $\leq (2t/n^2)^2 = O(1/n^2)$ when $t = cn/2$, so the expected number of multiple edges is $O(1)$. Multiple edges are a nuisance, but not a real problem. We have to be careful in Theorem 9.3 where the random variable of interest is also $O(1)$, but in the other cases the quantities of interest $\to \infty$, so multiple edges can be ignored.

It is easy to see that in order for two integers to be in the same cycle in the permutation, it is necessary that they are in the same component of the random graph. To estimate the difference between cycles and components, let F_t denote the event that a fragmentation occurs at time t and note that

$$D_t = N_t - 2 \sum_{s \leq t} 1_{F_s} \qquad (9.4)$$

A fragmentation occurs in the random permutation when a transposition occurs between two integers in the same cycle, so tree components in the random graph G_t^* correspond to unfragmented cycles in the random transposition random walk.

9.2.1 Phase transition

We are now ready to state our results of Berestycki and Durrett (2006) which explain Bourque and Pevzner's (2002) simlulation results. As the reader will see, it is convenient to write $t = cn/2$. There are three separate regimes based on the value of c.

Theorem 9.3. *Let $0 < c < 1$. The number of fragmentations*

$$Z_c := \sum_{s \leq cn/2} 1_{F_s} \Rightarrow \text{Poisson}(\kappa(c)) \tag{9.5}$$

where $\kappa(c) = (-\log(1-c) - c)/2$.

At $c = 0.8$, which corresponds to Bourque and Pevzner's $0.4n$ this says that the average number of fragmentations is $(\log(5)-0.8)/2 = 0.4047$. Translating back to discrete time $0.4n - E(D_{0.4n}) \to 0.4047$ as $n \to \infty$.
    In the critical case, $c = 1$, the discrepancy grows slowly with n.

Theorem 9.4. *Let χ have a standard normal distribution. As $n \to \infty$,*

$$\left(\frac{6}{\log n}\right)^{1/2} \left(\sum_{s \leq n/2} 1_{F_s} - \frac{1}{6}\log n\right) \Rightarrow \chi. \tag{9.6}$$

The supercritical regime, $c > 1$, is the most interesting case.

Theorem 9.5. *Let $c > 0$ be a fixed positive number. Then the number of cycles in the random permutation at time $cn/2$, $|D_{cn/2}|/n \to u(c)n$, where*

$$u(c) = 1 - \sum_{k=1}^{\infty} \frac{1}{c} \frac{k^{k-2}}{k!}(ce^{-c})^k \tag{9.7}$$

In addition, there is a central limit theorem

$$\frac{D_{cn/2} - u(c)}{\sqrt{n}} \Rightarrow \chi$$

Note that the last theorem is valid for all regimes. The large fluctuations here for $c < 1$ are due to the fact that the number of inversions N_t is a Poisson process and hence has $(N_t - t)/\sqrt{t} \Rightarrow \chi$. Although it is not obvious from formula (9.7), $u(c) = c/2$ for $c < 1$ and $u(c) < c/2$ when $c > 1$. Using

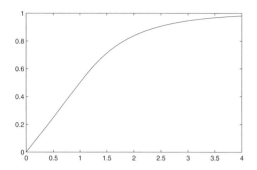

Fig. 9.2. Graph of the limit $u(c)$ in Theorem 9.5.

Stirling's formula, $k! \sim k^k e^{-k} \sqrt{2\pi k}$, it is easy to check that g' exists for all c and is continuous, but $g''(1)$ does not exist.

Why is this true? We begin with Cayley's result that there are k^{k-2} trees with k labeled vertices. For a proof of this fact and more on random graphs, see Bollobás (2001). At time $cn/2$ each edge is present with probability $1 - \exp(-c/n) \sim c/n$ so the expected number of trees of size k present is

$$
\sim \binom{n}{k} k^{k-2} \left(\frac{c}{n}\right)^{k-1} \left(1 - \frac{c}{n}\right)^{k(n-k)+\binom{k}{2}-k+1} \tag{9.8}
$$

since each of the $k-1$ edges needs to be present and there can be no edges connecting the k point set to its complement or any other edges connecting the k points other than the $k-1$ that make up the tree. For fixed k, (9.8) is asymptotic to

$$
n \frac{k^{k-2}}{k!} c^{k-1} \left(1 - \frac{c}{n}\right)^{kn}
$$

The quantity in parentheses at the end converges to e^{-ck} so we see that the sum in (9.7) gives the asymptotics for the number of tree components at time $cn/2$. The number of nontree components is $o(n)$, so recalling the formula for the distance we can guess that $|D_{cn/2}|/n \to u(c)n$. The central limit theorem can be proved using Pittel's (1990) central limit theorem for the number of components of the random graph. In the both cases, the main difficulty is showing that the result of the coagulation-fragmentation process of cycles in the random permutation is close to the random graph. □

Results for reversals

The law of large numbers for the distance in Theorem 9.5 extends easily to the reversal chain. Recall that the main difference lies in the fact that a reversal involving edges from different components in the breakpoint graph always yields a coagulation, but one involving two edges in the same component may or may not cause a fragmentation. The proof of Theorem 9.5 for transpositions is based on showing that fragmentations can be ignored, so this difference is unimportant and these results extend to reversals.

This is not true for the more precise results in Theorems 9.3 and 9.4. For example, the underlying data shows that up to $c = 1$, an average of 23% of the reversals have caused no change in the distance. For our purposes this is not important. In the subcritical and critical regimes, $t - D_t \geq 0$ is small for transpositions, and is even smaller for reversals.

9.2.2 Bayesian approach

As the previous results show, the actual number of inversions that have occurred in the history of two chromosomes may not be equal to the minimum number. York, Durrett, and Nielsen (2002) have developed a Bayesian approach. Assuming that the same N markers are present on both chromosomes, the arrangements are given as signed permutations, and that rearrangements are only due to inversions, they model the process of chromosome rearrangement as follows:

- The occurrence of inversions is a Poisson process with unknown mean λ; the probability of exactly L inversions is $e^{-\lambda}\lambda^L/L!$.
- A uniform prior distribution $u(\lambda)$ for $0 \leq \lambda \leq \lambda_{\max}$ is assumed.
- Each of the $N(N+1)/2$ inversions occurs with equal probability.

There are $(N(N+1)/2)^{L(X)}$ inversion sequences X of length $L(X)$, each of which has probability

$$P(X|\lambda) = e^{-\lambda}\frac{\lambda^{L(X)}}{L(X)!} \cdot \left(\frac{N(N+1)}{2}\right)^{L(X)}$$

Let Ω be the set of inversion sequences of any length, and let D be the data consisting of the marker order on the two chromosomes. In order to approximately calculate $P(X|D)$ and $P(\lambda|D)$, they define a Markov chain on $\Omega \times [0,\infty)$ with stationary distribution $P(X, \lambda|D)$.

They use a Metropolis-Hastings scheme in which λ and X are alternately updated. A Gibbs step is used to update λ, i.e., $P(\lambda|X, D) = ce^{-\lambda}\lambda^{L(X)}u(\lambda)$. The new path X' is generated as follows:

- Choose $\ell \leq L(X)$ with probability $q(\ell)$ and then a starting point uniformly at random from $0 \leq j \leq L - \ell$. In practice they use

$$q(\ell) = c\left[1 - \tanh\left(\xi\left(\frac{\ell}{\alpha N} - 1\right)\right)\right]$$

where e.g., $\alpha = 0.65$ and $\xi = 8$. This formula is used so that all lengths small with respect to N have roughly equal probability.
- Generate a new path to replace the segment from π_j to $\pi_{j+\ell}$ as follows: Use the breakpoint graph to choose inversions at random but so that each step decreases the distance to the target by 1 with high probability.

One must run the chain long enough to reach equilibrium before generating a sample points. York et al. (2002) use the method of Gelman and Rubin (1992) to decide when the Markov chain has converged. This involves running $m \geq 2$ chains for the same data starting from much different initial states. Let $X_{i,j}$ be the ith element of the jth Markov chain and $L_{i,j}$ its length. Define the between and within variances by

$$B = \frac{1}{m-1} \sum_j (<L>_j - <L>)^2$$

$$W = \frac{1}{m} \sum_j \frac{1}{n-1} \sum_i (L_{i,j} - <L>_j)^2$$

where $<L>_j = (1/n) \sum_i L_{i,j}$ and $<L> = 1/(mn) \sum_{i,j} L_{i,j}$. Convergence is indicated when

$$(\hat{R})^{1/2} = \sqrt{\frac{n-1}{n} + \frac{B}{W}} \approx 1$$

Analysis of two data sets

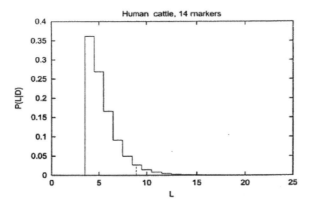

Fig. 9.3. Posterior distribution of X chromsome inversions for human-cattle comparison.

We begin by considering a data set with 14 markers that is a comparison of human and cattle X chromosomes. Assuming the human arrangement is the identity, the cattle data may be written as follows:

$$1\,2\,3\,4\,5\,7\,8 - 6\,9 - 11 - 10 - 14 - 13 - 12$$

In this case it is easy to see that the distance is 4: flip $-14 - 13 - 12$ then $-11 - 10$. At this point only -6 is out of place, which can be fixed by

$$7\,8 - 6 \rightarrow 6 - 8 - 7 \rightarrow 6\,7\,8$$

Figure 9.3 gives the posterior distribution of the number of inversions. Even in this simple case, the parsimony solution has probability ≈ 0.36.

Ranz, Casals, and Ruiz (2001) have constructed a comparative map between chromosome 2 of *D. repleta* and chromosome 3R of *D. melanogaster* with 79 markers. Numbering the markers on the *D. repleta* chromosome by their order on *D. melanogaster* we have

36	*37*	17	40	*16*	*15*	*14*	63	*10*	*9*	55	28
13	51	22	79	39	70	66	*5*	*6*	*7*	35	64
33	*32*	*60*	*61*	18	65	62	12	1	11	23	20
4	52	68	29	48	3	21	53	8	43	72	*58*
57	*56*	19	49	34	59	30	77	31	67	44	2
27	38	50	*26*	*25*	76	69	41	24	75	71	78
73	47	54	45	74	42	46					

Translocations and pericentric inversions (which contain the centromere) are rare so we can assume chromosome arms are conserved and compute the inversion distance. We do not know the signs of the markers so we let the starting permutation given above wander over the 2^{79} choices of sign by using moves that flip a single marker and accept the move with high probability if it reduces the distance.

By flipping markers as described in the previous sentence one can find a path that achieves a distance of 54, which provides an upper bound on the parsimony distance. The parsimony distance lies outside the 95% credible interval of $[71, 118]$ that comes from the Bayesian estimate. Indeed the posterior probability of 54 is so small that this value that it was never seen in the 258 million MCMC updates in the simulation run.

An alternative and simpler approach to estimating the number of inversions introduced by Durrett (2002) comes from considering $\phi(\eta) = -2 +$ the number of conserved adjacencies (places where two adjacent markers differ by 1). A simple calculation shows that ϕ is an eigenfunction of the random reversal Markov chain with eigenvalue $(n-1)/(n+1)$. That is, a randomly chosen inversion decreases the average value of this statistic by $(n-1)/(n+1)$. In our case the number of markers is $n = 26$ and the number of conserved adjacencies (indicated with italics) is 11 so our moment estimate is

$$m = \frac{\ln(9/80)}{\ln(78/80)} = 86.3$$

This agrees with the Bayesian analysis above where the mode of the posterior distribution is 87. However these two numbers differ drastically from the parsimony analyses. The breakpoint distance is $(80 - 11)/2 = 35$, while the parsimony distance is ≤ 54.

Unequal inversion rates

The analysis above assumes that all inversions occur at the same rate. However, inversions between an inverted and a wild type chromosome result in one with some genetic material omitted and some duplicated.

$$
\begin{array}{cccccccccccc}
1 & 2 & 3 & 4 & 5 & 6 & 7 & 8 & 9 & 10 & 11
\end{array}
$$

$$
\begin{array}{cccccccccccc}
1 & 2 & 3 & -9 & -8 & -7 & -6 & -5 & -4 & 10 & 11
\end{array}
$$

When a large amount of DNA is absent from a chromosome, the offspring is typically not viable, so we would expect longer inversions to occur at a lower rate. York, Durrett, and Nielsen (2007) have developed a Bayesian method that takes the lengths of the inversions into account. Analysis of data from *Drosophila melanogaster* and *D. yakuba* finds an inversion tract length of 4.8 megabytes with shorter inversions occuring more frequently.

9.3 Nadeau and Taylor's analysis

As mentioned earlier, genomes evolve by inversions within chromosomes, reciprocal translocation of genetic material between autosomes, i.e., the nonsex chromosomes, and fissions and fusions of chromosomes. In this section we give an account of a remarkable paper of Nadeau and Taylor (1984), who developed a method to estimate the number of breakpoints caused by these events.

In the mid 1980's when the paper was written, there was very little genetic data compared to what exists today. However, there were 54 loci in humans whose chromosomal locations were also known in mice. These data gave rise to 13 conserved segments where the genes were adjacent in both species. The genes involved, the length of the segment in centiMorgans, and the chromosome locations in mice and humans are given in the next table. The p's and q's in the Human column refer to the chromosome arms on which the genes are located.

Genes	length	Mouse	Human
B2m, Sdh-1	1	2	15q
Galt, Aco-1	5	4	9p
Pgm-2, Pgd, Gpd-1	24	4	1p
Pgm-1, Pep-7, Alb-1	12	5	4q
Gus, Mor-1	11	5	7q
Got-2, Prt-2	4	7	19
Ups, Es-17	10	8	16q
Mpi, Pk-3	6	9	15q
Pgm-3, Mod-1	3	9	6q
Acy-1, Trf, Bgl	10	9	3p
Igh, Pi	12	12	14q
Glo-1, H-2, C-4, Upg-1	9	17	6p

The lengths in the table above are the distances between the genes at the end of the segment. The actual conserved segment will be larger than this measurement. To estimate how much larger, we note that

Theorem 9.6. *If n genes are put randomly in a segment of length m, then the expected distance r between the leftmost and rightmost gene is*

$$r = m(n-1)/(n+1) \tag{9.9}$$

Here, and in what follows, we use the notation of Nadeau and Taylor (1984), even though in some cases it will look strange to probabilists. To explain (9.9) we will draw a picture:

Here the $n = 4$ genes define $n + 1 = 5$ segments. Symmetry implies that the segments have the same average length so the average distance between the leftmost and rightmost gene is $(n-1)/(n+1)$ times the total length m. For the reader who does not like the appeal to symmetry in the last argument, we give the following

Proof. We can suppose without loss of generality that $m = 1$. The probability that none of the genes fall in $[0, x]$ is $(1-x)^n$. Differentiating, we find that the location of the leftmost gene Y_1 has density function $P(Y_1 = x) = n(1-x)^{n-1}$. Integrating by parts, we have

$$EY_1 = \int_0^1 x \cdot n(1-x)^{n-1}\, dx = \int_0^1 (1-x)^n\, dx = -\left.\frac{(1-x)^{n+1}}{n+1}\right|_0^1 = \frac{1}{n+1}$$

Reflection symmetry implies that if Y_n is the location of the rightmost gene, then $E(1 - Y_n) = 1/(n+1)$. Combining these two facts, we have

$$E(Y_n - Y_1) = 1 - E(1 - Y_n) - EY_1 = \frac{n-1}{n+1} \qquad \square$$

Inverting the relationship in (9.9), we have

$$\hat{m} = r\frac{(n+1)}{n-1}$$

In words, the estimated length is the observed range multiplied by $(n+1)/(n-1)$. In the data above, segments with 2, 3, and 4 genes have their lengths multiplied by 3, 2, and 5/3. A little arithmetic then shows that the average length of conserved segments observed is 20.9 cM.

Because conserved segments were identified by the presence of two or more linked homologous markers, segments lacking identifiable markers and segments with a single identifiable marker were necessarily excluded in estimates of the conserved segment length. As a result, the estimate of the mean length of conserved segments is biased toward long segments. To correct for this, we prove

Theorem 9.7. *Let T be the total number of mapped homologous markers, G be the genome size in centiMorgans, $D = T/G$ be the density of markers, let L be the mean length of conserved segments, and x' be the average length of observed segments with at least two markers.*

$$Ex' = L \cdot \frac{LD + 3}{LD + 1}$$

Proof. If we assume that breakpoints are distributed "at random," then their locations will follow a Poisson process and the lengths will have an exponential distribution with mean L:

$$f(x) = \frac{1}{L} e^{-x/L}$$

We assume that markers are also distributed "at random," so the probability that a segment of length x will have at least two markers is

$$1 - e^{-Dx} - Dx e^{-Dx}$$

and the relative frequency of segments of length x in the sample is

$$S(x) = [1 - e^{-Dx} - Dx e^{-Dx}] f(x)$$
$$= \frac{1}{L} \left[e^{-x/L} - e^{-Bx} - Dx e^{-Bx} \right]$$

where $B = D + L^{-1}$. Normalizing this relative frequency to integrate to 1, we see that the expected value of observed lengths is

$$Ex' = \frac{\int_0^\infty x S(x) \, dx}{\int_0^\infty S(x) \, dx}$$

Recalling $\int_0^\infty e^{-\lambda x} \, dx = 1/\lambda$, $\int_0^\infty x e^{-\lambda x} \, dx = 1/\lambda^2$, and $\int_0^\infty x^2 e^{-\lambda x} \, dx = 2/\lambda^3$, we have

$$\int_0^\infty S(x) \, dx = 1 - \frac{1}{L(D + L^{-1})} - \frac{D}{L(D + L^{-1})^2}$$
$$\int_0^\infty x S(x) \, dx = L - \frac{1}{L(D + L^{-1})^2} - \frac{2D}{L(D + L^{-1})^3}$$

Combining the last three formulas, we have

$$Ex' = \frac{[L^2(D + L^{-1})^3 - (D + L^{-1}) - 2D]/(D + L^{-1})^3}{[L(D + L^{-1})^2 - (D + L^{-1}) - D]/(D + L^{-1})^2}$$

Expanding out the cube in the numerator and the square in the denominator, a lot of cancellation occurs, leaving us with

$$Ex' = \frac{L^2[D^3 + 3D^2 L^{-1}]}{LD^2 \cdot (D + L^{-1})} = L \cdot \frac{LD + 3}{LD + 1}$$

which is the desired result.   □

In the case under consideration, the total number of markers $T = 54$ and the genome size of mouse is $G = 1600$ centiMorgans, so $D = 0.338$ markers/cM. Setting

$$20.9 = \frac{L^2 D + 3L}{LD + 1}$$

and solving, we have $0.338L^2 + (3 - 20.9 \cdot 0.338)L - 20.9 = 0$, which gives $L = 8.1$. To calculate the number of segments, we divide $1600/8.1 = 197.53$. However, 18 of the breakpoints are caused by the ends of the 19 mouse chromosomes, so our estimate of the number of breakpoints caused by translocations is 179.53.

The final step in Nadeau and Taylor's (1984) analysis is to assign an uncertainty to the estimate. If the uncertainty in the estimation of x' were small, then the relationship between L and x' could be approximated by a linear function with slope dL/dx'. Using $\mathrm{var}\,(cx') = c^2\,\mathrm{var}\,(x')$, we have

$$\mathrm{var}\,(L) \approx \mathrm{var}\,(x') \cdot \left(\frac{dL}{dx'}\right)^2$$

The standard deviation of the transformed lengths of the 13 conserved segments in our sample is 12.8 cM. This leads to an estimate of $(12.8)^2/13 = 12.6$ for $\mathrm{var}\,(x')$. To evaluate the second term, we use calculus to conclude that

$$\frac{dx'}{dL} = \frac{(2DL + 3)(LD + 1) - D(L^2 D + 3L)}{(LD + 1)^2}$$

and $dL/dx' = 1/(dx'/dL)$. Substituting the numerical values of L, D, and $V(x')$, we estimate the standard deviation of our estimate of L to be 1.6 cM. Dividing this leads to an estimate of 178 ± 39 breakpoints or 89 ± 20 inversions and translocations between the genomes of man and mouse.

More data confirms the theory

Over time the number of markers in the comparative map between mouse and man has grown but the estimates of the average length of conserved segments (in centiMorgans) based on the theory in Nadeau and Taylor (1984) have remained roughly constant.

source	markers	segments	estimate
Nadeau and Taylor (1984)	54	13	8.1
Nadeau (1989)	157	27	10.1
Copeland et al. (1993)	917	101	8.8
DeBry and Seldin (1996)	1416	181	7.7

A more direct test of the theory is to look at the distribution of the number of genes per conserved segment. If breakpoints were a Poisson process with rate n and genes were a Poisson process with rate m, then the number of genes in a segment would have a shifted geometric distribution with success probability $n/(m + n)$. That is the probability of r genes would be

$$\left(\frac{m}{m+n}\right)^r \cdot \frac{n}{m+n} \qquad \text{for } r = 0, 1, 2, \ldots \qquad (9.10)$$

Taking a different approach to this question, Sankoff and Nadeau (1996) have shown:

Theorem 9.8. *Consider a linear interval of length 1, with $n > 0$ uniformly distributed breakpoints that partition the interval into $n+1$ segments. Suppose that there are m genes also uniformly distributed on the interval between 0 and 1, and independently of the breakpoints. For an arbitrary segment, the probability that it contains r genes, $0 \le r \le m$, is*

$$P(r) = \frac{nm!(n+m-r-1)!}{(n+m)!(m-r)!} \qquad (9.11)$$

To make the connection with (9.10) we can write (9.11) as

$$\frac{m}{n+m} \cdot \frac{m-1}{n+m-1} \cdots \frac{m-r+1}{n+m-r+1} \cdot \frac{n}{n+m-r}$$

Proof. The segment length between two adjacent breakpoints has probability density $f(x) = n(1-x)^{n-1}$. For a segment of length x, the probability that it has r genes is given by the binomial distribution. Thus

$$P(r) = \int_0^1 n(1-x)^{n-1} \binom{m}{r} x^r (1-x)^{m-r} \, dx$$

Repeated integration by parts shows that

$$\int_0^1 x^a (1-x)^{b-a} \, dx = \frac{b-a}{a+1} \int_0^1 x^{a+1}(1-x)^{b-a-1} \, dx$$

$$= \frac{(b-a) \cdot (b-a-1) \cdots 1}{(a+1) \cdot (a+2) \cdots b} \int_0^1 x^b \, dx = \frac{(b-a)!a!}{(b+1)!}$$

Using this result with $a = r$ and $b = n + m - 1$ shows

$$P(r) = n \cdot \frac{m!}{r!(m-r)!} \cdot \frac{(n+m-r-1)!r!}{(n+m)!}$$

which is the indicated result. $\square$

Sankoff, Parent, Marchand, and Ferretti (1997) applied this result to the human/mouse map that was available at the time their paper was written. Since one cannot observe n_0 the number of segments having no genes, they had to compare $f(r) = n_r / \sum_{k>0} n_k$ with $Q(r) = P(r)/(1 - P(0))$. They found that the fit was generally pretty good, but $f(1)$ was much larger than $Q(1)$.

Genome sequence data raises doubts

The draft human and mouse sequences revealed many undiscovered synteny blocks, a total of 281 of size at least 1 megabase and put the random breakage model to a new test, see Pevzner and Tesler (2003a). Although the number of synteny blocks is higher than the Nadeau-Taylor predictions, the lengths of these blocks still fit an exponential distribution. However the finer resolution data also revealed 190 short synteny blocks that are not consistent with the exponential distribution.

Pevzner and Tesler (2003b,c) argued that the surprisingly large number of breakpoints in clumps was an argument in favor of a different model of chromosome evolution that they called the *fragile breakage model*. This model postulates that the breakpoints mainly occur within relatively short hot spots of rearrangement. They observed that the existence of such fragile regions was supported by previous studies of cancer and infertility.

9.4 Genomic distance

As in the case of inversions, one can ask how many moves (inversions, translocations, fissions, and fusions) are needed to transform the autosomes of the mouse genome into those of the human genome. Hannenhalli and Pevzner (1995b) extended their earlier work on the reversal distance to compute the *genomic distance*, that is, the minimal number of moves needed to transform one genome into another one. The answer is in terms of seven parameters capturing different combinatorial properties of the sets of strings, so we will not go into the details here.

9.4.1 Graph distance

Instead, we will describe how to compute a lower bound on the distance from the breakpoint graph, which, in most applications, gives the right answer. To explain the method, we will consider an example

$$
\begin{array}{ll}
\text{genome } A & \text{genome } B \\
1, -6 & 1, 2, 3 \\
3, 5, -4 & 4, 5, 6 \\
2 &
\end{array}
$$

The first step is to double the markers. A signed marker $+x$, becomes $2x-1$, $2x$ and $-x$ becomes $2x$, $2x-1$. Doubling the markers in the A genome and adding A's at the ends (which we consider to be distinct vertices) gives edges

$$
A - 1, 2 - 12, 11 - A
$$
$$
A - 5, 6 - 9, 10 - 8, 7 - A
$$
$$
A - 3, 4 - A
$$

which are drawn as in Figure 9.4 as thick lines (black edges). Adding an empty chromosome to genome B to make the number of chromosomes equal, and drawing thin lines (gray edges) for adjacencies in the B genome gives the breakpoint graph. Note that, except for the special vertices A and B, which always have degree 1, all the other vertices in the breakpoint graph are incident with one gray and one black edge.

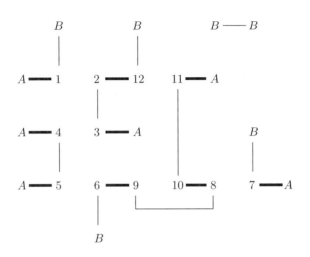

Fig. 9.4. Breakpoint graph for our simple example.

We have two types of connected components in the breakpoint graph, one without special vertices $x_1, x_2, \ldots x_{2r}$ that we call *cycles*, and a second type that begins and ends with special vertices that we call *paths*. We can write the breakpoint graph in a unique way a union of paths and cycles.

Let $c(A, B)$ be the number of components, i.e., paths and cycles, including empty chromosomes. Let $\#(A, A)$ be the number of paths that start with A and end with A. In the example drawn above the components are

$$A - 1 - B, \qquad A - 4 - 5 - A, \qquad A - 3 - 2 - 12 - B$$
$$A - 11 - 10 - 8 - 9 - 6 - B \qquad A - 7 - B \qquad B - B$$

the number of components $c(A, B) = 6$, and the number of same genome cycles $\#(A, A) = \#(B, B) = 1$.

Let n be the number of makers and k the number of chromosomes. The *graph distance* between two genomes A and B is defined as

$$d(A, B) = n + k - c(A, B) + \#(A, A). \tag{9.12}$$

The graph distance for our example is $6 + 3 - 6 + 1 = 4$. Again one can check that any move can only decrease $d(A, B)$ by at most 1, so the graph distance is a lower bound on the genomic distance. It is easy to check in the example we are considering that we can transform A into B in four rearrangements by using the following steps

$$
\begin{array}{ccccccccc}
1,\, -6 & & 1,\, 2 & & 1,\, 2 & & 1,\, 2,\, 3 & & 1,\, 2,\, 3 \\
3,\, 5,\, -4 & \rightarrow & 3,\, 5,\, -4 & \rightarrow & -6,\, 5,\, -4 & \rightarrow & 4,\, -5,\, 6 & \rightarrow & 4,\, 5,\, 6 \\
2 & & 6 & & 3 & & & &
\end{array}
$$

where in the third step, flipping the second chromosome does not count as a move.

9.4.2 Bayesian estimation

Following Durrett, Nielsen, and York (2003), we will assume that inversions and translocations arise in the genome at constant rates λ_I and λ_T, and that the process of inversions and translocations forms a Markov chain with state space Ω_O given by the set of all possible orderings of the markers on ordered chromosomes. Assuming all markers are signed indicating their orientation on the chromosome, there are

$$
|\Omega_O| = 2^N \frac{(M + N - 1)!}{(M - 1)!}
$$

orderings of N signed markers on M numbered chromosomes, allowing empty chromosomes, i.e., chromosomes containing no markers. In our representation, transitions between neighboring states (states differing by only a single inversion or a single translocation) in Ω_O occur at rate λ_I if the two states differ by an inversion and at rate λ_T if they differ by a translocation. Because of the symmetry of the transition rates, the Markov chain is reversible and has a uniform stationary distribution, i.e.,

$$
\pi_O(x) = 2^{-N} \frac{(M - 1)!}{(M + N - 1)!} \qquad \text{for all } x \in \Omega_O.
$$

Because the ordering of chromosomes is not of interest, we also consider a Markov chain with a collapsed state space (Ω_U) that ignores the numbering of chromosomes. An element of Ω_U, with M_0 empty chromosomes, corresponds to $2^{M-M_0} M!/M_0!$ elements of Ω_O. The stationary probability assigned to an element of Ω_U with M_0 empty chromosomes is then

$$
\pi_U(x) = 2^{-N+M-M_0} \frac{M!(M - 1)!}{M_0!(M + N - 1)!} \qquad \text{for all } x \in \Omega_U.
$$

Transitions between neighboring states in this collapsed Markov chain occur at rates λ_I if the two states differ by an inversion and λ_T if they differ by a translocation, except for events involving a chromosomal fission. When there are M_0 empty chromosomes, these transitions occur at rate $2M_0\lambda_T$, since they correspond to a translocation with an empty chromosome. With these rates the transition probabilities of the Markov chain obey the detailed balance equations and the Markov chain is reversible. In the following we will use the representation based on the Markov chain with state space on Ω_U.

Consider the genome of two organisms. The genomic marker data from one species x_1 can be transformed into the genomic marker data from another species x_2 through a sequence of inversions and translocations. Because the process we have defined is reversible we can write the sampling probability as

$$P(x_1, x_2|\Theta) = P(x_1)P(x_1 \rightarrow x_2|\Theta)$$

where Θ is the vector of parameters and $P(x_1 \rightarrow x_2|\Theta)$ is the transition probability for the transformation from x_1 to x_2. Because a model parameterized in terms of t, λ_T and λ_I is not identifiable, we arbitrarily set $t = 1$ and define $\Theta = (\lambda_T, \lambda_I)$. Therefore, λ_T and λ_I can be interpreted as the expected number of translocations and inversions per marker pair in the history of the two species. $P(x_1)$ does not depend on Θ, so the likelihood function is, therefore, simply given by

$$L(\Theta) = Pr(x_1 \rightarrow x_2|\Theta).$$

Let Ψ be the (countably infinite) set of all possible evolutionary paths from x_1 to x_2. We notice

$$P(x_1 \rightarrow x_2|\Theta) = \sum_{y \in \Psi} P(y|\Theta)$$

To estimate Θ we define a Markov chain with state space on $[0, \infty)^2 \times \Psi$ with stationary distribution given by the joint posterior distribution of parameters and evolutionary path, y,

$$\pi(y, \lambda_T, \lambda_I) = P(y, \lambda_T, \lambda_I|x_1, x_2)$$

To ensure the posterior is proper and biologically meaningful, the supports of λ_T and λ_I are restricted to the intervals $(0, \lambda_{Tmax})$ and $(0, \lambda_{Imax})$ respectively, and uniform priors u_T and u_I are used.

The Markov chain is simulated using the Metropolis-Hastings algorithm similarly to the method used in York et al. (2003). In brief, the posterior distribution is proportional to

$$P(y|\lambda_T, \lambda_I)u_T(\lambda_T)u_I(\lambda_I)$$

The Markov chain is then simulated by iteratively proposing new values of y, λ_T and λ_I, and accepting these new values according to the appropriate Metropolis-Hastings acceptance probabilities, see Section 9.2 or York et al. 2003 for more details.

One new complication is that if at time t there are $I_T(t)$ possible translocations and $I_I(t)$ possible inversions that can occur, the total rate of translocations and inversions is $I_T(t)\lambda_T + I_I(t)\lambda_I$. This rate is not constant in time and the probability of a path depends on the times between the events. To avoid keeping track of the interevent times, Durrett, Nielsen, and York (2003) used a method based on uniformization to ensure that the total rate of change is constant in time. In brief, they allowed pseudoevents of evolutionary change, which have no effect on marker order, to make the total rate of evolutionary change kept constant.

Analysis of three data sets

Tomato vs. eggplant. Doganlar et al. (2002) constructed a comparative map of tomato and eggplant consisting of 233 markers. Thinning their data set to choose one marker from each group in which the order was not resolved, leads to a data set with 170 markers. Analyzing the data by hand they concluded: "Overall, eggplant and tomato were differentiated by 28 rearrangements, which could be explained by 23 paracentric inversions and five translocations during evolution from the species' last common ancestor."

Six of the chromosomes are conserved between species, so the interesting part of the data can be written as

Eggplant		Tomato	
E3.	1, 2, 3, 4, 5, 6	T3.	1, −5, 2, 6
E4.	7, 8	T4.	21, −22, −20, 8
E5.	9, 10	T5.	−4, 14, 11, −15, 3, 9
E10.	11, 12, 13, 14, 15, 16, 17, 18	T10.	7, 16, −18, 17
E11.	19, 20, 21, 22	T11.	−19, 24 −26, 27, 25
E12.	23, 24, 25, 26, 27	T12.	−12, 23, 13, 10

Doubling the markers and defining the breakpoint graph, which we do not have to draw to compute the components, we find that there are seven shared ends in the two genomes which lead to short paths: $E - 1 - T$, $E - 12 - T$, $E - 13 - T$, $E - 16 - T$, and $E - 20 - T$. The other paths are

$$E - 17 - 6 - 7 - 27 - 26 - 19 - 18 - T$$
$$E - 21 - 28 - 29 - 5 - 4 - 11 - 10 - 2 - 3 - 9 - 8 - T$$
$$E - 36 - 32 - 33 - 35 - 34 - T$$
$$E - 37 - 47 - 46 - 25 - 24 - T$$
$$E - 44 - 42 - 43 - 40 - 41 - T$$
$$E - 45 - 23 - 22 - 30 - 31 - 14 - 15 - 39 - 38 - T$$
$$E - 54 - 49 - 48 - 52 - 53 - 51 - 50$$

The number of markers $n = 27$, chromosomes $k = 6$, components $c(E, T) = 12$, and the number of same genome paths is $\#(E, E) = \#(T, T) = 0$ so using (9.12) the distance

$$d(E, T) = 27 + 6 - 12 + 0 = 21$$

Adding to this seven inversions for the other six chromosomes we arrive at a distance of 28, in agreement with Doganlar et al. (2002).

Durrett, Nielsen, and York's Bayesian analysis produced 95% credible intervals of [5,7], [21,31], and [28,37] for the number of translocations, inversions, and total number of events (respectively) separating tomato and eggplant. Figure 9.5 gives the posterior distribution for the number of inversions.

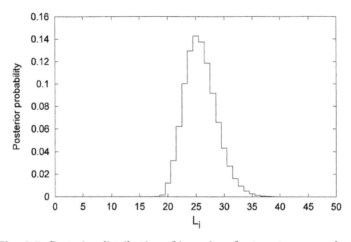

Fig. 9.5. Posterior distribution of inversions for tomato vs. eggplant.

Human vs. cat. Murphy et al. (2000) created a radiation hybrid map of the cat genome integrating 424 genes with 176 microsatellite markers. Using resources on the NCBI home page, http://www.ncbi.nlm.nih.gov, Durrett, Nielsen, and York (2003) were able to locate the position of 281 of the genes in the human genome. From this set they deleted 12 singleton disruptions, i.e., isolated genes that map to a chromosome different from their neighbors. In support of this practice they noted that none of the regions deleted are observed in the chromosome painting experiments of Weinberg et al (1997) and those techniques are thought to be capable of detecting segments as small as 5 megabases.

Parsimony analysis shows that the marker order in the human genome can be transformed into the cat genome in 78 moves (14 translocations and 64 inversions). Bayesian analysis gives 95% credible intervals of [12,15], [71,89], and [85,102] for the number translocation, inversions, and total number of events respectively. Note that the parsimony distance is not in the 95% credible interval for the total number of events. In fact, the posterior probability of a total of 78 events is approximately 2×10^{-5}. The posterior distribution for the number of translocations assigns probability

$$
\begin{array}{cccc}
12 & 13 & 14 & \geq 15 \\
0.5967 & 0.3451 & 0.0531 & 0.0050
\end{array}
$$

so this time the smallest value is the most likely. The posterior distribution of
the number of inversions is given in the next figure, with the mode being 79.

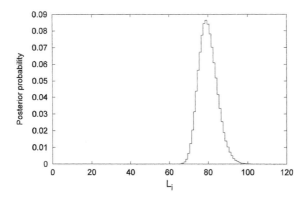

Fig. 9.6. Posterior distribution of inversions for human vs. cat.

Finally we would like to compare our estimates with those of Nadeau and
Taylor, described in Section 9.3. There are 83 conserved segments with 2 or
more markers with an average length of 16.3 megabases, which results in
an estimated average length of 18.2 megabases. Using 3.2 gigabases for the
length of the human genome yields an estimate of 175.8 segments. Subtracting
22 chromosomes in the human genome, we arrive at an estimate of 155.8
disruptions or an estimate of 77.9 events. This is almost exactly the parsimony
distance of 78, but is this lower than the Bayesian estimates. On the other
hand, the method of Nadeau and Taylor includes events that are not visible
with our set of markers.

Human vs. cattle. Band et al. (2000) constructed a radiation hybrid
map of cattle (Bos taurus) with a total of 638 genes. For the data see `http://`
`bos.cvm.tamu.edu/htmls`. Using resources on the NCBI home page, Durrett,
Nielsen, and York (2003) were able to locate the position of 448 of the genes in
the human genome. Deleting 24 singleton disruptions for the reasons indicated
above results in a map with 422 markers. Again the reduced map is consistent
with the results of chromosome painting, see Hayes 1995 and Chowdhary et
al (1996).

Parsimony analysis shows that the marker order in the human genome can
be transformed into the cattle genome in 155 moves (20 translocations and
135 inversions). The Bayesian approach experienced convergence problems in
this example. In the case of translocations there is a considerable difference
between the chains indicating convergence problems. The qualitative differences between chains in the number of inversions are not as great as in the case
of translocations. The modes are all in the range 185–191, but the variance
differs somewhat from run to run.

We cannot make statements with much confidence about the number of inversions and translocations. However two things are clear: (i) the number of events is roughly twice that in the human cat comparison even though the divergence times are similar and (ii) our conclusions differ considerably from those of Band et al. (2000), who say that their "comparative map suggests that 41 translocation events and a minimum of 54 internal rearrangements have occurred." They do not explain how they reached this conclusion. However, we would require a larger number of translocations if we had not deleted the singletons and they would underestimate the number of inversions if they used the breakpoint distance.

To estimate distances using the methods of Nadeau and Taylor we note that there are 125 conserved segments with two or more markers with an average length of 7.19 megabases, which results in an estimated average length of 7.57 megabases. Using 3.2 gigabases for the length of the human genome yields an estimate of 422.7 segments. Subtracting 22 chromosomes in the human genome gives an estimate of 400.7 disruptions, or an estimate of 200 events. This is somewhat larger than our Bayesian estimate, but that is consistent with the fact that our estimate is restricted to the events that can be detected by the 422 markers in our map.

9.5 Midpoint problem

Given that there are many paths that achieve the minimum distance, it is natural to use information about gene order for several species to identify the sequence of events that occurred. Here we are thinking of examples where the phylogeny is known. Simon and Larget (2001) and Larget et al. (2002) have used genome rearrangements to estimate phylogenetic relationships.

Hannenhalli et al. (1995) were the first to do this for three species: Herpes Simplex virus (H), Epstein Barr virus (E), and Cytomegalovirus (C). Comparing the gene orders in these viruses leads to the following segments:

$$H \quad 1, 2, 3, 4, 5, 6, 7$$
$$E \quad 1, 2, 3, -5, 4, 6, -7$$
$$C \quad 1, -2, -3, 7, -4, 5, 6$$

Constructing the breakpoint graph for the H-E comparison

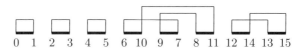

and using (9.1), we have $d(H, E) \geq 8 - 5 = 3$. For the other direction, we note that performing the reversal involving the black edges 6-10 and 8-11 leaves a situation where two more reversals will put the numbers in order.

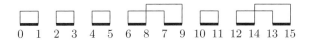

At the beginning of the section, we showed that $d(E, C) = 5$. The break-point graph for the H-C comparison is

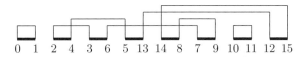

(9.1) implies that $d(H, C) \geq 8 - 3 = 5$. We leave it to the reader to show that $d(H, C) = 5$.

If we connect the three genomes in an unoriented tree, then there will be one intermediate genome M, the midpoint.

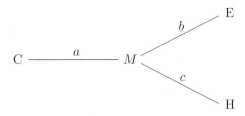

If we let a, b, and c be the number of changes on the indicated edges, then

$$a + b \geq d(C, E) = 5$$
$$a + c \geq d(C, H) = 5$$
$$b + c \geq d(E, H) = 3$$

Adding the three inequalities, we have

$$a + b + c \geq \frac{d(C, E) + d(C, H) + d(E, H)}{2} = \frac{13}{2} \qquad (9.13)$$

so $a + b + c \geq 7$. There are three triples (a, b, c) with sum 7 that satisfy these inequalities: $(4, 2, 1)$, $(4, 1, 2)$, and $(3, 2, 2)$. To check this note that (i) $b + c \geq 3$ implies $a \leq 4$, and (ii) $b, c \geq 5 - a$ implies $a + b + c \geq 10 - a$, so $a \geq 3$. If $a = 4$, $b \geq 1$, and $c \geq 1$, while if $a = 3$, $b \geq 2$, and $c \geq 2$.

For a given triple, say $(4, 2, 1)$, Hannenhalli et al. (1995) used a computer to generate all permutations that were distance 4 from C, distance 2 from E, and distance 1 from H. There is one that is in all three sets $1, 2, 3, -4, 5, 6, 7$, so it represents one possibility for the midpoint M. Considering $(4, 1, 2)$ leads to another possibility: $1, 2, 3, -4, 5, 6, -7$. The third $(3, 2, 2)$ does not lead to any scenario with seven inversions, so we have only two most parsimonious scenarios.

More recent approaches

Sankoff and Blanchette (1997, 1998a,b , 2000) have considered our problem for the "breakpoint" distance, which is $1/2$ the number of markers adjacent in one genome that fail to be adjacent in the other, rounded up to the next integer. That is, given n genomes $G_1, \ldots, G_n$ having a known phylogeny, one seeks genomes $H_1, \ldots, H_{n-2}$ for the internal nodes of the tree, so that the sum of the breakpoint distance between endpoints of edges of the tree is minimized. Blanchette et al. (1999) used BPANALYSIS, an implementation of the breakpoint analysis of Blanchette et al. (1997) on a problem with 11 genomes and 35 markers. More recently, an improvement of the BPANALYSIS, call GRAPPA has been developed by Moret et al. (2001).

Bourque and Pevzner (2002) developed an approach called the Multiple Genome Rearrangement Median (MGR-MEDIAN) algorithm based on the genomic distance which applies to n species. When $n = 3$ the algorithm works in two stages. In the first stage, rearrangement events in a genome that bring it closer to each of the other two of the three genomes are carried out "in a carefully selected order." In the second stage, moves are accepted if they bring two genomes closer together. For more on the computer implementation of this approach, see Tesler (2002).

Durrett and Interian (2006) have recently developed a new approach to the median problem. Given three genomes A, B, and C, they seek a genome M that minimizes $d(A, M) + d(B, M) + d(C, M)$, where d is the graph distance. To do this, they initialize the search process with a proposed midpoint and then proceed by iteratively making small changes in the midpoint. The proposed change is always accepted if it reduces the total number of moves, and with a fixed probability if it does not. This work is not yet published, but a paper and the computer code for the algorithm, MEDBYLS (Median by Local Search), can be found at `http://www.cam.cornell.edu/~interian/MEDbyLS_code.html`

Four data sets

Human-lemur-tree shrew. The first data set we will consider is a three way comparison of human (*Homo sapiens*) , lemur (*Eulemer macaco macaco*) and tree shrew (*Taupaia belangeri*). The reason for interest in this comparison is that the midpoint should provide an estimate for genome arrangement in the ancestor of all primates. Müller et al. (1997) did a reciprocal chromosome painting between human and lemur, and Müller et al. (1999) did a reciprocal painting between human and tree shrew, and a painting of lemur chromosomes with tree shrew paints. There are 37 segments in the comparison with lemur and 39 with tree shrew. Subdividing to obtain a common collection of segments, we arrive at the 41 homologous segments given in the table. Chromosome painting does not give information about orientation, so we have assigned signs to segments to minimize the distance.

Human	Lemur	Tree shrew	
1. 1,2,3	1. -22, -7,-6,-9,-8,39	1. 32,23,22	21. 15
2. 4,5	2. -5, -21, -31	2. -26, -38	22. 13
3. 6,7,8,9	3. 18, -14	3. 36	23. 11
4. 10,11,12,13	4. -29,-26,-25	4. 18,19	24. 7
5. 14,15	5. -24,27,-40	5. -37,-33	25. 10
6. 16,17	6. -38,10, 4	6. 9	26. 12
7. 18,19	7. 33,-36,30	7. -5,-39,8	27. 6
8. 20,21	8. -35,16,12	8. 17	28. -28,41
9. 22	9. 1	9. 27,-40	29. 16
10. 23,24	10. 32,-34	10. 1	30. 31
11. 25,26	11. 17	11. 30	
12. 27,28	12. -37,13	12. 4	
13. 29	13. -28,41	13. 20,21	
14/15. 30,31	14. 2	14. 14	
16. 32,33	15. 15	15. 35	
17. 34	16. 20	16. 24	
18. 35	17. 23	17. 29	
19. 36,37	18. 3	18. 34	
20. 38	19. 19	19. 2,3	
21. 39	20. -11	20. 25	
22. 40,41			

The distances between the three genomes are $d(L, H) = 21$, $d(L, T) = 19$, and $d(H, T) = 16$. Using (9.13), the total number of events

$$\geq \frac{d(L, H) + d(L, T) + d(H, T)}{2} = 28$$

The next table gives the midpoint M computed by local search, which has $d(L, M) = 12$, $d(TS, M) = 10$, $d(H, M) = 10$ for a total of 32 events. This is 4 more than the lower bound, but it is possible to show that 32 is the minimum. The numbering of the chromosomes is to facilitate comparison with the next midpoint.

1a. 1	10	7. 18, 19	12a/22a. 27, -40
1b. 2, 3	11	9. 22	12b/22b. -28, 41
2a. 4 .	-12, -16	10. 23, 24	13. 29
-5,-21,-20	-13, 37	11. 25, 26	14/15. 30, 31
3a. 6, 7	14		17. 34
3b/21. -9, -8, 39	15		18. 35
	17	16/19a. 32, 33, -36	20. 38

Müller et al. (1999) have proposed that the primitive primate karyotype consists of human autosomes 1a, 1b, 2a, 2b, 3/21, 4 to 11, 12a/22a, 12b/22b,

13, 14/15, 16a, 16b, 17, 18, 19a, 19b, and 20. Our interpretation of this mid-point N in terms of our segments is given in the next table. We have performed two inversions in human chromosome 3 since this improves the performance of their solution: $d(L, N) = 17$, $d(T, N) = 14$, $d(H, N) = 7$ for a total of 38 events, 6 more than the minimum.

1a. 1	4.10 11 12 13	7. 18 19	12a/22a. 27,-40
1b. 2 3	5. 14 15	9. 22	12b/22b. -28,41
2a. 4	6. 16 17	10. 23 24	13. 29
2b. 5		11. 25 26	14/15. 30 31
8. 20 21	19b. 37		17. 34
3/21.-7 -6 -9 -8 39		16a. 32	18. 35
		16b. 33	20. 38
		19a. 36	

Note that the expert solution has many fewer events in the human genome, while the one found by local search distributes the changes almost equally over the three lineages. It is interesting to note that the two solutions share many features in common even though the expert solution is informed by comparisons with other primate genomes while the computer generated solution only uses the three given as data. For an interesting expert analysis of rearrangements within primates, see Haig (1999).

   Human-cow-cat. Our second data set is a comparison of human, cat, and cattle constructed by Murphy et al. (2003). They have 300 markers on autosomes. We have deleted 12 markers whose position is in conflict with chromosome painting experiments of Hayes (1995) and Chowdhary et al. (1996) that compared human and cattle or work of Weinberg et al. (1997), and Murphy et al. (1999, 2000) that used chromosome painting results and a radiation hybrid map to compare humans and cats. In addition we deleted 3 markers to make block boundaries coincide and to allow the creation of smaller data set. The first step is to omit chromosomes that don't tangle with the others

I. Human 17	Cat E1	Cow 19
II. Human 14,15	Cat B3	Cow 10, 21
III. Human 6	Cat B2	Cow 9, 23
IV. Human: 11	Cat D1	Cow 15, 29.

The second is to reduce to syntenic segments, i.e., blocks of markers tha are the same but in a different order. This produces a data set small enough so that one can try to visualize the changes

Human	Cow	Cat
1: 1,2,3,4,5,6,7	1: -35, 11	A1: 28,-15,-14
2: 8,9	2: 9, -2	A2: 32,-10,16
3: 10,11	3: -4,-3	A3: -34,-8
4: 12,13	4:16	B1:18,-13,-12
5: 14,15	5: 26,-37,-25,38	B4: 22,25,26,37,38
7: 16,17	6:12	C1: 1,2,3,9
8: 18,19	7: -32,-15	C2:-35,11
9: 20,21	8:20	D2:-7,-23,24
10: 22,23,24	11:-8,21	D3: -27,36,31
12: 25,26,27	12:28	D4:20,21
13: 28	13:-22,34	E2:-33,30
16: 29,30	14:-19	E3:17,-29
18: 31	16:6,5,1	F1:5,-4,6
19: 32,33	17: -13,27,36	F2:19
20: 34	18:-30,33	
21: 35	20:14	
22: 36,37,38	22:10	
	24:31	
	25:29,-17	
	26:24	
	27:18	
	28:7,-23	

Since cow and cat both begin with C, we will use B for bovine and F for feline as the one letter abbreviations of those genomes. $d(F, H) = 51$, $d(F, B) = 82$, and $d(H, B) = 72$ so the lower bound from (9.13) is 103. The local search method produces a midpoint with $d(H, M) = 18$, $d(F, M) = 36$, $d(B, M) = 56$ for a total of 110 events. Murphy et al. (2003) have analyzed this data using the MGR-MEDIAN algorithm. This method has a parameter G that is a distance threshold used to filter out spurious markers that occur at isolated points. When $G = 4$ singletons are deleted, while increasing G allows for more complex microrearrangments. The solution they present in their Figure 2 has $G = 6$ uses 276 of the 300 markers. They find distances $d(H, M) = 16$, $d(F, M) = 21$, $d(B, M) = 27$ for a total of 64 events.

It is confusing to compare these answers, which assign a much different fraction of moves to the three branches. Murphy et al. (2003) presented their solution as the answer. However, it is now common to explore a wide range of alternative midpoints to see which features of the midpoint in which one can have confidence, see e.g., the supplementary material of Murphy et al. (2005).

Human-mouse-rat. The third data set is a comparison of human, mouse, and rat constructed by Colin Dewey and Lior Pachter which appeared in the April 1, 2004 issue of Nature devoted to the sequencing of the rat genome, see page 498. In our final example we will concentrate on the inference of the number of events rather than trying to reconstruct the changes. For reconstructions of the midpoint, see Bourque et al. (2004, 2005), or Murphy et al. (2005). The local search method produced 100 midpoints in which the total distance was 347 in 98 cases and 346 in 2. The average distance to the midpoint is 43.01 for mouse, 62.75 for rat, and 241.22 for human, in contrast to the distances of 50, 56 and 247 and a total of 353 reported by the Rat Genome Sequencing Project Consortium (2004). See Bourque, Pevzner, and Tesler (2004) for details of the computation.

If we use 15 million years for the divergence of mouse and rat, and 90 million years for their divergence from human, then the human to midpoint branch is 165 million years and the estimates of events per million years from local search are 3.01 for mouse, 4.33 for rat and 1.48 for human. This contrasts with the estimates of 2.25 for mouse and 1.25 for rat from Figure 3 of Murphy et al. (2005). It is interesting to note that the answer to the question: "do rats have more frequent rearrangements than mice?" is different. Murphy et al. (2005) give an estimate of 0.39 events per year for the 90 million years since divergence, and 2.11 per year on the branch from the divergence to the mouse-rat split. The local search method cannot estimate rates separately for the two branches but weighting the rates by the interval lengths $(0.39)(90/165) + (2.11)(75/165) = 1.17$, gives a rate smaller than 1.48.

Seven species comparison. The results of Murphy et al. (2005) quoted in the last two paragraphs come from a paper that compares human, mouse, rat, cat, dog, pig, and cow. Using sequenced genomes and radiation hybrid maps they constructed a map with 307 blocks. The pairwise distances between the species based on this map are

	Mouse	Rat	Cat	Cattle	Dog	Pig
Human	153	149	61	115	108	78
Mouse		60	169	209	184	172
Rat			164	203	180	170
Cat				130	114	87
Cattle					157	135
Dog						124

The most parsimonious solution involved 487 rearrangements. As the following recreation of their Figure 3 shows there is a great heterogeneity in the rates of rearrangements

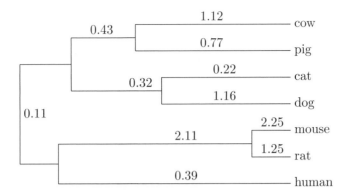

9.6 Genome duplication

In this section, we will consider three examples where DNA sequence data indicate that the whole genome has undergone duplication: yeast (*Saccharomyces cerevisiae*), maize, and *Arabidopsis thaliana*.

9.6.1 Yeast

Smith (1987) was the first to suggest that the yeast genome had undergone duplication based on the fact that the core histone genes occur as duplicate pairs. A few years later, Lalo et al. (1993) showed that there was a large duplication between chromosomes III and XIV covering the centromeric regions of these two chromosomes.

The completion of the sequencing of the yeast genome in 1997 made it possible for Wolfe and Shields (1997) to study this question in more detail. To look for duplicated blocks, they used the BLASTP to compare the amino acid sequences of all genes in the yeast genome. A BLASTP score of ≥ 200 (which will occur randomly with probability $p = 10^{-18}$ or less) was taken as evidence that two genes are homologues. The process of looking for duplicated blocks is complicated by the fact that after genes are duplicated, it is often the case that one member of the pair loses function and similarity is erased by the accumulation of mutations. For this reason, Wolfe and Shields defined a duplicate region when there were (i) at least three pairs of homologues with intergenic distances of at most 50 kb, and (ii) conservation of gene order and orientation (with allowance for small inversions within some blocks). With these criteria they found 55 duplicate regions containing 376 pairs of homologous genes. The duplicated regions were an average of 55 kb long and together they span about 50% of the yeast genome.

How did these 55 duplicated regions arise? They were formed either by many successive independent duplications (each involving dozens of kilobases) or simultaneously by a single duplication of the entire genome (tetraploidy) followed by reciprocal translocation between chromosomes to produce a mosaic of duplicated blocks. In support of the tetraploidy and translocation model, one can observe that for 50 of the 55 blocks, the orientation of the entire block with respect to the centromere is the same in the two copies. Block orientation is expected to be conserved if the blocks were formed by reciprocal translocation between chromosomes, whereas if each block was made by an independent duplication event, the orientation should be random. Since 50 heads or 50 tails in 55 tosses of a fair coin is extremely unlikely, this is evidence for the tetraploidy explanation.

To obtain more evidence for duplication and to estimate the age of the event, Wolfe and Shields (1987) compared 12 *S. cerevisiae* duplicate pairs with homologues in *Kluveromyces* and an outgroup. In 9 of these pairs, there was strong bootstrap support ($\geq 89\%$) for a branching order that places the two *S. cerevisiae* sequences together; in the others there was no strong support for any order. They then estimated the ages of the duplications in *S. cerevisiae* compared with the divergence from *Kluveromyces*. Three of the gene pairs yield very young ages, indicating that they have been involved in recent gene duplications. Of the five pairs for which there were sufficient data to calculate a confidence interval based on bootstrapping, the mean relative age of duplication is 0.74 (with a standard deviation of 0.12). Since the divergence of *Kluveromyces* and *S. cerevisiae* has been estimated at 1.5×10^8 years ago, this places the genome duplication in yeast at roughly 10^8 years ago. For more on the comparison of gene order in *Kluveromyces* and *S. cerevisiae*, see Keogh, Seoighe, and Wolfe (1998).

Seoighe and Wolfe (1998, 1999) followed up on this work by using simulation and the analytical methods of Nadeau and Taylor (1984) to investigate the extent of genome rearrangement after duplication in yeast. To simplify things they took the distance between two genes to be the number of genes between them rather than the distance in kilobases. In their simulations, an original genome with eight chromosomes was duplicated and genes were deleted at random until something resembling present-day yeast remained: 5790 genes on 16 chromosomes. Reciprocal translocations were then made between randomly chosen points in the genome, and blocks of duplicated genes were identified by using criteria similar to those in Wolfe and Shields (1997): three homologues with a maximum distance of 45 genes between them. Of all the parameter combinations they considered, a probability of 8% for retaining a duplicate pair and 75 reciprocal translocations produced the best fit to what was observed. It produced 62 duplicate blocks spanning 54% of the genome and containing 350 duplicate pairs. This was the only case in which all three statistics were within two standard deviations of their means.

To explain Seoighe and Wolfe's use of results of Nadeau and Taylor (1984), we need to distinguish between chromosomal regions demarcated by reciprocal translocations, called *segments*, and the number of duplicate regions that could be identified, called *blocks*. If we let L be the mean length of segments, then, as in Section 5.3, segment lengths have an exponential distribution:

$$\frac{1}{L}e^{-x/L}$$

The reader should not be too hasty to accept this generalization of the previous results since segment lengths are now numbers of genes rather than kb. However, our estimate of L will turn out to be 16.45, so we probably do not make much of an error approximating the geometric here by the exponential and the binomial in the next step by the Poisson distribution.

Having pointed out this slight inaccuracy, we now return to performing the computation as Seoighe and Wolfe (1998) have done and using their notation. The probability that a segment of length x contains three or more homologues is

$$1 - e^{-Dx} - Dxe^{-Dx} - \frac{(Dx)^2}{2}e^{-Dx}$$

where D is the density of homologous pairs in the whole genome. Thus the fraction of the genome covered by segments has expected value

$$F = N \int_0^\infty \left(1 - e^{-Dx} - Dxe^{-Dx} - \frac{(Dx)^2}{2}e^{-Dx}\right)\frac{1}{L}xe^{-x/L}\,dx$$

where N is the number of segments. Replacing N by $5790/L$ and recalling

$$\int_0^\infty x^k e^{-\lambda x}\,dx = k!/\lambda^{k+1} \tag{9.14}$$

where $0! = 1$, we have

$$F = \frac{5790}{L}\left(L - \frac{1}{L(D+L^{-1})^2} - \frac{2D}{L(D+L^{-1})^3} - \frac{3D^2}{L(D+L^{-1})^4}\right) \tag{9.15}$$

As in Nadeau and Taylor (1984), if m is the expected length of a segment that contains n paralogues separated by a total distance r then

$$m = r(n+1)/(n-1)$$

Using this, the fraction of the genome covered by blocks, 0.496, translates into a fraction 0.686 of the genome in identified segments. Using this in (9.14) we have $L = 5790/16.45 = 352$ segments organized as 176 pairs. Eight of these breakpoints are chromosome ends, yielding an estimate of 84 reciprocal translocations.

As a further check on the predictions of the calculations above we can observe that the probability that a segment of length x contains y homologues is $e^{-Dx}(Dx)^y/y!$, so the expected number of segments of length x with y homologues is

$$N \int_0^\infty e^{-Dx} \frac{(Dx)^y}{y!} \cdot \frac{1}{L} e^{-x/L} \, dx = \frac{ND^y}{L(D + L^{-1})^{y+1}} \qquad (9.16)$$

where we have used (9.14) to evaluate the integral. Comparing the expected values from (9.16) with the data gives the result in the following table. Here simulation refers to a model with 446 pairs of retained duplicates and 84 reciprocal translocations, and the intervals in this column are the mean $\pm$ 2 standard deviations. Since the simulation parameters are closely related to the parameters used in evaluating (9.16), it should come as no surprise that (9.16) agrees well with the simulation. The agreement between the data and simulation is not so good. In six cases (3, 5, 8, 10, 12, 13) the data lie outside the approximate 95% confidence interval based on the simulation. Since blocks are defined by the occurrence of 3 or more homologues, there are no data for 0, 1, or 2. The additional number of two-member blocks found in the real data is 34. This is larger than the theoretical prediction, but might reflect the fact that there have been some small-scale duplications since the whole genome duplication event.

y	data	(9.16)	simulation
0		49.6	49.4
1		35.6	35.9 ± 5.9
2		25.6	26.2 ± 5.2
3	10	18.4	18.5 ± 4.3
4	10	13.2	13.2 ± 3.5
5	6	9.5	9.6 ± 3.1
6	4	6.8	6.8 ± 2.5
7	6	4.9	4.8 ± 2.1
8	6	3.5	3.3 ± 1.7
9	1	2.5	2.4 ± 1.4
10	4	1.8	1.8 ± 1.2
11	2	1.3	1.3 ± 1.1
12	1	0.9	0.9 ± 1.0
13	4	0.7	0.6 ± 0.7
14	0	0.5	0.4 ± 0.6
15	0	0.3	0.3 ± 0.6
16–20	1	0.8	0.6 ± 2.6
21–25	0	0.1	0.0

Having estimated the number of reciprocal translocations since genome duplication, it is natural to inquire about the minimum number of translocations needed to rearrange the duplicated genes on the 16 chromosomes so that the yeast genome consists of two sets of 8 chromosomes with identical gene order. Seoighe and Wolfe found that after three initial inversions were performed to correct the orientation of the five blocks whose orientation relative to the centromere is opposite to their copies, this could be done in 41 steps. This is considerably smaller than the estimate of 84 given above. However, the reader should note that this only requires that the conserved blocks be brought into a symmetrical arrangement, while simulations in Seoighe and Wolfe (1998) show that the number of moves required to do this can be much smaller than the number performed to scramble the genome. El-Marbrouk, Bryant, and Sankoff (1999) attacked this problem using Hannenhalli's breakpoint graph. They did not use any inversions and found that a duplication followed by 45 translocations is necessary and sufficient to produce the current order of duplicated genes. For more on genome halving problems, see El-Marbrouk, Nadeau, and Sankoff (1999).

Recently, Friedman and Hughes (2000) have taken another look at duplicated genes in yeast. They used a cutoff score of $p = 10^{-50}$ to define homologous genes. By comparing the number of matches in windows of fixed size for the yeast genome with that of a randomly scrambled version, they determined that for their fixed window size 4 matches disregarding order were needed to define a block. With this criterion, they identified 39 blocks. To test the hypothesis of a single genome duplication they looked at the distribution of the fraction of synonymous substitutions, p_s. They found that the distribution of p_s was distinctly bimodal, indicating that some duplications were recent but a subset of 28 duplications occurred in the distant past. For the more ancient duplications they estimated an age of 200–300 million years in contrast to Wolfe and Shield's estimate of 100 million years.

9.6.2 Maize

Ahn and Tanksley (1993) constructed genetic linkage maps for rice and maize. They found that in some instances entire chromosomes or chromosome arms were nearly identical with respect to gene content and gene order. Their analysis also revealed that most (> 72%) of the single-copy loci in rice were duplicated in maize, suggesting the presence of a polyploidization event. This pattern extends to many other cereals. Moore et al. (1995) showed that the genomes of rice, wheat, maize, foxtail millet, sugar cane, and sorghum could be aligned by dissecting the individual chromosomes into segments and rearranging the linkage blocks into similar structures. Further work by Devos and Gale (1997) and Gale and Devos (1998) has brought more detail to the picture. However, their circular comparative maps are hard for us to interpret so we will stick with the simple linear picture in Moore et al. (1995). Using numbers and letters to indicate segments of rice chromosomes, the maize genome can be written as follows

M3 =	12a	1a	1b			M8 =	1a	5a	5b	1b
M6 =	6a	6b	5a	5b		M9 =	6a	6b	8	3c
M1 =	3c	8	10	3b	3a	M5 =	2	10	3b	3a
M4 =	11a	2				M7 =	11a	9	7	
M2 =	4a	4b	9	7		M10 =	4a	4b	12a	

The underlined groups are conserved between the two sets of chromosomes, so we will call them segments. Making choices about the orientation of the segments of length one that will minimize the distance leads to the following:

$$\begin{array}{lll}
M3 = +1 & +2 & +3 \\
M6 = +4 & +5 & \\
M1 = +6 & +7 & \\
M4 = +8 & +9 & \\
M2 = +10 & +11 &
\end{array}
\qquad
\begin{array}{lll}
M8 = -2 & +5 & +3 \\
M9 = +4 & -6 & \\
M5 = -9 & +7 & \\
M7 = +8 & +11 & \\
M10 = +10 & -1 &
\end{array}$$

If we invert $-2, 5$ in $M8$, then $M8 = -5 + 2 + 3$ and the resulting genomes are cotailed in the terminology of Hannenhalli. That is, the same blocks appear on the ends. The reader may notice that $+6$ is on the front end of $M1$ while -6 is on the back end of $M9$. However, this is exactly what we want. If we flip $M9$, then it has $+6$ on the front.

To construct the breakpoint graph, we follow the procedure in Section 9.1, but now we use 0's for the chromosome ends:

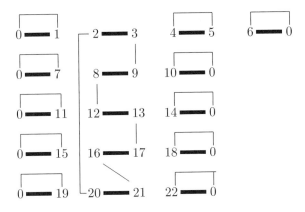

When M8, M9, M5, M7, and M10 are rearranged to match the order in the other five chromosomes, there will be 16 cycles. Now there are 12, so a minimum of 4 translocations is needed. It is easy to see that this is sufficient. We first make $M8$ match $M3$, then make $M9$ match $M6$, etc., and we are done in four moves.

The solution to undoubling the maize genome is elegant and parsimonious. Unfortunately, it is also wrong. The absence of duplicated blocks in millets, sorghum, and sugarcane locates the duplication at the indicated place in the phylogeny. Comparing with the haploid chromosome number x of closely related species makes it appear unlikely that the progenitor maize genome had five chromosomes. Wilson et al. (1999) have suggested that the progenitor maize genome had eight chromosomes, even though this requires six fusions after duplication to reduce the number to the current ten. Their suggestion for the makeup of the predoubling genome can be found in Figure 4 of their paper. It requires six inversions and one insertion to produce the current maize genome.

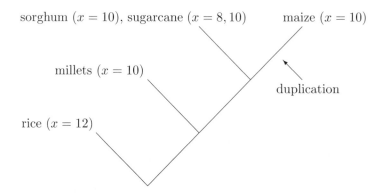

sorghum ($x = 10$), sugarcane ($x = 8, 10$)   maize ($x = 10$)

millets ($x = 10$)

duplication

rice ($x = 12$)

The tetraploid event in the evolution of maize could be an autotetraploid event (the doubling of one genome) or an allotetraploid event (the hybridization of two closely related species). Hexaploid bread wheat (*Triticum aestivum*) is an allopolyploid with genome construction AABBCC, formed through hybridization of *T. uratu* with a B genome of unknown origin, and subsequent hybridization only about 8000 years ago with C genome diploid, *T. tauschii*. In order to distinguish between these two scenarios, Gaut and Doebley (1997) examined the ages of 14 duplicate genes in maize.

loci	distance
orp1, orp2	0.298(1.44)
ant1, ant2	0.277(1.32)
ohp1, ohp2	0.254(1.19)
r, b	0.241(0.83)
cpna, cpnb	0.186(0.55)
cdc2a, cdc2b	0.177(1.04)
whp1, c2	0.169(0.66)
fer1, fer2	0.168(1.44)
cI, plI	0.159(1.05)
ibp1, ibp2	0.150(0.36)
tbp1, tbp2	0.147(1.20)
vpl4a, vpl4b	0.121(0.29)
obf1, obf2	0.104(0.48)
pgpa1, pgpa2	0.102(0.39)

In an autotetraploid, the divergence of the gene copies will all start at the time the species returns to being a diploid. In an allotetraploid, during the time of tetrasomic inheritance, genetic drift (or selection) could bring the alleles of one parent species to fixation. These genes would begin to diverge at the time of the shift to disomic inheritance, while loci that retain alleles from both parents would date to the time of the divergence of the two species that combined to make the allotetraploid. Thus the presence of two distinct age classes would be evidence for allotetraploidy. The previous table gives the synonymous distance between the duplicated loci with the variance times 1000 in parentheses. A χ^2 test soundly rejects ($p < 0.0001$) the hypothesis that all of these random variables have the same mean. The group above the line (group A) and the group below the line (group B) have 95% confidence intervals that don't overlap. A χ^2 homogeneity test was not significant for either group (group A, $p = 0.65$, and group B, $p = 0.08$).

Having demonstrated the existence of two groups of duplicated sequence pairs, Gaut and Doebley's next step was to explore how the duplication events are related to the divergence of maize and sorghum. To do this we first need an estimate of the rate of nucleotide substitutions. Gaut et al. (1996) estimated the synonymous substitution rate between the $Adh1$ and $Adh2$ loci of grasses at 6.5×10^{-9} substitutions per synonymous site per year. Given this rate, sequences in group A diverged roughly

$$\frac{0.267}{2 \cdot 6.5 \times 10^{-9}} = 20,500,000 \quad \text{years ago}$$

and pairs of duplicated sequences in group B diverged approximately 11.4 million years ago. Quite remarkably the average divergence between maize and sorghum falls between the group A distance and the group B distances, suggesting that the sorghum genome is more closely related to one of the maize sub genomes than to the other one. For more on this topic, see Gaut et al. (2000).

9.6.3 Arabidopsis thaliana

This plant was chosen as a model diploid plant species because of its compact genome size. Thus it was surprising when a 105 kilobase bacterial artificial chromosome clone from tomato sequenced by Ku et al. (2000) showed conservation of gene order and content with four different segments in chromosomes 2–5 of *Arabidopsis*. This pattern suggests that these four segments were derived from a common ancestral segment through two or more rounds of large scale genome duplication events, or by polyploidy. Comparison of the divergence of the gene copies suggested that one of the duplication events is ancient and may predate the divergence of *Arabidopsis* and tomato approximately 112 million years ago, while the other is more recent.

When the complete sequence of *Arabidopsis* was later published (see Nature 408, 796–815), dot-matrix plots of genic similarities showed that much of the genome fell into pairs. These plots provide compelling evidence in support for one polyploidy event, which the authors of *Arabidopsis* Genome Initiative took to be the more recent event proposed by Ku et al. (2000). A significantly different conclusion was made by Vision, Brown, and Tanksley (2000), who analyzed the duplications in the completed genome sequenced. Their results suggest that there were three rounds of duplications where 25, 36, and 23 blocks were duplicated 100, 140, and 170 years ago. An independent analysis of the duplicated genes by Lynch and Conrey (2000) suggested that there was one whole genome duplication 65 million years ago.

References

Adams, M.D., et al. (2000) The genome sequence of *Drosophila melanogaster*. *Science*. 287, 2185–2195

Aguadé, M., Miyashita, N., and Langley, C.H. (1992) Polymorphism and divergence in *Mst26A* male accessory gland gene region in *Drosophila*. *Genetics*. 132, 755–770

Ahn, S., and Tanksley, S.D. (1993) Comparative linkage maps of the rice and maize genomes. *Proc. Natl. Acad. Sci. USA*. 90, 7980–7984

Aldous, D.J. (1985) Exchangeability and related topics. Pages 1–198 in Lecture Notes in Math 1117. Springer-Verlag, New York

Aldous, D.J. (1989) Hitting times for random walks on vertex-transitive graphs. *Proc. Camb. Phil. Soc.* 106, 179–191

Anderson, E.C., and Slatkin, M. (2004) Population-genetic basis of haplotype blocks in the 5q31 region. *Am. J. Human Genetics*. 74, 40–49

Anderson, S., et al. (1981) Sequence and organization of the human mitochondrial genome. *Nature* 325, 457–465

Aquadro, C.F. (1991) Molecular population genetics of *Drosophila*. In *Molecular Approaches in Pure and Applied Entomology*. Edited by J. Oakeshott and M. Whitten. Springer-Verlag, New York

Aquadro, C.F., Begun, D.J., and Kindahl, E.C. (1994) Pages 46–56 in *Nonneutral Evolution: Theories and Molecular Data*. Edited by B. Golding. Chapman and Hall, New York.

Aquadro, C.F., and Greenberg, B.D. (1983) Human mitochondrial DNA variation and evolution: Analysis of nucleotide sequences from seven individuals. *Genetics*. 103, 287–312

Ardlie, K.G., Kruglyak, L., and Seielstad, M. (2002) Patterns of linkage disequilibrium in the human genome. *Nature Reviews Genetics*. 3, 299–309

Arnason, E. (2004) Mitochondrial cytochrome *b* variation in the high-fecundity Atlantic cod: trans-Atlantic clines and shallow gene genealogy. *Genetics.* 166, 1871–1885

Arratia, R., Barbour, A.D., and Tavaré, S. (1992) Poisson process approximations for the Ewens sampling formula. *Ann. Appl. Prob.* 2, 519–535

Arratia, R., Barbour, A.D., and Tavaré, S. (2003) *Logarithmic Combinatorial Structures : A Probabilistic Approach.* European Mathematical Society, Zurich

Arratia, R., Gordon, L., and Goldstein, L. (1989) Two moments suffice for Poisson approximation: the Chen-Stein method. *Ann. Prob.* 17, 9–25

Athreya, K.B., and Ney, P.E. (1972) *Branching Processes.* Springer, New York

Bader, D.A., Moret, B.M.E., and Yan, M. (2000) A linear time algorithm for computing inversion distance between signed permutations in linear time with an experimental study. Tech Report TR-CS-2000-42, Department of Computer Science, U. of New Mexico, available at `http://www.cs.unm.edu/`

Bafna, V., and Bansal, V. (2006) Inference about recombination from haplotype data: Lower bounds and recombination hot spots. *J. Comp. Biol.* 13, 501–521

Bafna, V. and Pevzner, P. (1995) Sorting by reversals: Genome rearrangement in plant organelles and evolutionary history of X chromosome. *Mol. Biol. Evol.* 12, 239–246

Bafna, V., and Pevzner, P. (1996) Genome rearrangements and sorting by reversals. *SIAM J. Computing.* 25, 272–289

Bailey, J.A., et al. (2002) Recent segmental duplications in the human genome. *Science.* 297, 1003–1007

Balding, D.J., Bishop, M., and Cannings, C. (2001) *Handbook of Statistical Genetics.* John Wiley and Sons, New York

Bamshad, M., and Wooding, S.P. (2003) Signatures of natural selection in the human genome. *Nature Reviews Genetics.* 4, 99-111

Band, M.J., et al. (2000) An ordered comparative map of the cattle and human genomes. *Genome Research.* 10, 1359–1368

Barton, N.H. (1995) Linkage and the limits to natural selection. *Genetics.* 140, 821–841

Begun, D.J., and Aquadro, C.F. (1992) Levels of naturally occurring DNA polymorphisms correlate with recombination rates in *D. melaongaster. Nature.* 356, 519–520

Berestycki, J., Berestycki, N., and Schweinsberg, J. (2006a) Beta-coalescents and continuous stable random trees. arXiv:math.PR/0602113 *Ann. Prob.*, to appear

Berestycki, J., Berestycki, N., and Schweinsberg, J. (2006b) Small time behavior of beta coalescents. arXiv:math.PR/0601032 *Ann. Inst. H. Poincaré. Probab. Statist.*, to appear

Berestycki, N. and Durrett, R. (2006) A phase transition in the random transposition random walk. *Prob. Theory. Rel. Fields.* 136, 203–233

Berry, A.J., Ajioka, J.W., and Kreitman, M. (1991) Lack of polymorphism on the *Drosophila* fourth chromosome resulting from selection. *Genetics.* 129, 1111–1117

Birkner, M., and Blath, J. (2007a) Inference for Λ-coalescents. WIAS preprint 1211. Available at http://www.wias-berlin.de/people/birkner/

Birkner, M., and Blath, J. (2007b) Computing likelihoods for coalescents with multiple collisions in the infinitely-many-sites model. WIAS preprint 1237. Available at http://www.wias-berlin.de/people/birkner/

Birkner, M., Blath, J., Capaldo, M., Etheridge, A., Möhle, M., Schweinsberg, J., and Wakolbinger, A. (2005) Alpha-stable branching and beta-coalescents. *Electronic Journal of Probability.* 10, 303–325

Blanchette, M., Bourque, G., and Sankoff, D. (1997) Breakpoint phylogenies. Pages 25-34 in *Genome Informatics 1997.* Edited by S. Miyano and T. Takagi, Universal Academy Press, Tokyo.

Blanchette, M., Kunisawa, T., and Sankoff, D. (1999) Gene order breakpoint evidence in animal mitochondrial phylogeny. *J. Mol. Evol.* 49, 193–203.

Bodmer, W.F., and Cavalli-Sforza, L.L. (1968) A migration matrix model for the study of random genetic drift. *Genetics.* 59, 565–592

Bollobás, B. (2001) *Random Graphs*, Second edition. Cambridge University Press

Boom, J.D.G., Boulding, E.G., and Beckenbach, A.T. (1994) Mitochondrial DNA variation in introduced populations of Pacific oyster, *Crassostrea gigas*, in British Columbia. *Can. J. Fish. Aquat. Sci.* 51, 1608–1614

Borodin, A.N., and Salminen, P. (2002) *Handbook of Brownian Motion – Facts and Formulae.* Birkhauser, Boston

Bowen, B.W., and Grant, W.S. (1997) Phylogeography of the sardines (*Sardinops* SPP.): Assessing biogeographic models and population histories in temperate upwelling zones. *Evolution.* 51, 1601–1610

Bourque, G. and Pevzner, P.A. (2002) Genome-scale evolution: reconstructing gene orders in the ancestral species. *Genome Research.* 12, 26–36

Bourque, G., Pevzner, P.A., and Tesler, G. (2002) Reconstructing the genomic architecture of ancestral mammals: Lessons from human, mouse, and rat genomes. *Genome Research.* 14, 507–516

Braverman, J.M., Hudson, R.R., Kaplan, N.L., Langley, C.H., and Stephan, W. (1995) The hitchhiking effect on the site frequency spectrum of DNA polymorphisms. *Genetics.* 140, 783–796

Bustamante, C.D., et al. (2005) Natural selection on protein coding genes in the human genome. *Nature.* 437, 1153–1157

Bustamante, C.D., Nielsen, R., Sawyer, S.A., Olsen, K.M., Purugganan, M.D., and Hartl, D.L. (2002) The cost of inbreeding in *Arabadopsis. Nature.* 416, 531–534

Bustamante, C.D., Wakeley, J., Sawyer, S., and Hartl, D.L. (2001) Directional selection and the site-frequency spectrum. *Genetics.* 159, 1779–1788

Cann, R.L., Stoneking, M., and Wilson, A.C. (1987) Mitochondrial DNA and human evolution. *Nature.* 325, 31–36

Caprara, A. (1997) Sorting by reversals is difficult. Pages 75–83 in the *Proceedings of RECOMB 97.* ACM, New York

Caprara, A. (1999a) Formulations and hardness of multiple sorting by reversals. *Proceedings of RECOMB '99.* ACM, New York

Caprara, A. (1999b) On the tightness of the alternating-cycle lower bound for sorting by reversals. *J. Combinatorial Optimization.* 3, 149–182

Cardon, L.R., and Bell, J.I. (2001) Association study designs for complex diseases. *Nature Reviews Genetics.* 2, 91–98

Casella, G., and Berger, R.L. (1990) *Statistical Inference.* Wadsworth & Brooks Cole, Pacific Grove, CA

Charlesworth, B., Morgan, M.T., and Charlesworth, D. (1993) The effects of deleterious mutations on neutral molecular variation. *Genetics.* 134, 1289–1303

Charlesworth, D., Charlesworth, B., and Morgan, M.T. (1995) The pattern of neutral molecular variation under the background selection model. *Genetics.* 141, 1619–1632

Chovnick, A., Gelbart, W., and McCarron, M. (1977) Organization of the *rosy* locus in *Drosophila melanogaster. Cell.* 11, 1–10

Chowdhary, B.P., Fronicke, L., Gustavsson, I. and Scherthan, H. (1996) Comparative analysis of cattle and human genomes: detection of ZOO-FISH and gene mapping-based chromosomal homologies. *Mammalian Genome.* 7, 297–302

Clark, A.G., et al. (1998) Haplotype structure and population genetic inferences from nucleotide sequence variation in human lipoprotein lipase. *Am. J. Human Genetics.* 63, 595–612

Cook, L.M., and Jones, D.A. (1996) The medionigra gene in the moth panaxia dominula: the case for selection. *Phil. Trans. Roy. Soc. London, B.* 351, 1623–1634

Copeland, N.G., et al. (1993) A genetic linkage map of mouse: Current applications and future prospects. *Science.* 262, 57–66

Cox, J.T. and Durrett, R. (2002) The stepping stone model: new formulas expose old myths. *Ann. Appl. Prob.* **12**, 1348-1377

Cox, J.T., and Griffeath, D. (1986) Diffusive clustering in the two dimensional voter model. *Ann. Probab.* 14, 347–370

Coyne, J.A. (1976) Lack of genic similarity between two sibling species of *Drosophila* as revealed by varied techniques. *Genetics.* 84, 593–607

Crow, J.F. and Aoki, K. (1984) Group selection for a polygenic trait: Estimating the degree of population subdivision. *Proc. Nat. Acad. Sci.* 81, 6073–6077

Crow, J.F., and Kimura, M. (1965) Evolution in sexual and asexual populations. *Amer. Naturalist.* 99, 439–450

Crow, J.F., and Kimura, M. (1969) Evolution in sexual and asexual populations: a reply. *Amer. Naturalist.* 103, 89–91

Crow, J.F., and Simmons, M.J. (1983) The mutation load in *Drosophila.* Pages 1–35 in *The Genetics and Biology of Drosophila.* Edited by M. Ashburner, H.L. Carson, and J.N. Thompson. Academic Press, London

Daly, M.J., Rioux, J.D., Schaffner, S.F., Hudson, T.J., and Lander, E.S. (2001) High-resolution haplotype structure in the human genome. *Nature Genetics.* 29, 229-232

Dawson, E., et al. (2002) A first generation linkage disequilibrium map of chromosome 22. *Nature.* 418, 544-548

De, A., and Durrett, R. (2007) Stepping stone spatial structure causes slow decay of linkage disequilibrium and shifts the site frequency spectrum. *Genetics.* 176, 969–981

DeBry, R.W., and Seldin, M.F. (1996) Human/mouse homology relationships. *Genomics.* 33, 337–351

Depaulis, F., Mousset, S., and Veuille, M. (1996) Neutrality tests based on the distribution of haplotypes under an infinite sites model. *Mol. Biol. Evol.* 15, 1788–1790

Deoaulis, F., and Veuille, M. (1998) Neutrality tests based on the distribution of haplotypes under the infinite sites model. *Mol. Biol. Evol.* 15, 1788-1790

Devos, K.M., and Gale, M.D. (1997) Comparative genetics in the grasses. *Plant Molecular Biology.* 35, 3–15

Diaconis, P., and Stroock, D. (1991) Geometric bounds on the eigenvalues of Markov chains. *Ann. Appl. Prob.* 12, 1348–1377

DiRienzo, A., et al. (1994) Mutational processes of simple sequence repeat loci in human populations. *Proc. Natl. Acad. Sci. USA* 91, 3166–3170

Dobzhansky, T., and Sturtevant, A. (1938), Inversions in the chromosomes of *Drosophila pseuodobscura*. *Genetics.* 23, 28–64

Doganlar, S., Frary, A., Daunay, M.C., Lester, R.N., and Tanksley, S.D. (2002) A comparative genetic linkage map of eggplant (*Solanum melongea*) and its implications for genome evolution in the Solanaceae. *Genetics.* 161, 1697–1711

Donnelly, P., and Tavaré, S. (1986) The ages of alleles and a coalescent. *Adv. Appl. Prob.* 18, 1–19

Dorit, R.L., Akashi, H., and Gilbert, W. (1995) Absence of polymorphism at the *ZFY* locus on the human Y chromosome. *Science* 268, 1183–1185

Drake, J.W., Charlesworth, B., Charlesworth, D., and Crow, J.F. (1998) Rates of spontaneous mutation. *Genetics.* 148, 1667–1686

Durrett, R. (1996) *Stochastic Calculus: A Practical Introduction.* CRC Press, Boca Raton, FL

Durrett, R. (2002) Shuffling chromosomes. *J. Theor. Prob.* 16, 725–750

Durrett, R. (2005) *Probability: Theory and Examples, 3rd Ed.* Duxbury Press, Pacific Grove, CA

Durrett, R. and Interian, Y. (2006) Genomic midpoints: computation and evolutionary implications. Unpublished manuscript

Durrett, R., and Limic, V. (2001) On the quantity and quality of single nucleotide polymorphisms in the human genome. *Stoch. Proc. Appl.* 93, 1–24

Durrett, R., Nielsen, R., and York, T.L. (2004) Bayesian estimation of genomic distance. *Genetics.* 166, 621–629

Durrett, R., and Popovic, L. (2007) Degenerate diffusions arising from gene duplication models. *Ann. Appl. Prob.*, to appear

Durrett, R., and Restreop, M. (2007) One dimensional stepping stone models, sardine genetics, and Brownian local time. *Annals of Applied Probability.* to appear

Durrett, R., and Schweinsberg, J. (2004) Approximating selective sweeps. *Theor. Pop. Biol.* 66, 129–138

Durrett, R., and Schweinsberg, J. (2005) A coalescent model for the effect of advantageous mutations on the genealogy of a population. *Stoch. Proc. Appl.* 115, 1628–1657

Dynkin, E.B. (1965) *Markov Processes.* Springer-Verlag, Berlin

Eanes, W.F., Kirchner, M., and Yoon, J. (1993) Evidence for adaptive evolution of the *G6pd* in the *Drosophila melanogaster* and *Drosophila simulans* lineages. *Proc. Natl. Acad. Sci. USA*. 90, 7475–7479

Eldon, B., and Wakeley, J. (2006) Coalescent processes when the distribution of offspring number among individuals is highly skewed. *Genetics*. 172, 2621–2633

El-Mabrouk, N., Bryant, D., and Sankoff, D. (1999) Reconstructing the pre-doubling genome. Pages 154–163 in the Proceedings of RECOMB 99. ACM, New York

El-Mabrouk, N., Nadeau, J.H., and Sankoff, D. (1999) Genome halving. Pages 235–250 in *Proceedings of the 9th Annual Symposium on Combinatorial Pattern Matching*. Lecture Notes in Computer Science 1448, Springer, New York

Eriksson, A., Fernström, P., Mehlig, B., and Sagitov, S. (2007) An accurate model for genetic hitchhiking. arXiv:0705.2286v1 [q-bio.PE]

Etheridge, A.M., Pfaffelhuber, P., and Wakolbinger, A. (2006) An approximate sampling formula under genetic hitchhiking. *Ann. Appl. Prob*. 16, 685–729

Ethier, S.N., and Griffiths, R.C. (1987) The infinitely-many-sites model as a measure-valued diffusion. *Ann. Prob*. 15, 515–545

Ethier, S.N., and Griffiths, R.C. (1990) On the two-locus sampling distribution. *J. Math. Biol*. 29, 131–159

Ethier, S.N., and Kurtz, T.G. (1986) *Markov Processes: Characterization and Convergence*. John Wiley and Sons, New York

Ewens, W.J. (1972) The sampling theory of selectively neutral alleles. *Theor. Pop. Biol*. 3, 87–112

Ewens, W.J. (1973) Conditional diffusion processes in population genetics. *Theor. Pop. Biol*. 4, 21–30

Ewens, W.J. (1979) *Mathematical Population Genetics*. Springer-Verlag, Berlin

Ewens, W. (2004) *Mathematical Population Genetics. I. Theoretical Introduction*. Springer, New York

Even, S. and Goldreich, O. (1981) The minimum-length generator sequence problem is NP-hard. *J. of Algorithms*. 2, 311–313

Eyre-Walker, A., Gaut, R.L., Hilton, H., Feldman, D.L., and Gaut, B.S. (1998) Ivestigation of the bottleneck leading to the domestication of maize. *Proc. Natl. Acad. Sci. USA*. 95, 4441–4446

Fay, J.C., and Wu, C.I. (1999) A human population bottleneck can account for the discordance between patterns of mitochondrial versus nuclear DNA variation. *Mol. Biol. Evol*. 16, 1003–1005

Fay, J.C., and Wu, C.I. (2000) Hitchhiking under positive Darwinian selection. *Genetics*. 155, 1405–1413

Fay, J.C., and Wu., C.I. (2001) The neutral theory in the genomic era. *Current Opinion in Genetics and Development*. 11, 642–646

Fearnhead, P. (2003) Consistency of estimators of the population-scaled recombination rate. *Theor. Pop., Biol.* 64, 67–79

Fearnhead, P., and Donnelly, P. (2001) Estimating recombination rates from population genetic data. *Genetics*. 159, 1299–1318

Fearnhead, P., and Donnelly, P. (2002) Approximate likelihood methods for estimating local recombination rates. *J. Roy. Stat. Soc. B* 64, 657–680

Fearnhead, P., Harding, R.M., Schneider, J.A., Myers, S., and Donnelly, P. (2004) Application of coalescent methods to reveal fine-scale rate variation and recombination hot spots. *Genetics*. 167, 2067–2081

Feller, W. (1951) Diffusion processes in genetics. Pages 227–246 in *Proc. Second Berkeley Symp. Math. Statist. Prob.*, U of California Berkeley Press

Felsenstein, J. (1974) The evolutionary advantage of recombination. *Genetics*. 78, 737–756

Felsenstein, J. (1976) The theoretical population genetics of variable selection and migration. *Annu. Rev. Genet.* 10, 253–280

Felsenstein, J. (1982) How can we infer geography and history from gene frequencies? *J. Theor. Biol.* 96, 9–20

Fisher, R.A. (1922) On the dominance ratio. *Proc. Roy. Soc. Edin.* 42, 321–341

Fisher, R.A. (1930) *The Genetical Theory of Natural Selection*. Reprinted in 1999 by Oxford University Press.

Fisher, R.A. (1935) The sheltering of lethals. *Amer. Naturalist*. 69, 446–455

Fisher, R.A., and Ford, E.B. (1947) The spread of a gene in natural conditions in a colony of the moth *Panaxia dominula*. *Heredity* 1, 143–174

Fisher, R.A., and Ford, E.B. (1950) The "Sewall Wright Effect." *Heredity*. 4, 117–119

Friedman, R., and Hughes, A.L. (2000) Gene duplication and the structure of eukaryotic genomes. *Genome Research*. 11, 373–381

Frisee, L., Hudson, R.R., Bartoszewicz, A., Wall, J.D., Donfack, J., and DiRienzo, A. (2001) Gene conversion and different population histories may explain the contrast between polymorphism and linkage disequilibrium levels. *Am. J. Human Genetics*. 69, 831–843

Fu, Y.X. (1995) Statistical properties of segregating sites. *Theor. Pop. Biol.* 48, 172–197

Fu, Y.X. and Li, W.H. (1993) Statistical tests of neutrality of mutations. *Genetics.* 133, 693–709

Gabriel, S.B., et al. (2002) The structure of haplotype blocks in the human genome. *Science.* 296, 2225-2229

Gale, M.D., and Devos, K.M. (1998) Comparative genetics in the grasses. *Proc. Natl. Acad. Sci. USA* 95, 1971–1974

Gaut, B.S., and Doebley, J.F. (1997) DNA sequence evidence for the segmental allotetraploid origin of maize. *Proc. Natl. Acad. Sci. USA.* 94, 6809–6814

Gaut, B.S., Le Thierry d'Ennequin, M., Peek, A.S., and Sawkins, M.C. (2000) Maize as a model for the evolution of plant nuclear genomes. *Proc. Natl. Acad. Sci. USA* 97, 7008–7015

Gaut, B.S., Morton, B.R., McCaig, B.M., and Clegg, M.T. (1996) Substitution rate comparisons between grasses and palms: Synonymous rate differences at the nuclear gene Adh parallel rate differences at the plastid gene rbcL. *Proc. Natl. Acad. Sci. USA.* 93, 10274–10279

Gelman, A., and Rubin, D.B. (1992) Inference from iterative simulation using multiple sequences (with discussion). *Statist. Sci.* 7, 457–511

Gillespie, J.H. (1991) *The Causes of Molecular Evolution.* Oxford University Press

Gillespie, J.H. (1998) *Population Genetics: A Concise Course.* John Hopkins University Press, Baltimore, MD

Golding, G.B. (1984) The sampling distribution of linkage disequilibrium. *Genetics.* 108, 257–274

Goldman, N., and Yang, Z. (1994) A codon-based model of nucleotide substitutions for protein-coding DNA sequences. *Mol. Biol. Evol.* 11, 725–736

Gonick, L., and Wheelis, M. (1991) *The Cartoon Guide to Genetics.* Harper-Collins, New York

Griffiths, R.C. (1981) Neutral two-locus multiple allele models with recombination. *Theor. Pop. Biol.* 19, 169–186

Griffiths, R.C. (1991) The two-locus ancestral graph. Pages 100–117 in *Selected Proceedings of the Symposium on Applied Probability, Sheffield.* Institute of Mathematical Statistics, Hayward, CA

Griffiths, R.C. (2003) The frequency spectrum of a mutation, and its age, in a general diffusion model. *Theor. Pop. Biol.* 64, 241–251

Griffiths, R.C., and Marjoram, P. (1996) Ancestral inference from samples of DNA sequences with recombination. *J. Comp. Biol.* 3, 479–502

Griffiths, R.C., and Marjoram, P. (1997) An ancestral recombination graph. Pages 257–270 in *Progress in Population Genetics and Human Evolution.* Edited by P. Donnelly and S. Tavaré. IMA Volumes in Mathematics and its Applications, Springer, New York

Griffiths, R.C., and Pakes, A.G. (1988) An infinite-alleles version of the simple branching process. *Adv. Appl. Prob.* 20, 489–524

Griffiths, R.C., and Tavaré, S. (1994a) Ancestral inference in population genetics. *Statist. Sci.* 9, 307–319

Griffiths, R.C., and Tavaré, S. (1994b) Sampling theory for neutral alleles in a varying environment. *Phil. Trans. Roy. Soc. London, B.* 344, 403–410

Haig, D. (1999) A brief history of human autosomes. *Phil. Trans. Roy. Soc. London, B.* 354, 1447–1470

Hamblin, M.T., and Aquadro, C.F. (1996) High nucleotide sequence variation in a region of low recombination in *Drosophila simulans* is consistent with the background selection model. *Mol. Biol. Evol.* 13, 1133–1140

Hamblin, M.T., and Veuille, M. (1999) Population structure among African and derived populations of *Drosophila simulans*; evidence for ancient subdivision and recent admixture. *Genetics.* 153, 305–317

Hannenhalli, S., Chappey, C., Koonin, E.V., and Pevzner, P.A. (1995) Genome sequence comparisons and scenarios for gene rearrangements: A test case. *Genomics.* 30, 299-311

Hannenhalli, S., and Pevzner, P.A. (1995a) Transforming cabbage into turnip (polynomial algorithm for sorting signed permutations by reversals). Pages 178–189 in *Proceedings of the 27th Annual ACM Symposium on the Theory of Computing.* Full version in the *Journal of the ACM.* 46, 1–27

Hannenhalli, S., and Pevzner, P. (1995b) Transforming men into mice (polynomial algorithm for the genomic distance problem. Pages 581-592 in *Proceedings of the 36th Annual IEEE Symposium on Foundations of Computer Science.* IEEE, New York

Harding, R.M., Fullerton, S.M., Griffiths, R.C., Bond, J., Cox, M.J., Schneider, J.A., Moulin, D.S., and Clegg, J.B. (1997) Archaic African and Asian lineages in the genetic ancestry of modern humans. *Am. J. Human Genetics.* 60, 772–789

Hartl, D., and Clark, A.G. (2007) *Principles of Population Genetics.* Fourth Edition. Sinauer, Sunderland, MA

Hayes, H. (1995) Chromosome painting with human chromosome-specific DNA libraries reveals the extent and distribution of conserved segments in bovine chromosomes. *Cytogenet. Cell. Genetics.* 71, 168–174

Hey, J. (1991) A multi-dimensional coalescent process applied to multi-allelic selection models and migration models. *Theor. Pop. Biol.* 39, 30–48

Hey, J., and Wakeley, J. (1997) A coalescent estimator of the population recombination rate. *Genetics.* 145, 833–846

Hill, W.G., and Robertson, A. (1966) The effect of linkage on limits to artificial selection. *Genet. Res.* 8, 269–294

Hill, W.G., and Robertson, A. (1968) Linkage disequilibrium in finite populations. *Theoret. Appl. Genetics* 38, 226–231

Hinds, D.A., et al. (2005) Whole genome patterns of common DNA variation in the human population. *Science.* 307, 1072–1079

Hoppe, F. (1984) Polya-like urns and the Ewens' sampling formula. *J. Math. Biol.* 20, 91–94

Horai, S., and Hayasaka, K. (1990) Intraspecific nucleotide sequence differences in the major noncoding region of human mitochondrial DNA. *Am. J. Human Genetics.* 46, 828–842

Hudson, R.R. (1983) Properties of a neutral allele model with intragenic recombination. *Theor. Pop. Biol.* 23, 183–201

Hudson, R.R. (1985) The sampling distribution of linkage disequilibrium under an infinite allele model without selection. *Genetics.* 109, 611–631

Hudson, R.R. (1987) Estimating the recombination parameter of a finite population model without selection. *Genet. Res.* 50, 245–250

Hudson, R.R. (1991) Gene genealogies and the coalescent process. Pages 1–44 in *Oxford Surveys in Evolutionary Biology.* Edited by D. Futuyama and J. Antonovics. Oxford University Press

Hudson, R.R. (1993) The how and why of generating gene genealogies. Pages 23–36 in N. Takahata and A.G. Clark. *Mechanisms of Molecular Evolution.* Sinauer, Sunderland, MA

Hudson, R.R. (2001) Two-locus sampling distributions and their applications. *Genetics.* 159, 1805–1817

Hudson, R.R., and Kaplan, N. (1985) Statistical properties of the number of recombination events in the history of a sample of DNA sequences. *Genetics.* 111, 147–164

Hudson, R.R., and Kaplan, N.L. (1988) The coalescent process in models with selection and recombination. *Genetics.* 120, 831–840

Hudson, R.R., and Kaplan, N.L. (1994) Gene trees with background selection. Pages 140–153 in *Nonneutral Evolution: Theories and Molecular Data.* Edited by G.B. Golding. Chapman and Hall, New York

Hudson, R.R., and Kaplan, N.L. (1995) Deleterious background selection with recombination. *Genetics.* 141, 1605–1617

Hudson, R.R., and Kaplan, N.L. (1996) The coalescent process and background selection. Pages 57–65 in *New Uses for New Phylogenies.* Edited by P. Harvey et al. Oxford University Press

Hudson, R.R., Kreitman, M., and Aguadé (1987) A test of neutral evolution based on nucleotide data. *Genetics.* 116, 153–159

Hudson, R.R., Slatkin, M., and Maddison, W.P. (1992) Estimation of levels of gene flow from DNA sequence data. *Genetics.* 132, 583–589

Huerta-Sanchez, E., Durrett, R., and Bustamante, C.D. (2007) Population genetics of polymorphism and divergence under fluctuating selection. *Genetics.* Available online as 10.1534/genetics.107.073361

Hughes, A.L., and Nei, M. (1998) Pattern of nucleotide substitution at major histocompatibility complex class I reveals overdominant selection. *Nature.* 335, 167–170

Hughes, M.K., and Hughes, A.L. (1993) Evolution of duplicate genes in the tetraploid animal *Xenopus laevis. Mol. Biol. Evol.* 10 (1993), 1360–1369

Itô, K., and McKean, H.P. (1974) *Diffusion Processes and their Sample Paths.* Springer-Verlag, New York

Jeffreys, A.J., Ritchie, A., and Neumann, R. (2000) High resolution analysis of haplotype diversity and meiotic crossover in the human TAP2 recombination hot spot. *Human. Mol. Genetics.* 9, 725–733

Jeffreys, A.J., Kauppi, L., and Neumann, R. (2001) Intensely punctuate meiotic recombination in the class II region of the major histocompatability complex. *Nature Genetics.* 29, 217–222

Jensen, L. (1973) Random selective advantage of genes and their probabilities of fixation. *Genet. Res.* 21, 215–219

Joyce, P., and Tavaré, S. (1987) Cycles, permutations and the structure of the Yule process with immigration. *Stoch. Proc. Appl.* 25, 309–314

Kaj, I., and Krone, S.M. (2003) The coalescent process in a population of varying size. *J. Appl. Prob.* 40, 33–48

Kaneko, M., Satta, Y., Matsura, E.T., and Chigusa, S. (1993) Evolution of the mitochondrial ATPase 6 gene in *Drosophila*: Unusually high level of polymorphism in *D. melanogaster. Genet. Res.* 61, 195–204

Kaplan, H., Shamir, R., and Tarjan, R.E. (2000) Faster and simpler algorithm for sorting signed permutations by reversal. *SIAM Journal on Computing.* 29, 880–892

Kaplan, N.L., Darden, T., and Hudson, R.R. (1988) The coalescent in models with selection. *Genetics.* 120, 819–829

Kaplan, N.L., and Hudson, R.R. (1985) The use of sample genealogies for studying a selectively neutral *m*-locus model with recombination. *Theor. Pop. Biol.* 28, 382–396

Kaplan, N.L., Hudson, R.R., and Langley, C.H. (1989) The "hitchhiking effect" revisited. *Genetics.* 123, 887–899

Karlin, S., and Levikson, B. (1974) Temporal fluctuations in selection intensities: Case of small population size. *Theor. Pop. Biol.* 6, 383–412

Kauppi, L., Jeffreys, A.J., and Keeney, S. (2004) Where the crossovers are: recombination distributions in mammals. *Nature Reviews Genetics.* 5, 413–424

Ke, X., et al. (2004) The impact of SNP density on fine-scale patterns of linakge disequilibrium. *Human Mol. Genetics.* 13, 577-588

Keightley, P.D. (1994) The distribution of mutation effects on viability in *Drosophila melanogaster. Genetics.* 138, 1315–1322

Keogh, R.S., Seoighe, C., and Wolfe, K.H. (1998) Evolution of gene order and chromosome number in *Saccharomyces, Kluyveromyces,* and related fungi. *Yeast.* 14, 443–457

Kim, Y., and Stephan, W. (2000) Joint effects of genetic hitchhiking and background selection on neutral variation. *Genetics.* 155, 1415–1427

Kim, Y., and Stephan, W. (2002) Detecting a local signature of genetic hitchhiking along a recombining chromosome. *Genetics.* 160, 765–777

Kimura, M. (1953) "Stepping stone" model of population. *Ann. Rep. Nat. Inst. Genetics Japan.* 3, 62–63

Kimura, M. (1954) Process leading to quasi-fixation of genes in natural populations due to random fluctuation of selection intensities. *Genetics.* 39, 280–295

Kimura, M. (1955) Solution of a proces of random genetic drift with a continuous model. *Proc. Natl. Acad. Sci. USA* 41, 144–150

Kimura, M. (1962) On the probability of fixation of mutant genes in a population. *Genetics.* 47, 713–719

Kimura, M. (1964) Diffusion models in population genetics. *J. Appl. Prob.* 1, 177–232

Kimura, M. (1969) The number of heterozygous nucleotide sites maintained in a finite population due to a steady flux of mutations. *Genetics.* 61, 893–903

Kimura, M. (1971) Theoretical foundations of population genetics at the molecular level. *Theor. Pop. Biol.* 2, 174–208

Kimura, M., and King, J.L. (1979) Fixation of a deleterious allele at one of two "duplicate" loci by mutation pressure and random drift. *Proc. Natl. Acad. Sci. USA* 76, 2858–2861

Kimura, M., and Maruyama, T. (1966) The mutational load with epistatic gene interactions in fitness. *Genetics.* 54, 1337–1351

Kimura, M., and Maruyama, T. (1971) Patterns of neutral polymorphism in a geographically structured population. *Genet. Res. Camb.* 18, 125–131

Kimura, M., and Ohta, T. (1969a) The average number of generations until the fixation of a mutant gene in a finite population. *Genetics.* 61, 763–771

Kimura, M., and Ohta, T. (1969b) The average number of generations until extinction of an individual mutant gene in a finite population. *Genetics.* 63, 701–709

Kimura, M., and Ohta, T. (1973) The age of a neutral mutant persisting in a finite population. *Genetics.* 75, 199–212

Kimura, M., and Ohta, T. (1974) On some principles governing molecular evolution. *Proc. Nat. Acad. Sci. USA,* 7, 2848–2852

Kimura, M., and Weiss, G.H. (1964) The stepping stone model of population structure and the decrease of genetic correlation with distance. *Genetics.* 49, 561–576

Kingman, J.F.C. (1978) The representation of partition structures. *J. London Math. Soc.* 18, 374–380

Kingman, J.F.C. (1982a) The coalescent. *Stoch. Proc. Appl.* 13, 235–248

Kingman, J.F.C. (1982b) Exchangeability and the evolution of large populations. Pages 97–112 in *Exchangeability in Probability and Statistics.* Edited by G. Koch and F. Spizzichino. North-Holland, Amsterdam

Kondrashov, A.S. (1988) Deleterious mutation and the evolution of sexual reproduction. *Nature.* 336, 435–440

Kondrashov, F., Rogozin, I.B., Wolf, Y.I., and Koonin, E.V. (2002) Selection in the evolution of gene duplications. *Genome Biology.* 3(2):research0008.1-9

Kreitman, M. (1983) Nucleotide polymorphism at the alcohol dehydrogenase locus of *Drosophila melanogaster. Nature.* 304, 412–417

Kreitman, M. (2000) Methods to detect selection in populations with applications to the human. *Annu. Rev. Genomics Hum. Genet.* 1, 539–559

Kreitman, M., and Aguadé, M. (1986a) Genetic uniformity in two populations of *Drosophila melanogaster* as revealed by filter hybridization of four-nucleotide-recognizing restriction enzyme digests. *Proc. Natl. Acad. Sci. USA* 83, 3562–3566

Kreitman, M., and Aguadé, M. (1986b) Excess polymorphism at the *Adh* locus in *Drosophila melanogaster*. *Genetics*. 114, 93–110

Krone, S.M., and Neuhauser, C. (1997) The genealogy of samples in models with selection. *Genetics*. 145, 519–534

Kruglyak, L. (1999) Prospects for whole-genome linkage disequilibrium mapping. *Nature Genetics*. 22, 139-144

Kruglyak, L., and Nickerson, D.A. (2001) Variation is the spice of life. *Nature Genetics*. 27, 234–236

Ku, H.M., Vision, T., Liu, J., and Tanksley, S.D. (2000) Comparing sequenced segments of the tomato and *Arabidopsis* genomes: Large scale duplication followed by selective gene loss creates a network of synteny. *Proc. Natl. Acad. Sci. USA*. 97, 9121–9126

Kuhner, M.K., Yamato, J., and Felsenstein, J. (2000) Maximum likelihood estimation of recombination rates from population data. *Genetics*. 156, 1393–1401

Lalo, D., Stettler, S., Mariotte, S., Slominski, P.P., and Thuriaux, P. (1993) Two yeast chromosomes are related by a fossil duplication of their centromeric regions. *C.R. Acad. Sci. III*. 316, 367–373

Larget, B., Simon, D.L., and Kadane, J.B. (2002) Bayesian phylogenetic inference from animal mitochondrial genome rearrangements. *J. Roy. Stat. Soc., B*. 64, 681–693

Lessard, S., and Wakeley, J. (2004) The two-locus ancestral graph in a subdivided population: convergence as the number of demes grows in the island model. *J. Math. Biol*. 48, 275–292

Lewontin, R.C. (1964) The interaction of selection and linkage. I. General considerations; heterotic models. *Genetics*. 49, 49–67

Li, W.H. (1977) Distribution of nucleotide differences between two randomly chosen cistrons in a finite population. *Genetics*. 85, 331–337

Li, W.H. (1997) *Molecular Evolution*. Sinauer, Sunderland, MA

Li, W.H., Gu, Z., Wang, H., and Nakrutencko, A. (2001) Evolutionary analyses of the human genome. *Nature*. 409, 847–849

Li, W.H., and Sadler, L.A. (1991) Low nucleotide diversity in man. *Genetics*. 129, 513–523

Li, N., and Stephens, M. (2003) Modeling linkage disequilibrium and identifying hot spots using single nucleotide polymorphism data. *Genetics*. 165, 2213–2233

Lieberman, E., Hauert, C., and Nowak, M.A. (2005) Evolutionary dynamics on graphs. *Nature*. 433, 312–316

Littler, R.A. (1975) Loss of variaibility at one locus in a finite population. *Math. Bio.* 25, 151–163

Lynch, M. (1987) The consequences of fluctuating selection for isozyme polymorphisms in daphnia. *Genetics.* 115, 657–669

Lynch, M., and Conrey, J.S. (2000) The evolutionary fate and consequences of duplicate genes. *Science.* 290, 1151–1155

Lynch, M., and Crease, T.J. (1990) The analysis of population survey data on DNA sequence variation. *Mol. Biol. Evol.* 7, 377–394

Lynch, M., and Force, A. (2000) the probability of duplicate gene preservation by subfunctionalization. *Genetics.* 154, 459–473

Malécot, G. (1969) *The Mathematics of Heredity.* Freeman, San Francisco

Marjoram, P., and Wall, J.D. (2006) Fast "coalescent" simulation. *BMC Genetics.* 7, paper 16

Markovstova, L., Marjoram, P., and Tavaré, S. (2001) On a test of Depaulis and Veuille. *Mol. Biol. Evol.* 18, 1132–1133

Maruyama, T. (1970a) On the fixation probability of mutant genes in a subdivided population. *Genet. Res.* 15, 221–225

Maruyama, T. (1970b) Effective number of alleles in a subdivided population. *Theoret. Pop. Biol.* 1, 273–306

Maruyama, T. (1970c) Stepping stone models of finite length. *Adv. Appl. Prob.* 2, 229–258

Maruyama, T. (1970d) Analysis of population structure. I. One-dimensional stepping stone models of finite length. *Ann. Hum. Genet.* 34, 201–219

Maruyama, T. (1971) The rate of decrease of heterozygosity in population occupying a circular or linear habitat. *Genetics.* **67**, 437–454

Maruyama, T. (1974) A simple proof that certain quantities are independent of the geographical structure of population. *Theor. Pop. Biol.* 5, 148–154

Maruyama, T., and Kimura, M. (1971) Some methods for treating continuous stochastic processes in population genetics. *Japanese Journal of Genetics.* 46, 407–410

Maruyama, T., and Kimura, M. (1974) A note on the speed of gene frequency changes in reverse directions in a finite population. *Evolution.* 28, 161–163

Matsen, E., and Wakeley, J. (2006) Convergence to the island model coalescent process in populations with restricted migration. *Genetics.* 172, 701–708

Maynard Smith, J. (1968) Evolution in sexual and asexual populations. *Amer. Naturalist.* 102, 469–473

Maynard Smith, J. and Haigh, J. (1974) The hitchhiking effect of a favorable gene. *Genet. Res.* 23, 23–35

McDonald, J.H., and Kreitman, M. (1991) Adaptive protein evolution at the *Adh* locus in *Drosophila. Nature.* 351, 652–654

McVean, G.A.T. (2002) A genealogical interpretation of linkage disequilibrium. *Genetics.* 162, 987–991

McVean, G.A.T., and Cardin, N.J. (2005) Approximating the coalescent with recombination. *Phil. Trans. Roy. Soc., B.* 360, 1387–1393

McVean, G.A.T., and Charlesworth, B. (2000) The effects of Hill-Robertson interference between weakly selected mutations on patterns of molecular evolution and variation. *Genetics.* 155, 929–944

McVean, G.A.T., Myers, S.R., Hunt, S., Deloukas, P., Bentley, D.R., and Donnelly, P. (2004) The fine scale structure of recombination rate variation in the human genome. *Science.* 304, 581–584

Modiano, G., Benerecetti, A.S., Gonano, F., Zei, G., Capldo, A., and Cavalli-Sforza, L.L. (1965) An anlaysis of ABO, MN, Rh, Hp, Tf and G6PD types in a sample from the human population in the Lecce province. *Ann. Human. Genet.* 29, 19–31

Möhle, M. (2000) Total variation distances and rates of convergence for ancestral coalescent processes in exchangeable population models. *Adv. Appl. Prob.* 32, 983–993

Möhle, M., and Sagitov, S. (2001) A classification of coalescent processes for haploid exchangeable population models. *Ann. Prob.* 29, 1547–1562

Moore, G., Devos, K.M., Wang, Z., and Dale, M.D. (1995) Grasses, line up and form a circle. *Current Biology.* 5, 737–739

Moran, P.A.P. (1958) Random processes in genetics. *Proc. Camb. Phil. Soc.* 54, 60–71

Moret, B., Wyman, S., Bader, D., Warnow, T., and Yan, M. (2001) A new implementation and detailed study of breakpoint analysis. Pages 583-594 in *Proceedings of the 6th Pacific Symposium on Biocomputing, Hawaii.*

Mueller, L.D., Barr, L.D., and Ayala, F.J. (1985) Natural selection vs. random drift: evolution from temporal variation in allele frequencies in nature. *Genetics.* 111, 517–554

Mukai, T. (1969) The genetic structure of natural populations of *Drosophila melanogaster.* VII. Synergistic interaction of spontaneous mutant polygenes controlling viability. *Genetics.* 61, 749–761

Muller, H.J. (1932) Some genetic aspects of sex. *Amer. Naturalist.* 66, 118–132

Muller, H.J. (1958) Evolution by mutation. *Bull. Amer. Math. Soc.* 64, 137–160

Muller, H.J. (1964) The relation of recombination to mutational advance. *Mutation Research.* 1, 2–9

Müller, S., O'Brien, P.C.M., Ferguson-Smith, M.A., and Weinberg, J. (1997) Reciprocal chromosome painting between human and prosimians (*Eulemer macaco macaco* and *E. fulvus mayottensis*). *Cytogenet. Cell Genet.* 78, 260–271

Müller, S., Stanyon, R., O'Brien, P.C.M., Ferguson-Smith, M.A., Plesker, R., and Weinberg, J. (1999) Defining the ancestral karyotype of all primates by multi-directional chromosome painting between tree shrews, lemurs, and humans. *Chromosoma.* 108, 393–400

Murphy, W.J., Bourque, G., Tesler, G., Pevzner, P., and O'Brien, S.J. (2003) Reconstructing the genomic architecture of mammalian ancestors using multispecies comparative maps. *Human Genomics.* 1, 30–40

Murphy, W.J., et al. (2005) Dynamics of mammalian chromosome evolution inferred from multispecies comparative maps. *Science.* 309, 613–617

Murphy, W.J., Sun, S., Chen, Z., Pecon-Slattery, J., and O'Brien, S.J. (1999) Extensive conservation of sex chromosome organization between cat and human revealed by parallel radiation hybrid mapping. *Genome Research.* 9, 1223–1230

Murphy, W.J., Sun, S., Chen, Z., Yuhki, N., Hirschman, D., Menotti-Raymond, M, O'Brien, S.J. (2000) A radiation-hybrid map of the cat genome: implications for comparative mapping. *Genome Research.* 10, 691–702

Muse, S.V., and Gaut, B.S. (1994) A likelihood approach for comparing synonymous and nonsynonymous nucleotide substitution rates, with applications to the chloroplast genome. *Mol. Biol. Evol.* 11, 715–724

Mustonen, V., and Lässig, M. (2007) Adaptations to fluctuating selection in *Drosophila. Proc. Natl. Acad. Sci. USA* 104, 2277–2282

Myers, S.R., and Griffiths, R.C. (2003) Bounds on the minimum number of recombination events in a sample history. *Genetics.* 163, 375–394

Myers, S., Bottolo, L., Freeman, C, McVean, G., and Donnelly, P. (2005) A fine scale map of recombination rates and hot spots across the human genome. *Science.* 310, 321–324

Nachman, M.W. (1998) Deleterious mutations in animal mitochondrial DNA. *Genetica.* 102/103, 61–69

Nadeau, J.H. (1989) Maps of linkage and synteny homologies between mouse and man. *Trends in Genetics.* 5, 82–86

Nadeau, J.H., and Sankoff, D. (1998) The lengths of undiscovered conserved segments in comparative maps. *Mammalian Genome.* 9, 491–495

Nadeau, J.H., and Taylor, B.A. (1984) Lengths of chromosomal segments conserved since divergence of man and mouse. *Proc. Natl. Acad. Sci. USA* 81, 814–818

Nagylaki, T. (1974) The decay of genetic variability in geographically structured populations. *Proc. Nat. Acad. Sci., USA* 71, 2932–2936

Nagylaki, T. (1980) The strong-migration limit in geographically structured populations. *J. Math. Biol.* 9, 101-114

Nei, M. (1975) *Molecular Population Genetics and Evolution.* American Elsevier, North-Holland, Amsterdam

Nei, M., Maruyama, T., and Chakraborty, R. (1975) The bottleneck effect and genetic variability in populations. *Evolution.* 29, 1–10

Nei, M., and Takahata, N. (1993) Effective population size, genetic diversity, and coalescence time in subdivided populations. *J. Mol. Evol.* 37, 240–244

Neuhauser, C., and Krone, S.M. (1997) Ancestral processes with selection. *Theor. Pop. Biol.* 51, 210–237

Nickerson, D.A., et al. (1998) DNA sequence diversity in a 9.7kb segment of the human lipoprotein lipase gene. *Nature Genetics.* 19, 233–240

Nielsen, R. (2000) Estimates of population parameters and recombination rates from single nucleotide polymorphisms. *Genetics.* 154, 931–942

Nielsen, R. (2001) Statistical tests of selective neutrality in the age of genomics. *Heredity.* 86, 641–647

Nielsen, R. (2005) Molecular signatures of natural selection. *Annu. Rev. Genet.* 39, 197–218

Nielsen, R., and Yang, Z. (1998) Likelihood methods for detecting positively selected amino acid sites and application to HIV-1 envelope gene. *Genetics.* 148, 929–936

Nordborg, M., Charlesworth, B., and Charlesworth, D. (1996) The effect of recombination on background selection. *Genet. Res.* 67, 159–174

Nordborg, M., and Krone, S. (2002) Separation of time scales and convergence to the coalescent in structured populations. Pages 194–232 in *Modern Developments in Theoretical Population Genetics*, edited by M. Slatkin and M. Veuille, Cambridge U. Press

Notohara, M. (1990) The coalescent and the genealogical process in geographically structured population. *J. Math. Biol.* 29, 59–75

Notohara, M. (1993) The strong-migration limit for the genealogical process in geographically structured populations. *J. Math. Biol.* 31, 115–122

Nurminsky, D.I. (2001) Genes in sweeping competition. *Cell. Mol. Life. Sci.* 58, 125–134

Oakeshott, J.G., et al. (1982) Alcohol dehydrogenase and glycerol-3-phosphate dehydrogenase clines in *Drosophila melanogaster* on different continents. *Evolution.* 36, 86–92

O'Connell, N. (1993). Yule process approximation for the skeleton of a branching process. *J. Appl. Prob.* 30, 725–729

O'Hara, R.B. (2005) Computing the effects of genetic drift an fluctuating selection on genotype frequency changes in the scarlet tiger moth. *Proc. Roy. Soc. B* 272, 211–217

Ohno, S. (1967) *Sex Chromosomes and Sex-linked Genes.* Springer, Berlin

Ohno, S. (1970) *Evolution by Gene Duplication.* Springer-Verlag

Ohta, T. (1972) Fixation probability of a mutant influenced by random fluctuation of selection intensity. *Genet. Res.* 19, 33–38

Ohta, T., and Kimura, M. (1969a) Linkage disequilibrium due to random genetic drift. *Genet. Res.* 13, 47–55

Ohta, T., and Kimura, M. (1969b) Linkage disequilibrium at steady state determined by random genetic drift and recurrent mutation. *Genetics.* 63, 229–238

Ohta, T., and Kimura, K. (1971) Linkage disequilibrium between two segregating nucleotide sites under the steady flux of mutations in a finite population. *Genetics.* 68, 571–580

Ohta, T. and Kimura, M. (1975) The effect of selected linked locus on the heterozygosity of neutral alleles (the hitch-hiking effect). *Genet. Res.* 25, 313–326

Palmer, J.D., and Herbon, L.A. (1988) Plant mitochondrial DNA evolves rapidly in structure but slowly in sequence. *J. Mol. Evol.* 28, 87–97

Patil, N., et al. (2001) Blocks of limited haplotype diversity revealed by high resolution scanning of human chromosome 21. *Science.* 294, 1719-1723

Pevzner, P.A. (2000) *Computational Molecular Biology: An Algorithmic Approach.* MIT Press, Cambridge

Pevzner, P.A. and Tesler, G. (2003a) Genome rearrangement in mammalian evolution: Lessons from human and mouse genomes. *Genome Research.* 13, 37–45

Pevzner, P.A., and Tesler, G. (2003b) Transforming men into mice: the Nadeau-Taylor chromosomal breakage model revisited. *Proceedings of RE-COMB '03.* Berlin, Germany

Pevzner, P.A., and Tesler, G. (2003c) Human and mouse genomic seqeunce reveal extensive breakpoint reuse in mammalian evolution. *Proc. Natl. Acad. Sci. USA* 100, 7672–7677

Pfaffelhuber, P., Haboud, B., and Wakolbinger, A. (2006) Approximate genealogies under genetic hitchhiking. *Genetics.* 174, 1995–2008

Pfaffelhuber, P., and Studeny, A. (2007) Approximating genealogies for partially linked neutral loci under a selective sweep. *J. Math. Biol.* 55, 299–330

Phillips, M.S., et al. (2003) Chromosome wide distribution of haplotype blocks and the role of recombination hot spots. *Nature Genetics.* 33, 382–387

Pitman, J. (1999) Coalescents with multiple collisions. *Ann. Prob.* 27, 1870–1902

Pittel, B. (1990) On tree census and the giant component in sparse random graphs, *Rand. Struct. Algor.*, 1, 311–342

Pluzhnikov, A., and Donnelly, P. (1996) Optimal sequencing strategies for surveying molecular genetic diversity. *Genetics.* 144, 1247–1262

Prince, V.E. and Pickett, F.B. (2002) Splitting pairs: the diverging fates of duplicated genes. *Nature Reviews Genetics.* 3, 827–837

Pritchard, J.K., and Przeworski, M. (2001) Linkage disequilibrium in humans: models and data. *Am. J. Human Genetics.* 69, 1-14

Ptak, S.E. et al. (2005) Absence of the TAP2 recombination hot spot in chimpanzees. *PLOS Biology.* 2, e155

Ranz, J.M., Casals, F., and Ruiz, A. (2001) How malleable is the eukaryotic genome? Extreme rate of chromosomal rearrangement in the genus *Drosophila*. *Genome Research.* 11, 230–239

Rat Genome Sequencing Consortium (2004) Genome sequence of the Brown Norway rat yields insights into mammalian evolution. *Nature.* 428, 493–521

Reich, D.E., Cargill, M., Bolk, S., Ireland, J., Sabeti, P.C., et al (2001) Linkage disequilibrium in the human genome. *Nature.* 411, 199–204

Reich, D.E., et al. (2002) Human genome sequence variation and the influence of gene history, mutation, and recombination. *Nature Genetics.* 32, 135–142

Risch, N., and Merikangas, K. (1996) The future of genetic studies of complex human diseases. *Science.* 273, 1516–1517

Rogers, A.R., and Harpending, H. (1992) Population growth makes waves in the distribution of pairwise genetic differences. *Mol. Biol. Evol.* 9, 552–569

Ross, M.T., et al. (2005) The finished sequence of the human *X* chromosome. *Nature.* 434, 325–337

Rousset, F. (1997) Genetic differentiation and estimation of gene flow from F-statistics under isolation by distance. *Genetics.* 145, 1219–1228

Sabeti, P.C., et al. (2002) Detecting recent positive selection in the human genome from haplotype structure. *Nature.* 419, 832-837

Sabeti, P.C., et al. (2006) Positive natural selection in the human lineage. *Science.* 312, 1614–1620

Sagitov, S. (1999) The general coalescent with asynchronous mergers of ancestral lines. *J. Appl. Prob.* 36, 1116–1125

Sankoff, D., and Blanchette, M. (1997) The median problem for breakpoints in comparative genomics. *Computing and Combinatorics, Proceedings of CO-COON '97.* Edited by T. Jiang and D.T. Lee. Springer Lecture Notes in Computer Science 1276, pages 251–263

Sankoff, D., and Blanchette, M. (1998a) Multiple genome rearrangement and breakpoint phylogeny. *J. Comp. Biol.* 5, 555–570

Sankoff, D., and Blanchette, M. (1998b) Multiple genome rearrangement. *Proc. RECOMB 98*, 243–247

Sankoff, D., and Blanchette, M. (2000) Multiple genome rearrangement and breakpoint phylogenies. *J. Comp. Biol.* 5, 555–570

Sankoff, D., and Nadeau, J.H. (1996) Conserved synteny as a measure of genomic distance. *Discrete Appl. Math.* 71, 247–257

Sankoff, D., Parent, M.N., Marchand, I., Ferretti, V. (1997) On the Nadeau-Taylor theory of conserved chromosome segments. Pages 262–274 in *Combinatorial Pattern Matching, Eighth Annual Symposium.* Edited by A. Apostolico and J. Hein. Lecture Notes in Computer Science 1264, Springer, Berlin

Sano, A., Shimizu, A., and Iizuka, M. (2004) Coalescent process with fluctuating population size and its effective size. *Theor. Pop. Biol.* 65, 39–48

Sawyer, S.A. (1977) Asymptotic properties of the equilibrium probability of identity in a geographically structured population. *Adv. Appl. Prob.* 9, 268–282

Sawyer, S.A., and Hartl, D.L. (1992) Population genetics of polymorphism and divergence. *Genetics.* 132, 1161–1176

Sawyer, S.A., Kulathinal, R.J., Bustamante, C.D., and Hartl, D.L. (2003) Bayesian analysis suggests that most amino acid replacements in *Drosophila* are driven by positive selection. *J. Mol. Evol.* 57, S154–S164

Schaeffer, S.W., and Miller, E.L. (1993) Estimation of linkage disequilibrium and the recombination parameter determined from the alcohol dehydrogenase region of *Drosophila pseudoobscura. Genetics.* 135, 541–552

Schweinsberg, J. (2003) Coalescent processes obtained from supercritical Galton-Watson processes. *Stoch. Proc. Appl.* 106, 107–139

Schweinsberg, J., and Durrett, R. (2005) Random partitions approximating the coalescence of lineages during a selective sweep. *Ann. Appl. Prob.* 15, 1591–1651

Seielstad, M.T., Minch, E., and Cavalli-Sforza, L.L. (1998) Genetic evidence for a higher migration rate in humans. *Nature Genetics.* 20, 278–280

Seoighe, C., and Wolfe, K.H. (1998) Extent of genome rearrangement after genome duplication in yeast. *Proc. Natl. Acad. Sci. USA.* 95, 4447–4452

Seoighe, C., and Wolfe, K.H. (1999) Updated map of duplicated regions in the yeast genome. *Gene.* 238, 253–261

Simon, D.L. and Larget, B. (2001) Phylogenetic inference from mitochondrial genome rearrangement data. *Springer Lecture Notes in Computer Science.* 2074, 1022-1028

Simonsen, K.L., Churchill, G.A., and Aquadro, C.F. (1995) Properties of statistical tests of neutrality for DNA polymorphism data. *Genetics.* 141, 413–429

Singh, R.S., Lewontin, R.C., and Felton, A.A. (1976) Genetic heterogeneity within electrophoretic "alleles" of xanthine dehydrogenase in *Drosophila pseudoobscura. Genetics.* 84, 609–629

Sjödin, P., Kaj, I., Krone, S., Lascoux, M., and Nordborg, M. (2005) On the meaning and existence of an effective population size. *Genetics.* 169, 1061–1070

Sakletsky, H., et al. (2003) The male-specific region of the human Y chromosome is a mosaic of discrete sequence classes. *Nature.* 423, 825–837

Slade, P.F. (2000a) Simulation of selected genealogies. *Theor. Pop. Biol.* 57, 35–49

Slade, P.F. (2000b) Most recent common ancestor probability distributions in gene genealogies under selection. *Theor. Pop. Biol.* 58, 291–305

Slatkin, M. (1991) Inbreeding coefficients and coalescence times. *Genet. Res. Camb.* 58, 167–175

Slatkin, M. (1993) Isolation by distance in equilibrium and non-equilibrium populations. *Evolution.* 47, 264–279

Slatkin, M., and Barton, N.H. (1989) A comparison of three indirect methods of estimating average levels of gene flow. *Evolution.* 43, 1349–1368

Slatkin, M., and Hudson, H. (1991) Pairwise comparison of mitochondrial DNA sequences in stable and exponentially growing populations. *Genetics.* 129, 555–562

Smith, M.M. (1987) Molecular evolution of *Saccharomyces cerevisiae* histone gene loci. *J. Mol. Evol.* 24, 252–259

Smith, N.G.C., and Fearnhead, P. (2005) A comparison of three estimators of the population-scaled recombination rate: Accuracy and robustness. *Genetics.* 171, 2051–2062

Song, Y.S., and Hein, J. (2004) On the minimum number of recombination events in the evolutionary history of a DNA sequence. *J. Math. Biol.* 48, 160–186

Song, Y.S., and Song, J.S. (2007) Analytic computation of the expectation of the linkage disequilibrium coefficient r^2. *Theor. Pop. Biol.* 71, 49–60

Song, Y.S., and Hein, J. (2005) Constructing minimal ancestral recombination graphs. *J. Comp. Biol.* 12, 147–169

Stephan, W. (1995) An improved method for estimating the rate of fixation of favorable mutations based on DNA polymorphism data. *Mol. Biol. Evol.* 12, 959–962

Stephan, W., Wiehe, T., and Lenz, M.W. (1992) The effect of strongly selected substitutions on neutral polymorphism: Analytical results based on diffusion theory. *Theor. Pop. Biol.* 41, 237–254

Stewart, F.M. (1976) Variability in the amount of heterozygosity maintained by neutral mutations. *Theor. Pop. Biol.* 9, 188–201

Stoneking, M., Bhatia, K., and Wilson, A.C. (1986) Rates of sequence divergence estimated from restriction maps of mitochondrial DNAs from Papua New Guinea. *Cold Spring Harbor Symposia on Quantitative Biology.* 51, 433–439

Strobeck, C. (1987) Average number of nucleotide differences in a sample from a single subpopulation: A test for population subdivision. *Genetics.* 117, 149–153

Strobeck, C., and Morgan, K. (1978) The effect of intragenic recombination on the number of alleles in a finite population. *Genetics.* 88, 829–844

Stumpf, M.P.H., and McVean, G.A.T. (2003) Estimating recombination rates from population-genetic data. *Nature Reviews Genetics.* 4, 959–968

Taillon-Miller, P. et al. (2000) Juxtaposed regions of extensive and minimal linkage disequilbrium in human Xq25 and Xq28. *Nature Genetics.* 25, 324–328

Tajima, F. (1983) Evolutionary relationship of DNA sequences in finite populations. *Genetics.* 105, 437–460

Tajima, F. (1989) Statistical method for testing the neutral mutation hypothesis by DNA polymorphism. *Genetics.* 123, 585–595

Takahata, N. (1983) Gene identity and genetic differentiation of populations in the finite island model. *Genetics.* 104, 497–512

Takahata, N. (1995) A genetic perspective on the origins and history of humans. *Annu. Rev. Ecol. Syst.* 26, 343–372

Takahata, N., Ishii, K., and Matsuda, H. (1975) Effect of temporal fluctuation of selection coefficient on gene frequency in a population. *Proc. Natl. Acad. Sci. USA* 72, 4541–4545

Takahata, N., and Nei, M. (1984) F_{ST} and G_{ST} statistics in the finite island model. *Genetics.* 107, 501–504

Tavaré, S. (1984) Line of descent and genealogical processes, and their applications in population genetics models. *Theor. Pop. Biol.* 46, 119–164

Tavaré, S. (1987) The birth process with immigration, and the genealogical structure of large populations. *J. Math. Biol.* 25, 161–168

Tavaré, S. (2004) Ancestral inference in population genetics. Pages 1-188 in *Ecole d'Eté de Probabilités de Saint-Flour XXXI*, Springer Lecture Notes in Math 1837

Tavaré, S., Balding, D.J., Griffiths, R.C., and Donnelly, P. (1997) Inferring coalescence times from DNA sequence data. *Genetics.* 145, 505–518

Taylor, C., Iwasa, Y., and Nowak, M.A. (2006) A symmetry of fixation times in evolutionary dynamics. *J. Theor. Biol.* 243, 245–251

Taylor, J.S., and Raes, J. (2004) Duplication and divergence: The evolution of new genes and old ideas. *Annu. Rev. Genet.* 38, 615–643

Tessler, G. (2002) GRIMM: genome rearrangements web server. *Bioinformatics.* 18, 492–493

Tsaur, S.C., Ting, C.T., and Wu, C.I. (1998) Positive selection driving the evolution of a gene of male reproduction, *Acp26Aa*, of *Drosophila*: II, Divergence versus polymorphism. *Mol. Biol. Evol.* 15, 1040–1046

Underhill, P.A. et al. (1997) Detection of numerous Y chromosome biallelic polymorhpisms by denaturing high performance liquid chromatography. *Genome Research.* 7, 996–1005

Vigilant, L., Pennington, R., Harpending, H., Kocher, T.D., and Wilson, A.C. (1989) Mitochondrial DNA sequences in single hairs from a southern African population. *Proc. Natl. Acad. Sci. USA.* 86, 9350–9354

Vision, T.J., Brown, D.G., and Tanksley, S.D. (2000) The origin of genomic duplications in *Arabidopsis*. *Science.* 290, 2114–2117

Wakeley, J. (1996) Distinguishing migration from isolation by distance using the variance of pairwise differences. *Theor. Pop. Biol.* 49, 369–386

Wakeley, J. (1997) Using the variance of pairwise differences to estimate the recombination rate. *Genet. Res.* 69, 45–48

Wakeley, J. (1998) Segregating sites in Wright's island model. *Theor. Pop. Biol.* 53, 166–174

Wakeley, J. (1999) Nonequilibrium migration in human history. *Genetics.* 153, 1863–1871

Wakeley, J. (2000) The effects of subdivision on the genetic divergence of populations and species. *Evolution.* 54, 1092–1101

Wakeley, J. and Aliacar, N. (2001) Gene genealogies in a metapopulation. *Genetics.* 159, 893–905

Wakeley, J., and Lessard, S. (2003) Theory of the effects of population structure and sampling on patterns of linkage disequilibrium applied to genetic data from humans. *Genetics.* 164, 1043–1053

Wall, J.D. (2000) A comparison of estimators of the population recombination rate. *Mol. Biol. Evol.* 17, 156–163

Wall, J.D., and Hudson, R.R. (2001) Coalescent simulations and statistical tests of neutrality. *Mol. Biol. Evol.* 18, 1134–1135

Wall, J.D. and Pritchard, J.K. (2003) Haplotype blocks and linkage disequilibrium in the human genome. *Nature Reviews Genetics.* 4, 587-597

Walsh, J.B. (1995) How often do duplicated genes evolve new functions? *Genetics.* 139, 421–428

Walsh, J.B. (2003) Population models of the fate of duplicated genes. *Genetica.* 118, 279–294

Wang, N., Akey, J.M., Zhang, K., Chakraborty, R., and Jin, L. (2002) Distribution of recombination crossovers and the origin of haplotype blocks: The interplay of population history, recombination, and mutation. *Am. J. Human Genetics.* 71, 1127–1234

Ward, R., and Durrett, R. (2004) Subfunctionalization: How often does it occur? How long does it take? *Theor. Pop. Biol.* 66, 93–100

Ward, R.H., Frazier, B.L., Dew-Jager, K., and Pääbo, S. (1991) Extensive mitochondrial diversity within a single Amerindian tribe. *Proc. Natl. Acad. Sci. USA.* 88, 8720–8724

Watson, J.D., and Crick, F.H.C. (1953a) Molecular structure of nucleic acids: A structure for deoxyribonucleic acids. *Nature.* 171, 737–738

Watson, J.D., and Crick, F.H.C. (1953b) The structure of DNA. *Cold Spring Harbor Symp. Quant. Biol.* 18, 123–131

Watson, J.D., and Crick, F.H.C. (1953c) Genetic implications of the structure of deoxyribonucleic acid. *Nature.* 171, 964–967

Watterson, G.A. (1975) On the number of segregating sites in genetical models without recombination. *Theor. Pop. Biol.* 7, 256–276

Watterson, G.A. (1977) Heterosis or neutrality? *Genetics.* 85, 789–814

Watterson, G.A. (1983) On the time for gene silencing at duplicate loci. *Genetics.* 105, 745–766

Weinberg, J., et al., (1997) Conservation of human vs. feline genome organization revealed by reciprocal chromosome painting. *Cytogent. Cell. Genet.* 72, 211–217

Weir, B.S. and Cockerham, C.C. (1984) Estimating F-statistics for the analysis of population structure. *Evolution.* 38, 1358–1370

Weiss, G.H., and Kimura, M. (1965) A mathematical analysis of the stepping stone model of genetic correlation. *J. Appl. Prob.* 2, 129–149

Whitlock, M.C., and McCauley, D.E. (1999) Indirect measures of gene flow and migration: $F_{ST} \neq 1/(4Nm + 1)$. *Heredity.* 82, 117–125

Wiehe, T., and Stephan, W. (1993) Analysis of genetic hitchhiking model and its application to DNA polymorphism data from *Drosophila melanogaster.* *Mol. Biol. Evol.* 10, 842–854

Wilkins, J.F. (2004) A separation-of-time scales approach to the coalescent in a continuous population. *Genetics.* 168, 2227–2244

Wilkins, J. F. and Wakeley, J. (2002) The coalescent in a continuous, finite, linear population, *Genetics.* **161**, 873–888

Wilkinson-Herbots, H.M. (1998) Genealogy and subpopulation differentiation under various models of population structure. *J. Math. Biol.* 37, 535–585

Wilson, W.A., Harrington, S.E., Woodman, W.L., Lee, M., Sorrells, M.E., and McCouch, S.R. (1999) Inferences on the genome structure of progenitor maize through comparative analysis of rice, maize, and the domesticated panicoids. *Genetics.* 153, 453–473

Winckler, W., et al. (2005) Comparison of fine scale recombination rates in humans and chimpanzees. *Science.* 308, 107–111

Wiuf, C., and Hein, J. (1999) Recombination as a point process along sequences. *Theor. Pop. Biol.* 55, 248–259

Wolfe, K.H., and Shields, D.C. (1997) Molecular evidence for an ancient duplication of the entire yeast genome. *Nature.* 387, 708–713

Wright, S. (1931) Evolution in Mendelian populations. *Genetics.* 16, 97–159

Wright, S. (1938) The distribution of gene frequencies under irreversible mutation. *Proc. Natl. Acad. Sci. USA* 24, 253–259

Wright, S. (1939) Statistical genetics in relation to evolution. *Actualités Scientifique et Industrielles.* No. 802, Hermann et Cie, Paris.

Wright, S. (1942) Statistical genetics and evolution. *Bull. Amer. Math. Soc.* 48, 223–246

Wright, S. (1943) Isolation by distance. *Genetics* 28, 114–156

Wright, S. (1945) The differential equation of the distribution of gene frequencies. *Proc. Natl. Acad. Sci. USA* 31, 382–389

Wright, S. (1948) On the roles of directed and random changes in gene frequency in the genetics of populations. *Evolution.* 2, 279–294

Wright, S. (1951) The genetical structure of populations. *Ann. Eugen.* 15, 323–354

Wright, S. (1951) Fisher and Ford on "The Sewall Wright Effect." *Amer. Scientist.* 39, 452–479

Wright, S. (1986) *Evolution: Selected Papers.* U of Chicago Press.

Yang, Z., Nielsen, R., Goldman, N., and Pedersen, A.-M. K. (2000) Codon-substitution models for variable selection pressure at amino acid sites. *Genetics.* 155, 431–449

York, T.L., Durrett, R., and Nielsen, R. (2002) Bayesian estimation of inversions in the history of two chromosomes. *J. Comp. Biol.* 9,808–818

York, T.L., Durrett, R., and Nielsen, R. (2007) Dependence of paracentric inversion rate on tract length. *BMC Bioinformatics.* 8, paper 115

Zähle, I., Cox, J.T., and Durrett, R. (2005) The stepping stone model, II. Genealogies and the infinite sites model. *Ann. Appl. Prob.* 15, 671–699

Zeng, K., Shi, S., Fu., Y.X., and Wu, C.I. (2006) Statistical tests for detecting positive selection by utilizing high frequency SNPs. *Genetics.* 174, 1431–1439

Zhang, J. (2003) Evolution by gene duplication: an update. *Trends in Ecology and Evolution.* 18, 292–298

Zhang, K., Akey, J.M., Wang, N., Xiong, M., Chakraborty, R., and Jin, L. (2003) Randomly distributed crossovers may generate block-like patterns of linkage disequilibrium. *Human Genetics.* 113, 51–59

Index